DYNAMIC GRAPHICS
FOR STATISTICS

The Wadsworth & Brooks/Cole Statistics/Probability Series

Series Editors

Ole Barndorff-Nielsen, Århus University
Peter J. Bickel, University of California, Berkeley
William S. Cleveland, AT&T Bell Laboratories
Richard M. Dudley, Massachusetts Institute of Technology

Richard A. Becker, John M. Chambers, and Allan R. Wilks, *The New S Language: A Programming Environment for Data Analysis and Graphics*

Peter J. Bickel, Kjell Doksum, and John L. Hodges, Jr., *A Festschrift for Erich L. Lehmann*

George E. P. Box, *The Collected Works of George E. P. Box*, Volumes I and II, edited by George C. Tiao

Leo Breiman, Jerome H. Friedman, Richard A. Olshen, and Charles J. Stone, *Classification and Regression Trees*

John M. Chambers, William S. Cleveland, Beat Kleiner, and Paul A. Tukey, *Graphical Methods for Data Analysis*

William S. Cleveland and Marylyn E. McGill, *Dynamic Graphics for Statistics*

Khosrow Dehnad, *Quality Control, Robust Design, and the Taguchi Method*

Richard Durrett, *Lecture Notes on Particle Systems and Percolation*

Franklin A. Graybill, *Matrices with Applications in Statistics*, Second Edition

Lucien M. Le Cam and Richard A. Olshen, *Proceedings of the Berkeley Conference in Honor of Jerzy Neyman and Jack Kiefer*, Volumes I and II

P. A. W. Lewis and E. J. Orav, *Simulation Methodology for Statisticians, Operations Analysts, and Engineers*

H. Joseph Newton, *TIMESLAB: A Time Series Analysis Laboratory*

John Rawlings, *Applied Regression Analysis: A Research Tool*

John A. Rice, *Mathematical Statistics & Data Analysis*

Joseph P. Romano and Andrew F. Siegel, *Counterexamples in Probability and Statistics*

Judith M. Tanur, Frederick Mosteller, William H. Kruskal, Erich L. Lehmann, Richard F. Link, Richard S. Pieters, and Gerald R. Rising, *Statistics: A Guide to the Unknown*, Third Edition

John W. Tukey
The Collected Works of John W. Tukey
Volume I: *Time Series: 1949-1964*, edited by David R. Brillinger
Volume II: *Time Series: 1965-1984*, edited by David R. Brillinger
Volume III: *Philosophy and Principles of Data Analysis: 1949-1964*,
　　　　edited by Lyle V. Jones
Volume IV: *Philosophy and Principles of Data Analysis: 1965-1986*,
　　　　edited by Lyle V. Jones
Volume V: *Graphics: 1965-1985*, edited by William S. Cleveland

DYNAMIC GRAPHICS
FOR STATISTICS

Edited by

William S. Cleveland
AT&T Bell Laboratories

Marylyn E. McGill
MEM Research, Inc.

Wadsworth & Brooks/Cole Advanced Books & Software
A Division of Wadsworth, Inc.

© 1988 by Wadsworth, Inc., Belmont, California 94002.
All rights reserved. No part of this book may be reproduced,
stored in a retrieval system, or transcribed, in any form or
by any means—electronic, mechanical, photocopying, recording,
or otherwise—without the prior written permission of the
publisher, Wadsworth & Brooks/Cole Advanced Books & Software,
Pacific Grove, California 93950, a division of
Wadsworth, Inc.

Printed in the United States of America

10 9 8 7 6 5 4 3 2 1

Library of Congress Cataloging-in-Publication Data

Dynamic graphics for statistics / edited by William S. Cleveland.
  Marylyn E. McGill.
    p. cm. — (The Wadsworth & Brooks/Cole statistics/probability series)
  ISBN 0-534-09144-X
  1. Statistics—Graphic methods. 2. Computer graphics.
  I. Cleveland, William S., 1943-   II. McGill, Marylyn E., 1932-
    III. Series.
QA276.3.D96 1988
519.5′028′566—dc19   88-26017
                                                          CIP

Sponsoring Editor: *John Kimmel*
Editorial Assistant: *Maria Tarantino*
Production Coordinator: *Dorothy Bell*
Cover Design: *Flora Pomeroy*
Art Coordinator: *Lisa Torri*
Interior Color Separation: *Vec-Tron Data Graphics, Schiller Park, Illinois*
Printing and Binding: *Arcata Graphics, Fairfield, Pennsylvania*

Trademark Acknowledgements

Adage is a registered trademark of Adage, Inc.

Apollo is a registered trademark of Apollo Computer, Inc.

Mac-II, Macintosh, and MacPaint are registered trademarks of Apple Computer, Inc.

Teletype and UNIX are registered trademarks of AT&T.

BMD is a registered trademark of BMDP Statistical Software, Inc.

Consumer Reports is a registered trademark of Consumers Union of United States, Inc.

Cray is a registered trademark of Cray Research, Inc.

MacSpin is a trademark of D^2 Software, Inc.

Data Desk is a registered trademark of Data Description, Inc.

DEC, PDP, VAX-11/750, VAX-11/780, and VAX/VMS are registered trademarks of Digital Equipment Corporation.

Honeywell is a registered trademark of Honeywell, Inc.

MULTIBUS is a registered trademark of Intel Corporation.

IBM, ASP, HASP, IBM 360/91, IBM 370/168, IBM 2701, IBM 3081, IBM 3277, IBM AT, IBM PC, IBM PGA, IBM PS/2, IBM XT, OS/MVT, System/360, and VS/2 R1.6 are registered trademarks of International Business Machines Corporation.

Lexidata is a registered trademark of Lexidata Corporation.

Minitab is a registered trademark of Minitab, Inc.

Motorola is a registered trademark of Motorola, Inc.

Polaroid is a registered trademark of Polaroid, Inc.

DI-3000 is a registered trademark of Precision Visuals, Inc.

Ramtek is a registered trademark of Ramtek, Inc.

SAS is a registered trademark of SAS Institute, Inc.

Geometry Engine is a registered trademark of Silicon Graphics, Inc.

NeWS and SUN are registered trademarks of Sun Microsystems, Inc.

Symbolics and Symbolics-LISP are registered trademarks of Symbolics, Inc.

Tektronix is a registered trademark of Tektronix, Inc.

Varian is a registered trademark of Varian Associates.

ETHERNET, LAMPS, and Smalltalk-80 are registered trademarks of Xerox Corporation.

Contents

Contributors ix
Preface xi

1. **Dynamic Graphics for Data Analysis** (1987)
   Richard A. Becker, William S. Cleveland and Allan R. Wilks   1

   Comments
   John W. Tukey, Peter J. Huber, William F. Eddy, Howard Wainer and Edward R. Tufte   50

   Rejoinder
   Richard A. Becker, William S. Cleveland and Allan R. Wilks   66

2. **Some Approaches to Interactive Statistical Graphics** (1969)
   D. F. Andrews, E. B. Fowlkes and P. A. Tukey   73

3. **PRIM-9: An Interactive Multidimensional Data Display and Analysis System** (1974)
   Mary Anne Fisherkeller, Jerome H. Friedman and John W. Tukey   91

4. **Kinematic Display of Multivariate Data** (1982)
   D. L. Donoho, P. J. Huber, E. Ramos and H. M. Thoma   111

5. **An Introduction to Real Time Graphical Techniques for Analyzing Multivariate Data** (1982)
   Jerome H. Friedman, John Alan McDonald and Werner Stuetzle   121

6. **Control and Stash Philosophy for Two-Handed, Flexible, and Immediate Control of a Graphic Display** (1982)
   John W. Tukey   133

7. **Orion I: Interactive Graphics for Data Analysis** (1983)
   John Alan McDonald   179

8. **Brushing Scatterplots** (1987)
   Richard A. Becker and William S. Cleveland   201

9. **Plot Windows** (1987)
   Werner Stuetzle   225

10. **The Use of Brushing and Rotation for Data Analysis** (1988)
    Richard A. Becker, William S. Cleveland and Gerald Weil   247

11. **Elements of a Viewing Pipeline for Data Analysis** (1988)
    Andreas Buja, Daniel Asimov, Catherine Hurley and John A. McDonald   277

12. **EXPLOR4: A Program for Exploring Four-Dimensional Data Using Stereo-Ray Glyphs, Dimensional Constraints, Rotation, and Masking** (1988)
    D. B. Carr and W. L. Nicholson   309

13. **MACSPIN: Dynamic Graphics on a Desktop Computer** (1988)
    Andrew W. Donoho, David L. Donoho and Miriam Gasko   331

14. **The Application of Depth Separation to the Display of Large Data Sets** (1988)
    Thomas V. Papathomas and Bela Julesz   353

15. **A High Performance Color Graphics Facility for Exploring Multivariate Data** (1988)
    C. Ming Wang and Harold W. Gugel   379

16. **Dynamic Graphics for Exploring Multivariate Data** (1988)
    Forrest W. Young, Douglas P. Kent and Warren F. Kuhfeld   391

Contributors

David F. Andrews
Department of Statistics
University of Toronto
Toronto, Ontario
Canada

Daniel Asimov
Department of Statistics
University of California
Berkeley, California

Richard A. Becker
AT&T Bell Laboratories
Murray Hill, New Jersey

Andreas Buja
Statistical Research Group
Bell Communications Research
Morristown, New Jersey
and
University of Washington
Department of Statistics
Seattle, Washington

Daniel B. Carr
Battelle Pacific Northwest
  Laboratories
Richland, Washington

William S. Cleveland
AT&T Bell Laboratories
Murray Hill, New Jersey

Andrew W. Donoho
D^2 Software, Incorporated
Austin, Texas

David L. Donoho
Department of Statistics
University of California
Berkeley, California

Mary Anne Fisherkeller
Department of Statistics
Stanford University
Stanford, California

Edward B. Fowlkes
Statistical Research Group
Bell Communications Research
Morristown, New Jersey

Jerome H. Friedman
Department of Statistics
Stanford University
Stanford, California

Miriam Gasko
Department of Statistics
University of California
Berkeley, California

Harold W. Gugel
Computer Science Department
General Motors Research
  Laboratories
Warren, Michigan

Peter J. Huber
Department of Statistics
Harvard University
Cambridge, Massachusetts

Catherine Hurley
Department of Statistics
University of Waterloo
Waterloo, Ontario
Canada

Bela Julesz
AT&T Bell Laboratories
Murray Hill, New Jersey

Douglas P. Kent
SAS Institute Incorporated
Cary, North Carolina

Warren F. Kuhfield
SAS Institute Incorporated
Cary, North Carolina

John A. McDonald
Department of Statistics
University of Washington
Seattle, Washington

Wesley L. Nicholson
Battelle Pacific Northwest
  Laboratories
Richland, Washington

Thomas V. Papathomas
AT&T Bell Laboratories
Murray Hill, New Jersey

Ernesto Ramos
Department of Statistics
Harvard University
Cambridge, Massachusetts

Werner Stuetzle
Department of Statistics
University of Washington
Seattle, Washington

H. Mathis Thoma
Department of Statistics
Harvard University
Cambridge, Massachusetts

John W. Tukey
Princeton University
Princeton, New Jersey

Paul A. Tukey
Statistical Research Group
Bell Communications Research
Morristown, New Jersey

C. Ming Wang
Computer Science Department
General Motors Research
  Laboratories
Warren, Michigan

Gerald Weil
AT&T Bell Laboratories
Murray Hill, New Jersey

Allan R. Wilks
AT&T Bell Laboratories
Murray Hill, New Jersey

Forrest W. Young
Psychometric Laboratory
University of North Carolina
Chapel Hill, North Carolina

Preface

The essential characteristic of a dynamic graphical method is the direct manipulation of elements of a graph on a computer screen; the manipulation is carried out through the use of an input device such as a mouse, and in high-performance implementations, the elements change virtually instantaneously on the screen. In the future, most graphical data analysis will be carried out in a dynamic graphics environment. There are two reasons. First, the addition of dynamic capabilities to the methodology of traditional static data display provides an enormous increase in the power of graphical methods to convey information about data - wholly new methods become available and many capabilities that are cumbersome and time-consuming in a static environment become simple and fast in a dynamic one. Second, the price and availability of powerful statistical-computing environments are rapidly evolving in a direction that will permit the use of dynamic graphics.

This book contains a collection of papers about dynamic graphics. Some papers are new; others are technical reports from the past and re-publications of journal articles. The original date of writing or issue of each paper is shown in the table of contents. The writings cover a time period of about two decades, ranging from the late 1960's to the present. Technology has raced forward during this time period so that implementation of dynamic methods in computer systems is now vastly different; but fundamental ideas about basic graphical principles and data-analytic goals that are discussed in the early papers are relevant today. It is both instructive and interesting to see how these ideas have evolved through time.

There are a number of people to commend. John Kimmel's editorial mentoring made this collection possible. The authors were surprisingly cooperative in getting things done quickly; perhaps everyone realized that speed was essential because research in dynamic graphics and commercial development are now moving forward at a furious pace.

Finally, the authors and editors owe Susan A. Tarczynski much gratitude. Sue, who started out with an assortment of manuscripts in hardcopy and electronic form, used her considerable word-processing skills to produce a book whose typography is beautifully designed and executed.

William S. Cleveland
Marylyn E. McGill

1

DYNAMIC GRAPHICS
FOR DATA ANALYSIS

Richard A. Becker
William S. Cleveland
Allan R. Wilks
AT&T Bell Laboratories

ABSTRACT

Dynamic graphical methods have two important properties: direct manipulation of graphical elements on a computer graphics screen and virtually instantaneous change of the elements. The data analyst takes an action through manual manipulation of an input device and something happens in real time on the screen. These computing capabilities provide a new medium for the invention of graphical methods for data analysis. A collection of such methods—identification, deletion, linking, brushing, scaling, rotation, and dynamic parameter control—is reviewed. Those who develop dynamic methods must deal with a number of computer hardware and software issues, because for a dynamic method to work there must be a sufficiently fast flow of information along the channel that starts with the analyst's input and ends with the changed graph. Furthermore, for a dynamic method to be useful, the visual and manual tasks must be easy to perform. Several computing issues dealing with speed and ease of use—system bandwidth, input and output devices, low-level algorithms, device independent graphics, and data analysis environments—are reviewed.

Reprinted from (1987) *Statistical Science*, **2**: 355-395, by permission of the Institute of Mathematical Statistics.
From *Dynamic Graphics for Statistics*, William S. Cleveland and Marylyn E. McGill, eds., copyright © 1988 by Wadsworth, Inc. All rights reserved.

1. INTRODUCTION

Dynamic graphical methods have two important properties: direct manipulation and instantaneous change. The data analyst takes an action through manual manipulation of an input device and something happens, virtually instantaneously, on a computer graphics screen. Figure 1 shows an example in which a dynamic method is used to turn point labels on and off. The data analyst moves a rectangle over the scatterplot by moving a mouse; the figure shows the rectangle in a sequence of positions. When the rectangle covers a point, its label appears and when the rectangle no longer covers the point, its label disappears.

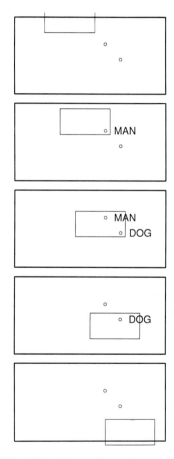

Figure 1. Dynamic Graphics. In a dynamic graphic method, the data analyst directly manipulates graphic elements on the screen, causing virtually instantaneous changes of the elements. The figure shows an example. The analyst moves a rectangle around the screen by moving a mouse; when points enter the rectangle their labels appear and when the points leave the rectangle their labels disappear. The method used is brushing, which is discussed later.

1.1 The Importance of Dynamic Methods

In the future, dynamic graphical methods will be ubiquitous. There are two reasons. One is that the addition of dynamic capabilities to the methodology of traditional static data display provides an enormous increase in the power of graphical methods to convey information about data—wholly new methods become possible and many capabilities that are cumbersome and time consuming in a static environment become simple and fast [Tukey and Tukey, 1985]. Huber [1983] aptly describes the importance of the dynamic environment: "We see more when we interact with the picture—especially if it reacts instantaneously—than when we merely watch." This does not mean that current static methods will be discarded, but rather that there will be a much richer collection of methods. The second reason is that the price and availability of powerful statistical computing environments are rapidly evolving in a direction that will permit the use of dynamic graphics [McDonald and Pedersen, 1985a and 1985b]. Thus, it seems likely that the methods described in this paper will be standard methodology in the future. Furthermore, because the number of people that have so far been involved in research in dynamic methods is relatively small, the development of new dynamic methods should accelerate as the appropriate computing environments become more widely available.

1.2 Two Early Systems

A recognition of the potential of direct manipulation, real-time graphics for data analysis goes back as far as the early 1960s when computer graphics systems first began appearing. Tukey and Wilk [1965] write: "Visual presentation of $z = f(x,y)$, is far from easy, yet badly needed. Of three possibilities—contours, families of cross sections, and isometric views—the first seems, so far, most likely to be effective, though direct-interaction graphical consoles may offer other possibilities."

In the late 1960s, Fowlkes [1971] developed a dynamic system for making probability plots. By turning a knob, the user could continuously change a shape parameter for the reference distribution on a probability plot displayed on the screen; the knob could be turned to make the pattern of points as nearly straight as possible. The data could be transformed by a power transformation, $(y + c)^p$, with c and p each changed by a knob. Points could be deleted by positioning a cursor on them. The system demonstrated that dynamic graphical methods had the potential to be important tools for data analysis.

Another early system was PRIM-9 [Fisherkeller, Friedman, and Tukey, 1975], a set of dynamic tools for projecting, rotating, isolating, and masking multidimensional data in up to nine dimensions. Rotation was the central operation; this dynamic method allows the data analyst to study three-dimensional data by showing the points rotating on the computer screen. Isolation and masking were features that allowed point deletion in a lasting or in a transient way. PRIM-9 was an influential system; many subsequent systems were modeled after it and during the 1970s dynamic graphics and PRIM operations were nearly synonymous. (In fact, in the rush to implement PRIM systems, Fowlkes' ideas were nearly forgotten.) As the reader will see from the descriptions to follow and their origins, it was not until the early 1980s that significantly different methods would begin appearing; this was stimulated in large measure by new computing techniques coming from computer scientists.

1.3 Contents of the Paper

A variety of dynamic graphical methods are described and illustrated in Section 2 of this paper. Sections 2.1 to 2.6 cover identification, deletion, linking, brushing, scaling, and rotation. Section 2.7 describes in a general way what many of the methods are doing—providing dynamic parameter control—and thereby opens the door to a large collection of potential methods. Computing issues are discussed in Section 3 of the paper; hardware and software considerations tend to be much more tightly bound to the success of dynamic methodologies than is the case for static graphical methods. Section 4 of the paper is a brief summary and discussion.

2. METHODS

2.1 Identification of Labeled Data

Identification has two directions. Suppose we have a collection of elements on a graph (e.g., points) and each element has a name or label. In one direction of identification we select a particular element and then find out what its label is; we will call this *labeling*. In the other direction, we select a label and then find the location on the graph of the element corresponding to this label. We will call this *locating*. Identification tasks, although seemingly mundane, are so all-pervasive that simple ways of performing them are of enormous help to a data analyst.

Labeling Points

Suppose x_i and y_i for $i = 1$ to n are measurements of two variables that have labels. Figure 2 shows an example. The data are measurements of the brain weights and body weights of a collection of animal species [Crile and Quiring, 1940]. Biologists study the relationship between these two variables because the ratio (brain weight)/(body weight)$^{2/3}$ is a rough measure of intelligence [Gould, 1979, Jerison, 1955]. In Figure 2, the data are graphed on a log scale and the axes are scaled so that $y = 2x/3$ is a 45° line; thus, 45° lines are contours of constant intelligence under this measure. Each point on the graph has a label: the name of the species.

In analyzing bivariate data with labels, we almost always want to know the labels for all or some of the points of the scatterplot. For example, in Figure 2 the point graphed by a *filled circle* is interesting because it

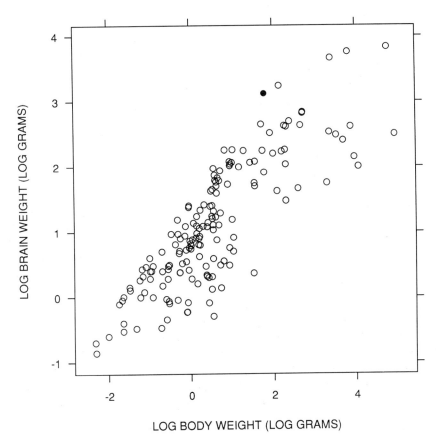

Figure 2. Identification. Log brain weight is graphed against log body weight for a collection of animal species. Each point has a name, the species name. As we saw in the previous figure, dynamic graphics provides easy methods for labeling: making the name for a point appear on the screen. We can also locate: find the data point with a particular name. In this figure, the species modern man was located by selecting it from a pop-up menu, which resulted in the circle for modern man being highlighted.

lies on a 45° line that is above the lines for all other points, which means that this species has a higher intelligence measure than all others. Which is it? With static displays, finding out the labels of points on scatterplots is a cumbersome task. It is not possible to routinely show the labels of all points because overplotting frequently causes an uninterpretable mess; this would be the case in Figure 2. With dynamic displays, getting label information is simple because the data analyst can turn labels on and off very rapidly. One method for doing this has been illustrated in Figure 1. It turns out that the label for the interesting point in Figure 2 is, not surprisingly, "modern man."

Locating Points

In analyzing labeled points we often want to go in the other direction and locate a point with a certain name. For example, we might have asked where modern man is in Figure 2. This task also can be accomplished in a particularly efficient way by using dynamic graphics. The data analyst can press a button on a mouse to bring up a menu on the screen that shows the labels, move the mouse to select the label, and then release the button [Becker and Cleveland, 1987]; this can result in the point being highlighted, as modern man is highlighted in Figure 2.

Locating Different Data Sets: Alternagraphics

In graphing two-variable data we often encounter another type of identification problem: the quantitative information is partitioned into subsets, and we want to locate the different subsets on the plot. For example, in Figure 2 the data can be categorized into five subsets: primates, nonprimate mammals, fish, birds, and dinosaurs. The subsets are shown in Figure 3. Another way to think of this is that there is a third variable, a categorical one, that we also want to show.

Subsets of the quantitative information on a graph can be enormously varied. Here are just three examples: 1) each subset is a set of points, as in Figure 3; 2) each subset is a collection of contours of a third variable as a function of the two axis variables; 3) the first subset is a scatterplot of points and the remaining subsets are regression curves of y on x resulting from several different models.

Comparing subsets can be surprisingly tricky when using static graphics. Part of the problem is that we want to be able to do more than just figure out to which subset each element of the graph belongs; we want to perceive each subset as a whole, mentally filtering out the others. With static graphics, many methods have been suggested [Cleveland, 1985]—use different plotting symbols, use different colors, or connect points of the same subset by lines. Often, none of these methods works, because the overlap of elements of the graph makes it impossible to distinguish the different subsets. In such cases the only solution is to use juxtaposed panels as illustrated in Figure 3. The drawback to such a display is that we cannot compare the locations of different subsets as effectively as when all of the data are on the same panel.

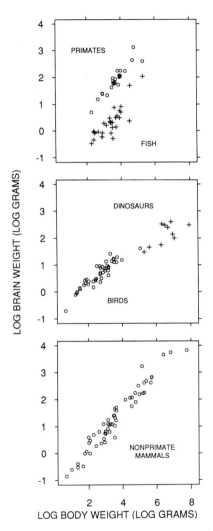

Figure 3. Alternagraphics. Another identification task for which dynamic methods are particularly well suited is locating different data subsets on a graph. For example, the brain and body data consist of five groups of species. In alternagraphics, data sets are identified by highlighting each in turn in rapid succession or by showing each set by itself and rapidly cycling through the sets.

A dynamic method that is often effective for identifying subsets is alternagraphics [Tukey, 1973]. At a given moment in time the viewer can identify some of the subsets and the selection of identified subsets can be changed quickly. There are many ways of implementing this idea. One is to cycle through the subsets showing each for a short time period. More explicitly, a graph of the first subset appears on the screen for, say, 1 sec, then it is replaced for 1 sec by a graph of the second subset, and so forth until we get to the last subset. Then the process repeats. Of course, the subsets are all shown on common axes so that the scale of the pictures remains identical as various subsets are shown. Another technique is to show all of the data at all times and have the cycling consist of a highlighting of one subset at each stage. A third technique is to provide the data analyst with the capability of turning any subset on or off with a simple and rapid action such as the use of a mouse to point to a subset name on a menu and then clicking a button; with such a technique the analyst can rapidly get any panel of Figure 3 [Donoho, Donoho, and Gasko, 1986].

2.2 Deletion

Another fundamental operation that can be easily carried out by using dynamic graphics is deleting points from a graph. Figure 4 illustrates one simple use of deletion: A scatterplot is made and there is an outlier that causes the remaining points on the graph to be crammed into such a small region that their resolution is ruined; the analyst removes the point by touching it with a cursor, and after the deletion the graph is automatically rescaled and redrawn on the screen. For example, Fowlkes [1971] used this dynamic deletion of outliers for probability plots; after points were selected for deletion, the expected order statistics of the reduced sample were recomputed automatically and the graph redrawn.

Deletion is actually a very general concept that can enter dynamic graphical methods in many ways. Its basic purpose is to eliminate certain graphical elements so that we can better study the remaining elements. For example, the outlier deletion lets us focus more incisively on the remaining data, and in alternagraphics, subsets can be temporarily deleted to allow better study of the remaining subsets. Other applications of deletion will be given in later sections.

2.3 Linking

Suppose we have n measurements on p variables and that scatterplots of certain pairs of the variables are made. A *linking* method enables us to visually link corresponding points on different scatterplots. For example, suppose there are four variables—x, y, w, and z—and that we graph y against x and z against w. To link points on the two scatterplots means to see by some visual method that the point (x_k, y_k) on the first plot corresponds to (w_k, z_k) on the other plot. To illustrate this, consider the Anderson [1935] iris data made famous by Fisher [1936]. There are 150 measurements of four variables: sepal length, sepal width, petal length, and petal width. Two scatterplots are shown in Figure 5. The data have been jittered, that is, small amounts of noise added, to avoid the overlap of plotted symbols on the graph. Each scatterplot has two clusters, and we immediately find ourselves wanting to know if there is some correspondence between the clusters of separate plots.

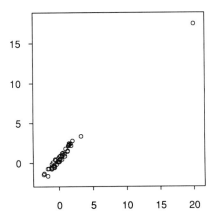

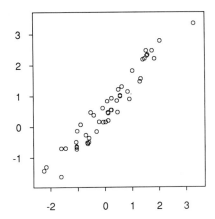

Figure 4. Deletion. In a dynamic graphics environment, points can be deleted by direct manipulation, for example, touching them with a cursor. In this figure, the outlier in the top panel is deleted, producing the graph of the bottom panel.

Linking is a concept that has long existed in the development of static displays [Chambers *et al.*, 1983; Diaconis and Friedman, 1980; Tufte, 1983]. One method for linking is the M and N plot of Diaconis and Friedman [1980]; lines are drawn between corresponding points on the two scatterplots. Another method is to use a unique plotting symbol for each point [Chambers *et al.*, 1983] on a particular plot and to use the same symbol for corresponding observations on different plots.

A third method is the scatterplot matrix, all pairwise scatterplots arranged in a rectangular array, which arose, in part, because it provides a certain amount of linking. An example is Figure 6, which shows the iris data. To maximize the resolution of the plotted points, scale information is put inside the panels of the off-diagonal of the matrix; the labels are the

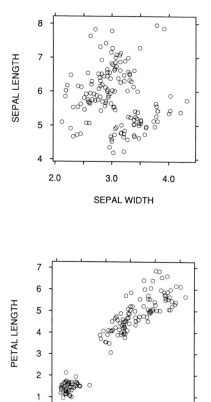

Figure 5. Linking. The scatterplots show the four variables of the Anderson iris data. Linking points on different scatterplots means visually connecting corresponding points.

variable names and the numbers show the ranges of the variables. Consider the cluster to the northwest in the sepal length and width plot of the (1,2) panel. Does this cluster correspond to one of the two clusters in the petal length and width scatterplot of panel (4,3)? By scanning horizontally from the (1,2) panel to the (1,3) panel and then vertically to the (4,3) panel we can see that the top half or so of the northwest sepal cluster corresponds to the top half or so of the northeast petal cluster. By scanning vertically from the (1,2) panel to the (4,2) panel and then horizontally to the (4,3) panel we can see that the left half or so of the northwest sepal cluster corresponds to the bottom half or so of the northeast petal cluster. The union of these two scans shows that most of the northwest sepal cluster corresponds to most of the northeast petal cluster; it is a good guess that the remaining pieces of the clusters also correspond.

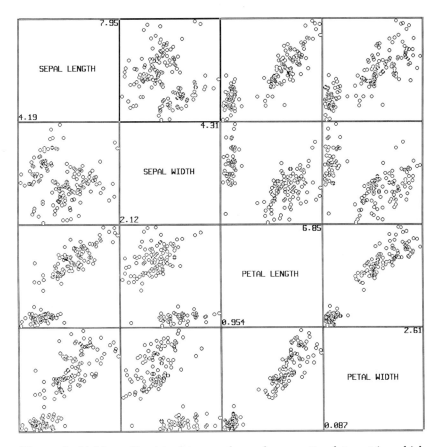

Figure 6. Linking. The iris data are shown by a scatterplot matrix, which allows a partial linking of points. overlapping points on this screen dump sometimes give unusual effects, because of the XOR style of graphics being used (see Section 3.3).

The earliest dynamic method for linking of which we are aware is described by Newton [1978]. The analyst used a light pen to draw a polygon bounded by a set of points on one plot; the enclosed points, together with all corresponding points on other scatterplots, were highlighted. For example, in Figure 7, the points in the northwest sepal cluster of the (1,2) panel, as well as all corresponding points on other panels, have been highlighted. Now we can see easily that the northwest sepal cluster corresponds to the northeast petal cluster.

McDonald [1982] developed another dynamic system for linking. In his method a cursor is positioned on one plot. All points close to the cursor on the plot are red, all points that are intermediate distances from the point are purple, and all far points are blue. All points corresponding to

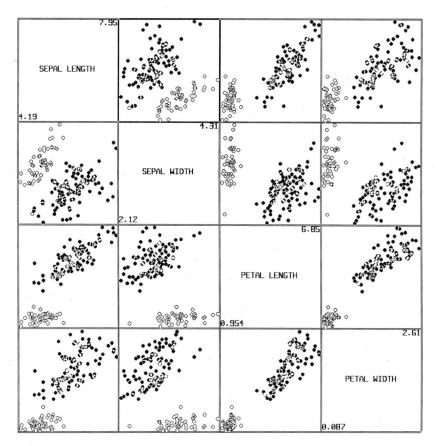

Figure 7. Linking. The northwest cluster in the (1,2) panel has been highlighted together with all corresponding points on the other panels. Now we can see that the northwest cluster in (1,2) corresponds to the northeast cluster of (1,2).

the same observation on the collection of scatterplots have the same color. When the cursor is moved, the coloring changes to reflect the change in the cursor position.

2.4 Linking, Deleting, and Labeling by Brushing

All of the methods for linking discussed in the previous section have certain limitations. The scatterplot matrix only allows partial linking. The M and N plot quickly becomes cluttered by the connecting lines. The unique plotting symbol method does not allow effective linking of groups of points and the overplotting of symbols obscures them. McDonald's method gives quite interesting results in some cases but has a limited domain of application because the coloring scheme does not provide a sufficiently incisive specification of regions of interest. Polygon drawing can provide incisive specification, but it is manually awkward and does not lead directly to a sufficiently rapid way of changing the region that is highlighted. In this section we will see how brushing overcomes these limitations.

Brushing is a dynamic method in which the data analyst moves a rectangle around the screen by moving a mouse [Becker and Cleveland, 1984 and 1987]. The rectangle, which is called the *brush*, is shown in the (3,2) panel of Figures 8 and 9. Brushing allows linking in such a way that a wide variety of data analytic tasks can be carried out. Also, brushing can be used to delete and label points. In the next sections we will describe the features of brushing: movement of the brush, the ability to change the shape of the brush, three *paint* modes (transient, lasting, and undone), and four *operations* (highlight, shadow highlight, delete, and label).

The Scatterplot Matrix

Brushing can, in principle, be done on any subset of the collection of $p(p-1)$ scatterplots of the p variables [Stuetzle, 1987]. However, if we are interested in all p variables and in their relationships, it is usually best to operate on the entire scatterplot matrix. This matrix has been a successful static graphical tool because it provides a highly integrated view of all of the data. As we saw earlier, it is possible to carry out, purely visually, a partial linking of points on different scatterplots, even without brushing. Furthermore, by scanning a row or a column, we can see one variable graphed against all others in a convenient and efficient way. Of course, if p or n are too large it may not be possible to put the full matrix on the screen and have good resolution; in such a case we have to be satisfied with looking at subsets of the variables.

Highlighting, Brush Shape, and the Three Paint Modes of Brushing

When the paint mode is *transient*, brushing with the *highlight* operation works as follows. A panel of the scatterplot matrix is selected as the active panel. All points on the active panel that are inside the brush are highlighted, as well as the points on other panels that correspond to these points. This is illustrated in Figure 8. The active panel is (3,2) and the brush is on this panel; it is covering the outlier on the sepal width and petal length scatterplot and so we can distinguish the points on other plots that correspond to the outlier. When the brush is moved, points that are no longer inside the brush are no longer highlighted and the new

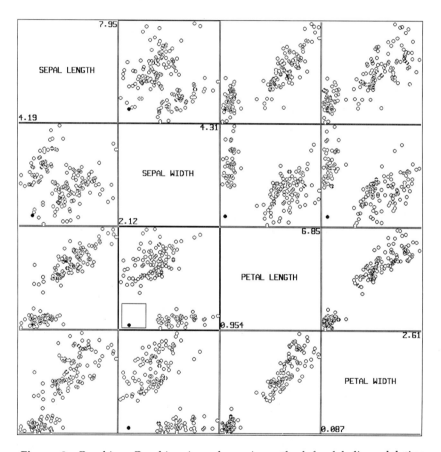

Figure 8. Brushing. Brushing is a dynamic method for labeling, deleting and linking by highlighting. The rectangle in the (3,2) panel is the brush, which the analyst moves around the screen by moving a mouse. When the operation is highlight, all points inside the rectangle on the active panel and all corresponding points on other panels are highlighted.

points inside the brush become highlighted; that is, the linked highlighting is transient. For example, in Figure 9 the brush has been moved to the right on the (3,2) panel and other points are highlighted.

Consider, on one of the panels of a scatterplot matrix, a subset of the points that cannot be enclosed in a rectangle without enclosing points not in the subset. With the transient highlighting just described, we cannot simultaneously highlight the points of the subset without also highlighting points not in the subset. An example is the northwest cluster in panel (1,2) of Figure 9. The problem is solved by the *lasting* paint mode: when one button of the mouse is pressed, points that are highlighted remain highlighted even after the brush no longer covers

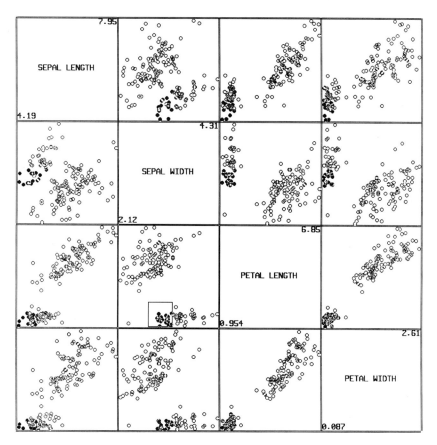

Figure 9. Transient Paint Mode. When the paint mode is transient, points previously inside the rectangle are no longer highlighted. In this figure the rectangle has been moved to the right from its position in the previous figure, and new points are highlighted.

16 DYNAMIC GRAPHICS FOR DATA ANALYSIS

them. Thus, an irregular cluster can be highlighted by using a small brush and moving it over all sections of the cluster as if it were being painted. This is illustrated by the sequence in the first column of Figure 10. Lasting highlight can be removed with the *undo* paint mode; when another button of the mouse is depressed, the effect of the brush is to reverse the highlighting of the points inside it. This is illustrated by the sequence in the second column of Figure 10.

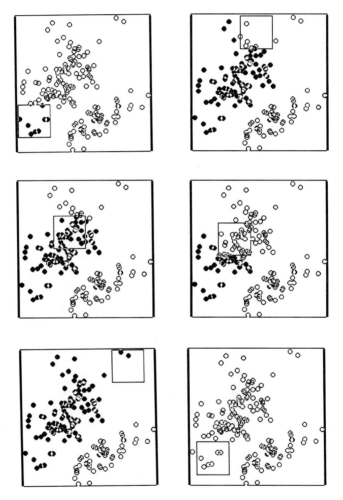

Figure 10. Lasting and Undo Paint Modes. When the paint mode is lasting, points remain highlighted. This enables highlighting of irregular clusters by painting, as illustrated by the sequence of panels in the first column of the figure. When the paint mode is undo, lasting highlight is removed. This enables removal of lasting highlight by painting, as illustrated by the sequence of panels in the second column of the figure.

Usually, lasting highlight is most conveniently carried out when the brush is in the form of a small square or small rectangle; this helps prevent the slopping of highlight onto unwanted points (although undo makes it easy to fix mistakes). A small brush is also useful for highlighting a single point, as in Figure 8. Other brush shapes are useful in other situations, as we shall now see.

Conditioning on a Single Variable

By making the brush long and narrow, we can highlight points for which the values of a particular variable are in a narrow range; thus, we can study the relationships of other variables for the selected variable held fixed to this range. This is illustrated in Figure 11. The data are from an experiment in which 30 rubber specimens were rubbed with an

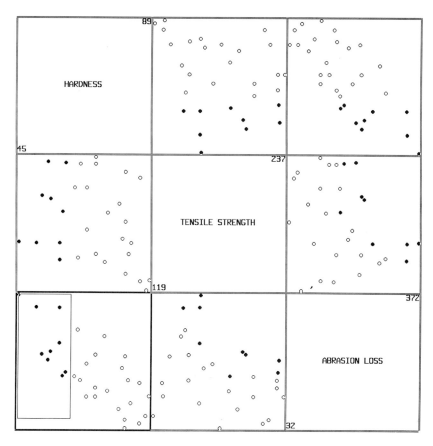

Figure 11. Conditioning. The brush is highlighting points with low values of hardness. The (3,2) panel shows the dependence of abrasion loss on tensile strength for hardness held fixed to low values.

abrasive material [Davies, 1957]. The abrasion loss (the amount of rubber rubbed off), the hardness, and the tensile strength of each specimen were measured. The purpose of the experiment was to see how abrasion loss depends on hardness and tensile strength. In the original analysis of these data, a linear regression model was fit. In Figure 11 the brush is selecting low values of hardness; from the highlighting in the (3,2) panel we can see that there is a *nonlinear* dependence of abrasion loss on tensile strength for hardness held fixed to low values. Figure 12 shows a sequence consisting of just the (3,1) and (3,2) panels of the matrix. As we go from top to bottom in the figure, the brush is moving from left to right in the (3,1) panel, conditioning on higher and higher values of hardness.

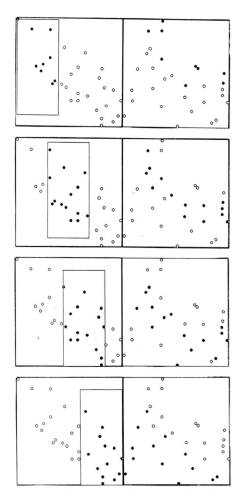

Figure 12. Conditioning. Each of the four pairs of graphs are the (3,1) and (3,2) panels of the scatterplot matrix in the previous figure. The brush is being moved from left to right on the (3,1) panel, conditioning on different ranges of hardness.

The highlighted points of the (3,2) panels generally form nonlinear patterns that appear to be just vertical translations of one another. Figure 13 is similar except that now we are conditioning on different ranges of tensile strength; the highlighted points of the panels on the left, which show the dependence of abrasion loss on hardness for tensile strength held fixed to the various ranges, also appear to be vertical translations of one another, but unlike those in Figure 12 they have a linear pattern.

Figures 12 and 13 suggest three things about the data. First, there is a nonlinear dependence of abrasion loss on tensile strength; the linear model used in the original analysis [Davies, 1957] appears incorrect. Second, there is a linear dependence of abrasion loss on hardness. Third,

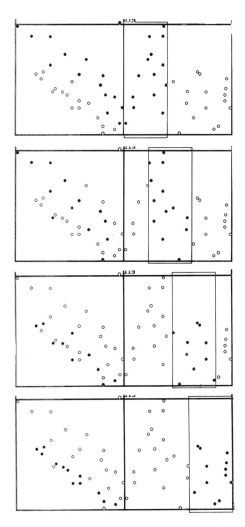

Figure 13. Conditioning. In this figure the brush is being moved on the (3,2) panel, conditioning on different ranges of tensile strength.

because of the vertical translation of patterns in Figures 12 and 13, a model for the data does not need a hardness and tensile strength interaction term. These conclusions have been confirmed by regression modeling of these data [Cleveland and Devlin, 1986].

Conditioning on Two Variables and the Shadow Highlight Operation

Figure 14 shows a scatterplot matrix with four variables: wind speed, temperature, solar radiation, and the cube roots of concentrations of the air pollutant, ozone. The original data, from a study of the dependence of ozone on meteorological conditions [Bruntz et al., 1974], were measurements of the four variables on 111 days from May to September of 1973 at sites in the New York City metropolitan region. There was one measurement of each variable on each day so the data consist of 111 points in a four-dimensional space. The analysis began with the cube root transformation of ozone because of the success of power transformations in previous analyses of air pollution data. Then a nonparametric regression curve was fit using locally weighted multiple regression, a local fitting procedure [Cleveland and Devlin, June 1988]; that is, cube root ozone was smoothed as a function of the meteorological variables.

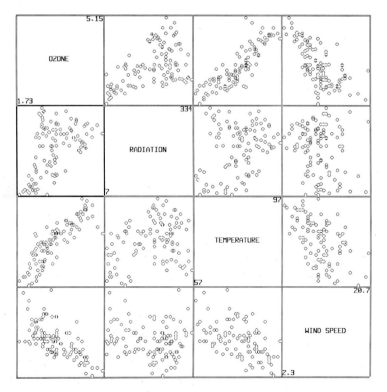

Figure 14. Scatterplot Matrix. Four variables are shown.

The data of the scatterplot matrix in Figure 14 are the three meteorological variables and the fitted, or smoothed, values of cube root ozone. Brushing was applied to these data to explore the fitted regression function, and to determine if there is appreciable nonlinearity in the surface; whether the nonlinearity exists is not immediately clear from the scatterplot matrix alone.

In Figure 15 highlighting is being carried out, but in a different way from that for the abrasion loss data that were analyzed in the previous section. First, although all data appear on the active panel, a scatterplot of radiation against temperature, only points that are highlighted appear on the other panels. This *shadow highlight* operation is useful in situations, such as this one, where the amount of overlap of the plotting symbols is so great that the highlighted symbols are hard to distinguish from those not highlighted. A second difference is that a square brush is being used in the active panel; thus, we are conditioning on values of radiation and temperature fixed to certain ranges. The points of the (1,4) panel

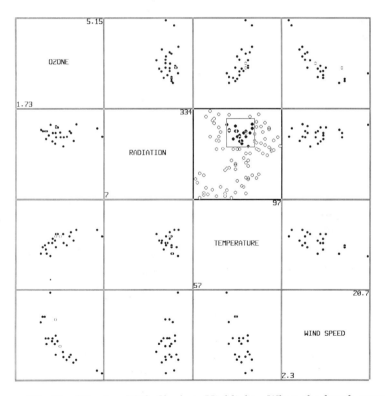

Figure 15. Conditioning With Shadow Highlight. When the brush operation is shadow highlight, all points on the active panel are shown, but on the other panels only highlighted points are shown. The points of the (1,4) panel show how ozone depends on wind speed for high values of radiation and middle values of temperature.

show us how the smoothed cube root ozone values depend on wind speed for the other two independent variables held fixed to these ranges. This two-variable conditioning can be done anywhere in the (2,3) panel to condition on other ranges of radiation and temperature, or can be done in the (2,4) panel or the (3,4) panel to condition on radiation and wind speed or temperature and wind speed.

When brushing is used in this way, conditioning on two variables, it becomes quite clear that the fitted surface is highly nonlinear for the ozone data. For example, in Figure 15, where we are conditioning on middle values of temperature and high values of radiation, the dependence of ozone on wind speed has substantial curvature.

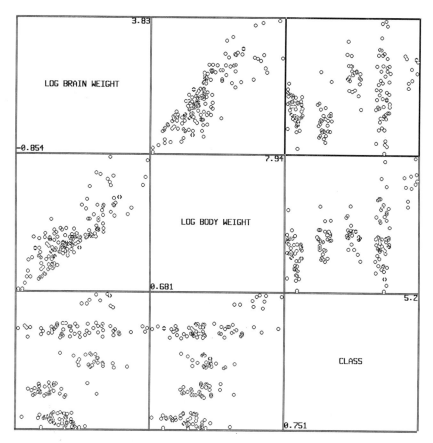

Figure 16. Alternagraphics by Brushing. Brushing can be used to show different data sets by adding a categorical variable that indicates to which set each observation belongs. This figure shows a scatterplot matrix for the brain and body weight data, including a categorical variable identifying each species as belonging to one of five categories of animals. From left to right on the (1,3) and (2,3) panels the categories are birds, fish, primates, nonprimate mammals and dinosaurs.

Brushing and Alternagraphics

Brushing in highlight or shadow highlight mode can also be used to perform alternagraphics. A categorical variable is defined that describes the subsets—for example, it takes on the value 1 if a point is in subset 1, 2 if it is in subset 2, and so on—and is plotted along with the other variables. Then, a highlighting of the different subsets can be gotten by applying conditioning to the categorical variable. It is also helpful to jitter the values of the categorical variable—that is, add small amounts of random noise—to reduce overplotting. This method is illustrated in Figures 16 and 17 for the brain and body weights discussed earlier. Figure 16 shows the full scatterplot matrix of the three variables: log brain weight, log body weight, and a categorical variable with jitter added that describes the species class. (From left to right, the classes are birds, fish,

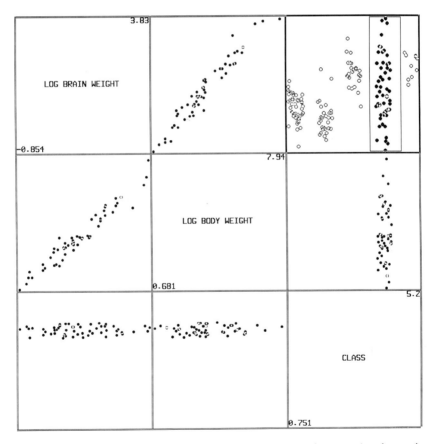

Figure 17. Alternagraphics by Brushing. Here one of categories (nonprimate mammals) of Figure 16 has been highlighted in shadow highlight mode. By choosing a tall narrow brush and moving it across the (1,3) panel, each category is momentarily visible on all of the plots.

primates, nonprimate mammals, and dinosaurs.) In Figure 17 the mode is shadow highlight, and the brush in the (1,3) panel is selecting the value of the categorical variable corresponding to nonprimate mammals, so the scatterplot in the (1,2) panel is the brain and body weights for this group of species. By moving the brush on the (1,3) panel we can quickly select any other group.

The Delete Operation

The *delete* operation is carried out in the same way as highlighting. With no buttons depressed, points inside the brush on the active panel are deleted in a transient way along with corresponding points on other

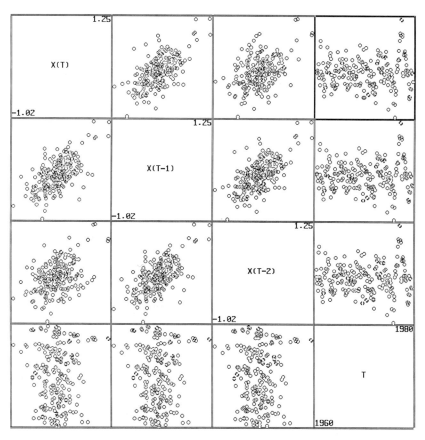

Figure 18. Time Series. A time series, $x(t)$, is explored by the scatterplot matrix. The (1,3) and (2,3) panels suggest that there is significant correlation at lags 1 and 2. However, panel (1,4) shows there is a peculiar excursion of the data over a short time interval toward the end of the data record.

panels. Pressing a mouse button causes the deletion to be lasting, and pressing another mouse button causes the deletion to be undone, that is, deleted points reappear. Figures 18 and 19 illustrate the use of brush deletion in the analysis of a time series. The figures are based on the residuals, $x(t)$, of monthly atmospheric concentrations of CO_2 [Keeling, Bacastow, and Whort, 1982] after trend and seasonal components were removed. The four variables in the scatterplot matrix are $x(t)$, $x(t-1)$, $x(t-2)$, and t. The (1,2) and (1,3) panels of Figure 18 suggest there are nonzero autocorrelations at lags 1 and 2. However, the (1,3) panel shows a peculiar excursion of the data near the end of the observation period. It is natural to ask what effect this excursion has on the autocorrelation. In Figure 19 a brush has been positioned in the (1,4) panel so that the points

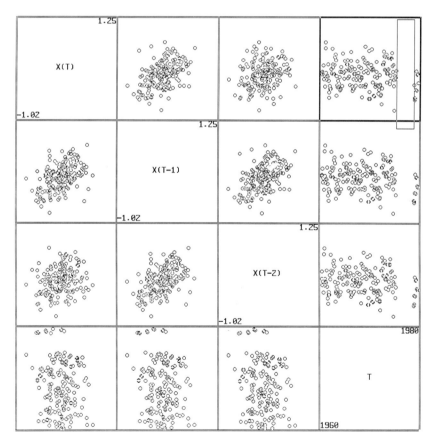

Figure 19. Deletion by Brushing. Points can be deleted with brushing in the transient or lasting paint mode, and lasting deletion can be undone by the undo mode. In this figure, the brush is removing the data of the short excursion. The autocorrelation at lag 2 has disappeared and the autocorrelation at lag 1 has been substantially reduced.

of the excursion are deleted. The autocorrelation at lag 1 is reduced and the autocorrelation at lag 2 disappears altogether; thus, much of the observed autocorrelation at these lags is induced by the data of the short, final time interval.

The Label Operation

The *label* operation is carried out in the same way as highlighting and deleting, that is, labeling can be transient, lasting, or undone. Figure 1, discussed at the beginning of the paper, illustrates the use of transient labeling.

2.5 Scaling

An important factor on a two-variable graph is the aspect ratio: the physical length of the vertical axis divided by the length of the horizontal axis. Changing the aspect ratio of a graph can have a radical effect on our ability to visually decode the quantitative information on the graph. Figures 20 and 21 are an example. The data are the yearly sunspot numbers from 1749 to 1924 that Yule [1927] analyzed in his landmark paper on autoregressive processes. In Figure 20 the aspect ratio is one and in Figure 21 it is 0.065. In Figure 20 we cannot see an important feature of the sunspot data that is apparent in Figure 21—the sunspot numbers rise more rapidly than they fall; that is, the absolute rate of change is greater when sunspot numbers are increasing than when they are decreasing.

In studying a two-variable graph, we often judge the slopes of line segments to determine the rate of change of y as a function of x. For example, in Figure 21 we judge the slopes of line segments connecting *successive* data points to determine that the sunspots rise more rapidly than they fall. In general the judgment of slope is enhanced when the angles of line segments with positive slopes are centered on 45° and when the angles of line segments with negative slopes are centered on −45° [Cleveland and McGill, 1987]. When slopes tend away from ±45°, either toward the vertical or horizontal, relative judgments of slope decrease in accuracy. Changing the aspect ratio of a graph changes the angles of line segments on the graph and therefore our ability to judge slopes. In Figure 20 we lose our ability to judge local slopes because the angles of the line segments connecting successive points are too close to ±90°.

For static graphs the aspect ratio can be chosen by selecting a set of line segments whose slopes are of interest and then selecting the aspect ratio so that the median of the absolute values of the angles of the segments is 45° [Cleveland and McGill, 1987]. But there are problems with this solution. One is that this optimum aspect ratio for one set of slopes

METHODS 27

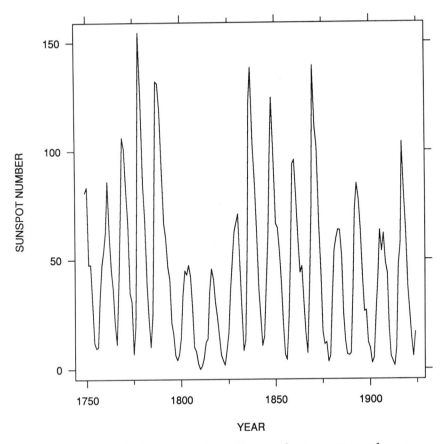

Figure 20. Dynamic Shape Control. The yearly sunspot numbers are graphed. The shape parameter of the graph—the vertical distance spanned by the data divided by the horizontal distance spanned by the data—is equal to one. From this graph we get a good portrayal of the variation in the peaks of the sunspot numbers but a poor portrayal of the shapes of the cycles.

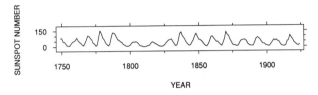

Figure 21. Dynamic Shape Control. The shape parameter of the graph is now 0.065 and we get a good portrayal of the shapes of the cycles—for example, we can see that the data tend to rise more rapidly than they fall—but a poor portrayal of peak-to-peak variation. Dynamic graphics can be used to continuously vary the shape parameter of a graph, quickly moving from this figure to the previous one.

might be quite different from that for another set of interesting slopes on the graph. Furthermore, optimum aspect ratios for many graphs, particularly those of long time series, are so small that for the usual page width (about 18 cm in this journal), the optimally scaled graph is too small in the vertical direction to have sufficient resolution along the vertical axis.

Dynamic scaling, in which the data analyst controls the length and width of the two axes, provides a convenient solution to these problems. The data analyst controls the aspect ratio and the computer makes the graph on the screen vary in a seemingly continuous way; the rapid and easy change from one aspect ratio to another allows accurate judgments of different sets of slopes. One particularly nice implementation of Buja et al., [1988] even solves the problem of small aspect ratios interfering with vertical resolution by using horizontal scrolling. To get a small aspect ratio without ruining the vertical resolution, the width of the graph is expanded beyond the width of the screen; the analyst sees only a portion of the graph at any moment and can scroll back and forth, bringing different parts of the graph into view in order to see all of it.

2.6 Rotation

Suppose x_i, y_i, and z_i, for $i = 1$ to n, are measurements of three variables. The data configuration is a point cloud: n points in three dimensions. The computer screen is two-dimensional, so we must be satisfied with inferring the three-dimensional structure from two-dimensional views. This, however, is not as limiting as it might seem because one of the wonders of the human visual system is that its algorithms for perceiving the three-dimensional world have as their input two-dimensional views falling on the retinas. Just as a motion picture gives a three-dimensional sense from pictures shown on a flat screen, so can a computer display convey three-dimensional point clouds by a rapidly changing sequence of two-dimensional views [Fisherkeller, Friedman and Tukey, 1975]. The point cloud is projected onto the screen while the viewing angle changes through time; the effect is an apparent rotation of the cloud that allows us to see its three-dimensional structure. Rotation is portrayed in Figure 22.

Control

There are many possibilities for selecting an axis about which the cloud rotates [Fisherkeller, Friedman and Tukey, 1975; McDonald, 1982; Tukey, 1987]. One is to use one of the three coordinate axes of the data. Rotation about one of them changes the orientation of the others from our fixed viewpoint—that is, the projection plane of the screen—so the apparent motion of the cloud about a given coordinate axis will vary according to the position of the other two axes. A second possibility for axis selection is to have three fixed screen axes, that is, axes that are fixed relative to the projection plane; the most sensible are screen horizontal, screen vertical, and perpendicular to the screen. Once one of these six axes has been selected, the data analyst can be given control of the sign and speed of rotation.

A third possibility for axis selection can lead to very delicate control of the cloud. The data analyst selects a particular direction in the projection plane by moving a mouse or a tracker ball. The motion of the cloud is what we would get if we reached out and spun the cloud by pushing on the front of it in the selected direction. If the amount of rotation depends on how far the device is moved, the data analyst has the feeling of very direct control of the cloud. The cloud can also be given inertia, so that once it is set in motion it continues until the analyst changes the motion.

It is entirely reasonable to implement all three of the above rotation-control methods in a dynamic graphics system and allow each data analyst using the system to discover his most comfortable control method.

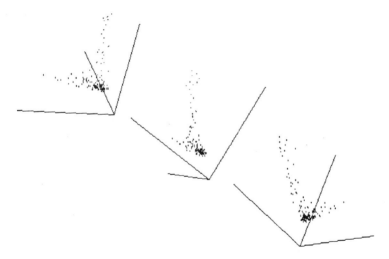

Figure 22. Rotation. Rotation on a computer graphics screen of measurements of three variables allows us to see the point cloud in depth, which provides a three-dimensional scatterplot. Rotation is a particularly effective method when we are searching for trivariate structure, such as clusters or manifolds, or when we want to study the local dependence of one variable on the other two. When the goal is to study conditional dependence, brushing is typically more effective. The portrayal of a rotating point cloud in this figure is from the MACSPIN manual [Donoho, Donoho and Gasko, 1985]. Reprinted with permission of D^2 Software, Inc.

Enhancements

Motion is only one factor that the human visual system uses to construct scenes; consequently, rotation by itself does not give as strong a three-dimensional view as we get of the real world. Remember that points in three-space are not typical of the real world—real objects are composed of surfaces, not points. However, certain enhancements to rotation can increase the perception of depth. First, we can add stereo vision [Becker, Cleveland, and Weil, 1988; Donoho et al., 1986; Littlefield, 1984]. One way to do this is to show two superimposed views of the cloud from two slightly different viewpoints, one view in red, the other in green, with overlap in yellow, and then view the display with glasses that have one red lens and one green. Each eye gets one view as in real-world vision. (There are other methods of generating stereo views that allow color in the pictures.)

Stereo vision enhances the three-dimensional effect of the rotating cloud but, even more importantly, the three-dimensional effect remains even when the motion stops. This is important for reasons that will be given shortly. Because our visual systems also use perspective to see depth, we can enhance point cloud rotation by having the sizes or intensities of the plotting symbols obey the rules for perspective. Another way to enhance the three-dimensional effect is to enclose the cloud in a rectangular box whose edges are the axes of the three variables; the box provides perspective, which enhances the depth effect, and also helps us perceive the axis directions.

Limitations

It should not be supposed that rotation and its enhancements give us for three variables exactly what scatterplots give us for two. On a scatterplot, we make a variety of judgments of horizontal and vertical positions to visually decode information [Cleveland, 1985]. Point cloud rotation and its enhancements add depth, but our judgment of a position in depth is not nearly as acute as our judgment of horizontal and vertical positions. Furthermore, when a point cloud is rotating it is easy to lose a strong cognitive fix on the axes and on which variable goes with which axis, even when the axes are shown. It is almost always the case in analyzing a set of data that recognizing some pattern in the data is not enough; we need to be able to interpret the pattern in terms of the measured variables. This is illustrated in the (1,2) panel of Figure 6. We can readily perceive from this scatterplot that there are two clusters of points, and we can almost instantaneously conclude that the cluster in the upper left consists of sepals that are shorter and wider than those in the lower right. For a scatterplot this process of interpreting a perceived pattern in terms of the measured variables is done so effortlessly that we are hardly aware that it is distinct from perceiving the pattern. For a rotating three-dimensional point cloud we are often painfully aware of the separation.

The problem is alleviated somewhat by having stereo; then the motion can be stopped and a particular view contemplated. But even with this enhancement, our cognitive processes are considerably less adept than they are for two dimensions.

Rotation With More Than Three Variables: Advanced Strategies

Suppose there are p variables where $p > 3$. We can study any three at a time by rotation. If we do this, without saving the results of a rotation in any way, we can effect rotation only about axes that are a linear combination of three variables. We can, however, adopt a replacement strategy, similar to one used in PRIM-9 [Fisherkeller, Friedman, and Tukey, 1975], that allows arbitrary rotations. Think of the procedure as a recursion with a succession of rotation sessions. After the kth session, suppose the data are described by a coordinate system with variables $z_1, \ldots, z_p$. Three of the variables, say z_1 to z_3, are selected for the next session. The data are rotated to get a final view for the session. The variables z_1, z_2 and z_3 are now replaced by a new set of three variables: the three that describe the screen axes. At the beginning of the sessions, the z_i are just the original data variables. Needless to say we rapidly lose a graphical appreciation of the meaning of identified structures in terms of the original variables. However, the system can keep track of the coefficients of the linear transformation of the z_i back to the original variables as way of helping us to infer meaning.

During these rotation sessions we can also condition on certain ranges of the z_i [Donoho, Donoho and Gasko, 1986; Fisherkeller, Friedman and Tukey, 1975]. For example, we might rotate with z_1, z_2 and z_3 and show only points for which $a < z_4 < b$, which is a form of deletion, or we might highlight points in this range. This allows us to study the relationship of z_1, z_2 and z_3 conditional on z_4 in the range a to b. In addition, we can hold the rotation fixed to a particular view and vary a and b.

Far more experimentation is needed with these advanced strategies. In applications, it is easy to lose all sense of where one is and what identified structures mean. Carefully worked-out examples, complete with strategies and what they show about the phenomenon under investigation, are needed.

Brushing vs. Rotation

Brushing and rotation can both be used for studying multidimensional data. Becker, Cleveland and Weil [1987] compared the two methods by implementing both on a high performance graphics workstation and using both to analyze a large number of data sets. We will not repeat the details of the investigation here, but will describe the basic conclusions.

Suppose the data are three-dimensional and the goal is to examine the trivariate structure. Simply because we can see directly the three-dimensional structure, rotation is particularly useful for exploring three-dimensional properties such as clusters and collections of points that lie close to one-dimensional and two-dimensional manifolds. Brushing tends to be less useful here, except in special cases.

Suppose the data are three-dimensional and the goal is to explore the local dependence of one variable on the two others. Again, rotation is a useful tool and brushing tends to be of less help.

Suppose the data are three-dimensional and the goal is to explore the conditional dependence of one variable on another for the third held fixed. Here brushing is particularly appropriate although rotation tends to perform less well. The reason is that the conditioning techniques of brushing allow us to incisively study conditional dependence, the usual goal, for example, when we do a regression analysis.

Once the data are in four dimensions or more the picture becomes somewhat less clear. If the goal is to study dependence of one variable on the others, brushing typically leads to more insight. For multivariate studies where understanding the p-dimensional structure is the goal, both methods can lead to separate and important insights. For example, when a cluster is definable by its projection onto the plane of two original variables, it can be explored through highlighting the cluster by brushing the scatterplot of the two variables. However, if the cluster is definable by a projection onto the space of three original variables, but not onto any plane of two of the variables, it is detectable only through rotation.

One strength of brushing for four or more variables is that the scatterplot matrix provides a single integrated view of the data unless, of course, a large value of n or p causes resolution problems. Even when several rotating clouds are on the screen, the same level of integration is not achieved. The integration allows a number of data analytic tasks to be carried out in a straightforward manner. An example is provided by the data of Figure 23, which are the logarithms of the weights of six body organs for 73 cardiomyopathic hamsters [Ottenweller, Tapp and Natelson, 1986]. The outlier in the (3,4) panel, which is being highlighted by the brush, suggests that one hamster has an unusually large spleen or an unusually small liver. By scanning the spleen plots in row 4 and the liver plots in row 3, we can see that the hamster has an enlarged spleen.

Thus, the overall conclusion is that neither brushing nor rotation uniformly dominates the other and it is quite reasonable to have both tools available. In fact, not only do brushing and rotation complement one another, they can be combined to create more than the sum of their parts [Becker, Cleveland and Weil, 1988; Stuetzle, 1987]. A subset of three of the variables is selected and rotation is applied to them on a panel adjacent to the scatterplot matrix. This panel takes part in the brushing operations just like the panels of the scatterplot matrix. When a matrix panel is brushed, the results are also shown on the rotation panel; furthermore, the rotation can be stopped at a particular view and the rotation panel selected as the master panel for the brushing operations.

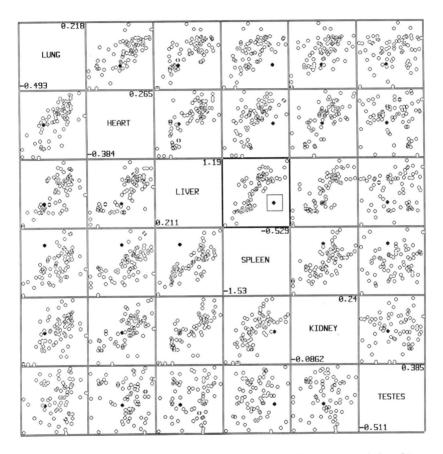

Figure 23. Four or More Variables. It is quite sensible to use both brushing and rotation when there are four or more variables. One reason that brushing is effective is the single integrated view provided by the scatterplot matrix. In this figure the logarithms of the weights of hamster organs are graphed. There is an outlier in the (3,4) panel, which is being highlighted by the brush. By scanning rows 3 and 4, we can see the hamster has an enlarged spleen.

2.7 The General Case: Parameter Control

Many of the methods described in the previous sections provide the data analyst with a method for rapidly changing a parameter that affects a graphical display. The following are examples: alternagraphics—the value of the categorical variable that breaks the data up into subsets; scaling—the height or width of the graph; rotation—angle of view. Dynamic methodology can in principle be used to control any parameter, discrete or continuous, that can affect a graphical display. Two further examples are the powers of power transformations of variables on a graph [Becker and McGill, 1985; Fowlkes, 1971] and the smoothness parameter of a nonparametric regression curve superimposed on a two variable graph [Friedman, McDonald and Stuetzle, 1982]. Thus, there are a vast number of methods that can potentially be implemented in a dynamic graphics environment.

3. COMPUTING ISSUES

This section focuses on some of the computing issues involved in *developing* a dynamic method. The discussion is needed because of two circumstances: current hardware must be pushed nearly to its limits to achieve the requisite speed for dynamic methods, and present software environments leave the programmer with many steps between the conception of a new idea and an easy to use implementation of it. With present technology, developing easy to use, responsive methods requires some knowledge of hardware and software issues.

Two aspects of the hardware for a graphics computing system have a large impact on the speed and ease of use of dynamic techniques: the bandwidth of the complete system and the specific input/output devices used in the system. These are discussed in Sections 3.1 and 3.2.

There are several levels of software used in the implementation of dynamic displays—low level software that operates on bits in frame buffers, intermediate level software that provides basic graphics library functions and high level software that provides an *environment* where data analysts can make good use of dynamic graphics. The levels of software needed to generate a particular dynamic display are determined by many factors including the complexity of the display and the hardware on which it will run. A simple dynamic method may be trivial to implement on very powerful hardware. Conversely, a graphical method requiring fast display of large amounts of data on an inexpensive computer system will probably force the programmer to use many low level graphics algorithms in order to get adequate performance.

Low level graphics techniques are described in Section 3.3. These are the techniques typically used by programmers who need maximal control over the graphics display because higher level software either does not perform the right task or does not perform the task with enough speed. We then discuss higher level, device-independent techniques in Section 3.4 that are easy to use and are accessible to programmers and to data analysts with programming experience. Finally, in Section 3.5, we take up the data analysis environments that make dynamic graphics displays available to data analysts.

3.1 System Bandwidth

Bandwidth is the speed with which user input, such as mouse moves and button presses, can be translated into picture updates on the screen. There are several steps in this translation, any of which can be the bottleneck that determines the overall bandwidth. Composite speed may be limited by the speed of numerical or graphics computations, by the speed of the display hardware or by the speed of passing information from input device to computer or from computer to display device. It is hard to quantify the required overall speed, but as a general rule of thumb, the bandwidth should be sufficiently high that the *perceived* response of the display to the input device is virtually instantaneous. McDonald and Pedersen [1985a] attempt to quantify hardware bandwidths of the various components of modern workstations, and also provide a general background on hardware from a data analyst's viewpoint.

Roughly speaking, the hardware necessary for doing dynamic graphics consists of computing horsepower sandwiched between input and output devices. The ways in which such a complete system is typically arranged can be broadly characterized as either the *time-shared* model, the *workstation* model or the *distributed processing* model, which is often a hybrid of the other two. Each organization has implications for dynamic graphics.

In the time-shared model (Figure 24a), the user has a terminal that is attached to a host computer. The terminal's display and input device are controlled by a fixed *terminal program* that runs in a processor inside the terminal. Because the terminal program is fixed, graphics capabilities will be controlled from the host computer by relatively primitive commands. This means that dynamic graphics techniques are limited in this model by the amount of information that can be exchanged between the host and the terminal; typical host-terminal communication speeds are relatively low. For example, to display a thousand points on a terminal might typically require sending four bytes per point. The communication channel between terminal and host rarely transmits more than 2000 bytes per sec, so that each display of a thousand points would take a minimum of 2 sec to send; this is much too slow for dynamic graphics. Furthermore, although host computers in a time-shared environment may have a fast processor, the response time for a particular user can be slow because of other users employing the system. Because of these speed problems, the time-shared model is generally unsatisfactory for dynamic graphics.

In the workstation model (Figure 24b), the data analyst has the sole use of a computer. In order to be usable for dynamic graphics this computer should have, as a rough minimum, a processor that can perform one million instructions per second, a main memory capacity of one megabyte, a minimum screen resolution of about 500 by 500 and a high speed network connection to access a disk or other workstations—this connection should have a minimum speed of about one million bits per

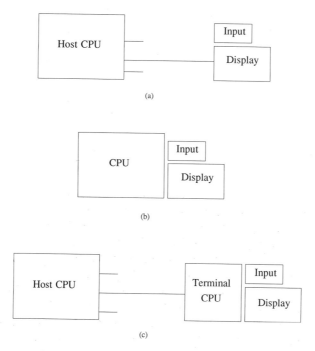

Figure 24. Models for Graphics Systems. Three models that are used for computing systems: (a) the time-shared model, in which a host computer is shared among several users, each with a terminal that has a fixed, relatively simple terminal program; (b) the workstation model, in which the user has the exclusive use of an entire computer (but perhaps sharing resources such as disk); and (c) the distributed computing model, in which the graphics computations are split between a shared host computer and a dedicated, programmable local computer.

sec. These requirements are discussed in more detail by Morris *et al.* [1986]. The great advantage of the workstation model is that because the processor, the graphics display and the input device are all housed together, they can be connected by high speed communications channels. Also, because the computer is dedicated to one person, there are no time-sharing delays. Furthermore, workstations are becoming affordable for most data analysts.

The distributed processing model (Figure 24c) is often a hybrid of the time-shared and workstation models. The user has a programmable computer with a graphics display and this local computer is attached to a time-shared host computer. Examples of distributed configurations include the AT&T Teletype 5620 programmable graphics terminal and Sun Microsystems' NeWS window system. In this model, the graphics computations are carried out in several processors and the distribution of the computational tasks is under control of the software rather than the hardware. The information flow between the host computer and the local graphics computer may contain programs as well as data. Typically, extensive computations are carried out at the more powerful host machine and time-critical graphics operations are performed by the local graphics machine.

There are several advantages of the distributed processing model over the workstation model. One is that the local graphics computer can be totally dedicated to a particular graphics task, although a workstation will also typically have to perform other tasks associated with a sophisticated operating system. Because the cost of computing is shared among a number of users on a time-shared host, more computing power may be available to the person with the large host and local graphics processor than the person with the workstation. This has an impact on any preprocessing that needs to be done for a dynamic technique. Also, the host machine gives the user the advantages associated with a shared environment, such as access to the files of others (although, with the advent of network file systems, this is also available in the workstation model).

The chief disadvantage of the distributed processor model is that software design is more complicated. The implementation requires part of the software to be running on the host and part to be running in the local machine; deciding how to do this partitioning can be delicate [Pike, Guibas and Ingalls, 1984].

3.2 Graphics Input and Output

Most current systems use a *mouse* for graphics input and a *raster scan display* for graphics output. A mouse is a small hand-held device that keeps track of relative changes in its position on a flat surface. These changes are continuously sent to the computer, which is then responsible for keeping track of the position of the mouse. The mouse is typically equipped with one, two or three buttons and the state of those buttons, pressed or not, is also continuously available to the computer.

A raster scan display has much in common with a television. It has an electron gun behind the screen that is continuously re-exciting spots on the screen with short persistence, i.e., that quickly fade if not re-excited. The electron gun follows a fixed scanning pattern from left to right, top to bottom. The value for each spot on the screen (on/off or a color) is stored elsewhere in fast memory (called a *frame buffer*) and these values are read in synchrony with the scanning of the electron gun.

An asset of raster scan devices is their low cost. Raster scan hardware has two parts, the drawing hardware (screen, electron gun, etc.) and the associated frame buffer that describes the image. The drawing hardware is inexpensive because it is little different than television technology. Raster scan was once expensive because of the cost of the associated memory. Plummeting memory prices, however, have changed this. For example, a medium resolution color screen might have a 1024 by 1024 array of *pixels* (a pixel, or picture element, is a frame buffer location,

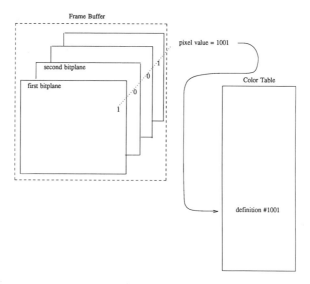

Figure 25. Color Displays. Relationships between the frame buffer, bitplanes and color map in a typical color device are shown here. The frame buffer may be viewed as a stack of bitplanes, here four. Reading the bits in the same position for each plane gives the value of the screen pixel at that position. The pixel value is used as an index into a table of color definitions, each of which may be set independently of the values in the frame buffer.

corresponding to an individual spot on the screen) with 8 bits of information stored for each pixel, and this requires one megabyte of memory. In the past 5 years, the cost of memory has decreased 10-fold. This trend, more than any other, has accounted for the recent upsurge in the popularity and availability of raster scan devices.

Raster scan devices also have added flexibility in the frame buffer for color displays. In these displays, each screen pixel is associated with a frame buffer location where several (usually from 4 to 32) bits of information are stored. Most modern color displays use these bit patterns as indices into a *color map*: a table of color definitions. The arrangement of the frame buffer—several bits for each pixel—can be viewed alternatively by considering, say, the first bit for each pixel as a collection; this rectangular collection of bits is called the first *bitplane*, and the other bitplanes are similarly defined (see Figure 25). Most devices, under control of a numerical *read mask*, have the ability to display any subset of the bitplanes at any time.

Some hardware offers the ability to *pan* and *zoom*. When the display hardware is reading the frame buffer, it can be set up to begin its scan anywhere in the frame buffer—the visual effect is to shift the picture horizontally or vertically (or both) with wraparound. In addition, the hardware can be instructed to replicate the values it reads from the frame buffer so that what would have been one pixel becomes a block of identical pixels on the screen. The visual effect is to zoom in on a part of the picture.

All of these features—color maps, bitplanes, zoom and pan—can be used separately or together to make rapid display changes, and are thus good grist for the dynamic graphics mill. There are some subtle tradeoffs involved between them, however. For a good discussion of the techniques and tradeoffs, see Heckbert [1984].

3.3 Low-Level Graphics Algorithms

It is often necessary to customize software to particular hardware in order to get maximum performance from a graphics device. Performance is essential to the success of dynamic methods and often the hardware used for dynamic graphics is just barely capable of achieving adequate performance levels. Even with very powerful hardware, it usually takes little time before data analysts push the limits of the hardware by wanting to increase the number of points that are displayed, to add more complex numerical methods to the display computations or to complicate the methodology in other ways. It is easy to push even the most powerful hardware to its limit. The remainder of this section describes some of the low level algorithms useful for squeezing graphics performance out of various types of hardware.

Integer Arithmetic

One technique applicable to many display devices is the use of integer arithmetic for time-critical computations. This is especially useful on devices without floating point hardware that would otherwise need to use (slow) software floating point routines. Another advantage of integer arithmetic is that it tends to match the resolution available on graphics devices. Display devices often have vertical and horizontal resolutions of fewer than 1000 pixels. This means that computations accurate to 10 bits are often sufficient for graphics; these can be done with hardware supported 16-bit integer computations. This integer arithmetic can be fast and can also save space. On the other hand, more programming care is required in translating ideas that are naturally expressed as floating point computations into equivalent integer computations.

Precomputation

Tradeoffs of speed versus memory usage are generally as applicable to graphics as they are to other computer algorithms. Time versus memory tradeoffs can be made by using precomputations; for example, sines and cosines can be precomputed and stored in a table so that trigonometric computations can be avoided during the display. Other tradeoffs involve precomputed pictures.

Suppose we would like to produce an alternagraphic display [Tukey, 1982]. Sometimes, precomputing the pictures is the easiest method of accomplishing this; in this case, the length of time it takes to generate the pictures is then not linked to the speed at which they can be displayed. When the displays are precomputed and then shown, the procedure is usually referred to as an *animation sequence*. Such a sequence might be used to show the contours of a surface varying in time or to show successive slices through a three-dimensional set of data. In these cases the sequence can rapidly show (say from 3 to 30 pictures per sec) a series of pictures that change slightly from one to the next, giving the viewer the impression of a continuously varying display. Alternagraphic displays also cycle through a set of pictures, although typically more slowly (a second or so per picture), and in this case, the various pictures would usually change substantially from frame to frame.

One of the major advantages of using precomputed pictures is that there is no need to write special purpose software to create the pictures to be displayed. All that is necessary is to augment the available graphics software so that it can draw pictures in portions of selected bitplanes. Once the pictures are drawn, then it is a simple matter to write a program to display them alternately on the screen. The precomputed pictures can also be stored offscreen and then rapidly copied into the active memory when they are to be displayed. Later, we will discuss operations for copying bitmaps. More complicated schemes are possible on devices with color maps, bitplanes, zoom and pan [Heckbert, 1984].

An example of the use of precomputation for dynamic graphics involves producing a power transformation display. Suppose we have paired observations x_i and y_i for $i = 1, \ldots, n$. A power transformation can often be used to linearize a curved plot, to stabilize variance, or to enhance symmetry. It is sometimes difficult to tell what will happen if we transform both variables simultaneously. Therefore, we want to allow two-dimensional motion of an input device to select transformation powers α and β and to show the changing scatterplot of y^β against x^α. In order to prevent scale changes from affecting the plot, we show $y_i^\beta/(\max(y)^\beta - \min(y)^\beta)$ against $x_i^\alpha/(\max(x)^\alpha - \min(x)^\alpha)$. We also need to make corrections for the sign of α and β as well as use a logarithmic transformation for $\alpha = 0$ or $\beta = 0$.

Suppose the hardware configuration is one in which computation is not enormously fast but there is a fair amount of memory. Straightforward computations would require floating point arithmetic to compute $\exp(\alpha \log(x_i))$ and $\exp(\beta \log(y_i))$ for each data point each time α and β changed. This is a lot of computation to do at a rate of approximately 10 times per sec, even for a moderate amount of data, say several hundred observations.

We could use tables of $\log(x_i)$ and $\log(y_i)$ to speed up this computation. However, by carrying precomputation just one step further, all floating point arithmetic can be avoided during display. Select a set of reasonable α values; for example, the 81 values $\alpha_j = -4$ (.1) 4. Next, compute a matrix with n rows and 81 columns where the i, j entry is $x_i^{\alpha_j}/(\max(x)^{\alpha_j} - \min(x)^{\alpha_j})$. Compute a similar matrix for y and β values. (We cannot compute just one table for grouped x and y values because the denominators differ for x and y.) To display the transformed scatterplot, it is simple to find the column of the x matrix corresponding to the current value of α and the column of the y matrix corresponding to the current value of β. All that the device has to do is quickly produce the scatterplot from the precomputed values. Notice that it would probably be infeasible to precompute all possible 81^2 pictures in advance; instead we have stopped just short of producing pictures during the precomputation. The important feature of this problem that allows us to succeed is that the power transformation can be decomposed into separate operations on the x and y coordinates.

Raster-Op

Recently, bitmapped display devices have become very popular. Most graphics on these devices is carried out by one very powerful operator, known as *raster-op* or *bitblt* [Newman and Sproull, 1979; Pike, Guibas and Ingalls, 1984]. Many bitmapped display devices provide special hardware to assist raster-op.

The raster-op operator takes two rectangular arrays of bits, the *source* and the *destination*, and combines them using various functions. The function *store* copies the source rectangle to the destination (ignoring any bits that were present in the destination); the *or* function produces a one-bit wherever a one-bit appears in either the source or the destination; *exclusive-or* (XOR) produces a one bit whenever the source and destination bits are not the same. See Figure 26.

Raster-op using XOR is widely used in bitmap graphics because it is its own inverse. If a source image has been copied onto the destination using XOR, the image will be erased completely if it is copied a second time using XOR. Of course, XOR is not without its drawbacks. When a picture is drawn using XOR to position a number of objects on the plot, any overlapping portions of different objects are invisible. This leads to strange effects in scatterplots in which there is a lot of overlap. In fact, if an even number of points in a plot have exactly the same coordinates, they completely obscure each other. This makes it important to use jittering techniques in which small amounts of random noise are added to data points to avoid exact overlap [Chambers *et al.*, 1983] prior to making displays that are done with XOR mode.

To see how a dynamic display may be implemented on a bitmapped device, suppose we want to display three-dimensional data by means of rotation. Suppose, too, that the bitmapped display device has only moderate computational power and no rotation hardware. What programming techniques can we use to produce an effective display?

First, we display the points on the screen in their initial position by repeatedly using raster-op to copy an offscreen image of a plotting symbol onto the screen at plotting positions of the points. Next, we must compute the new position of each point as the point cloud turns through a small angle. This can be done by computation of a rotation matrix using trigonometric functions, composition of the rotation with the previous rotation matrix and multiplication of the data by the composite matrix. However, an approximation to these computations can be done very quickly in integer arithmetic using only shift and add instructions. This technique was used in a version of point cloud rotation by Andrews [1981] to achieve point cloud rotation on a small and slow 8-bit personal computer.

We can move each point from its present position to its new position (after rotation) by means of raster-op. For each observation we use raster-op with XOR to copy an offscreen image of the plotting symbol once again into its current position (thus deleting the point) and then use XOR again to copy the image into its new position. Thus, we delete the first point and then redraw it, delete the second point and redraw it, and so forth through all of the points. This allows the display to move without ever visibly erasing and redrawing, although the rotating point cloud may not appear perfectly rigid when the number of points gets large.

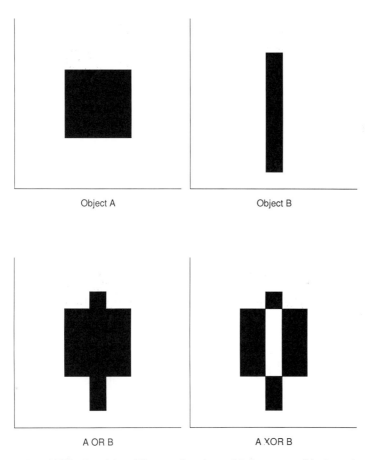

Figure 26. XOR Graphics. The result of combining two objects using the OR and XOR raster-op operations. Black represent one-bits and white represents 0-bits. Object A OR B has black wherever either A is black or B is black. Object A XOR B has black wherever either A or B but not both, is black.

Raster-op can also be used with extra memory to get what appears to be an instantaneous change from one rotation angle to another. Memory is allocated offscreen for a bitmap the same size as the onscreen bitmap. The points are drawn into the offscreen bitmap in their new position after rotation, and when all of the points have been drawn, the offscreen bitmap is copied into the onscreen bitmap using the raster-op with the STORE function. This may be faster than incremental plotting techniques when there are a large number of points on the display, because it is only necessary to draw each point once. Incremental techniques undraw the point and then redraw it in a new position, operating on each point twice. The drawbacks to using offscreen memory are that it requires a potentially large offscreen bitmap and it may be somewhat slower for small numbers of points because the offscreen memory must be cleared and then copied into the frame buffer.

3.4 Device-Independent Graphics: Graphics Standards

It may take specialized programming techniques to get adequate performance for dynamic displays on commonly available graphics devices. Also, although some of the low level operations we described in that section are common to a number of graphics devices, actual implementations usually vary slightly from machine to machine. This means that dynamic methods whose algorithms use low level routines are usually difficult to "port" from one variety of hardware to another.

The Graphical Kernel System, known as GKS [GKS, 1985], is now an international graphics standard. There are also a number of implementations of the ACM-SIGGRAPH CORE proposal [CORE, 1977]. These standards describe device-independent low level support for a variety of graphics input and output devices. They provide device independent ways of drawing lines, markers (point symbols), polygonal areas and characters. They provide a mechanism to translate from two-dimensional or three-dimensional *world coordinate systems* into coordinate systems appropriate for the graphics device. That is, they allow the programmer to specify how the natural units of measurement for the data are to be mapped onto the surface of the graphics device. The lines, markers, polygons and characters can be specialized by means of various graphical attributes such as line texture, width, color and character size, rotation and font.

Dynamic aspects of graphics are provided in the standards by a mechanism known as *segments*: a segment contains descriptions of graphical objects and also has attributes of its own such as visibility and a transformation matrix that describes how its coordinate system is to be mapped onto the screen. Graphic input is handled by a number of symbolic input types: button, valuator, locator, stroke and pick. Buttons can be pushed, valuators provide a single (continuously variable) value, locators provide single (x, y) coordinate pairs, strokes give a stream of coordinate pairs (corresponding, for example, to the positions of a mouse over time) and a pick identifies a particular segment that is visible on the display.

Data analysts should not think of graphics standards as a panacea, for there are many issues in data graphics that are not addressed by any of the proposed standards. In order to have a well defined scope, the current graphics standards have decided that they will not include applications software. This means that there is no explicit recognition of an applications area, such as data display, within the graphics standards, although the standards provide a foundation upon which good data graphics can be built. Much of the impetus for graphics standards comes from high visibility, high demand applications such as computer-aided design. Therefore, most of the effort of the computer graphics community is directed toward the realistic display of three-dimensional objects and scenes. The graphics standards contain many hooks for the display of such objects (three-dimensional perspective display of shaded surfaces, hidden surface elimination, etc). These are nontrivial problems and having good solutions in standard graphics software is quite valuable for typical applications, but the value of these features for data display applications is often much less.

Earlier we discussed how rotation could be implemented on a bit-mapped display device using raster-op as the primary graphical operation. Let us try to see how graphics standards could help us to implement rotation. Using the CORE system (GKS does yet not have three-dimensional primitives), this is easy. We create a segment that uses a three-dimensional *polymarker* (points in three-space) subroutine to display our points, and then, in a loop, we change the transformation matrix to give different views of our points. That is all there is to it.

On present hardware, the results range from beautiful to disastrous. The beautiful results come on devices where the hardware provides assistance in rotating objects. On these machines, the rotation is convincing even with a large number of data points. The disappointing results come on lower cost display devices that do not have special hardware for rotation. On these devices, the program may draw the points in one position, then erase the screen, then redraw the points in a new position. A cycle like this can take a second or two for even just a few data points. There is no illusion of a rotating three-dimensional point cloud. On the same device, the bitmapped approach can give very good results; the problem is that the operations available in the graphics standard do not map into very efficient operations on the low cost display devices. Of course, as hardware costs continue to decline, the performance of inexpensive graphics devices using standard graphics software will continue to improve. Nevertheless, it will be several years before adequate dynamic graphics is commonly available through the use of standard software.

3.5 Data Analysis Environments: Windows to the World of Data

Dynamic graphics routines, to be useful, must be embedded in a data analysis environment, which must include much more than the dynamic graphics routines. A data analyst at any time may require data management, basic graphics, statistical modeling, computations for transformations and a variety of other functions. The process of data analysis consists of jumping rapidly from one task to another as the patterns revealed in the data by one method suggest the use of another method. Thus, it is critical that dynamic methods be fully integrated with the other tools that the data analyst requires. Anything that makes it difficult to perform these rapid jumps impedes the progress of the analysis. In the past, much of the software for dynamic displays of data was implemented as standalone packages. This is appropriate for research into dynamic displays, but it is quite inadequate for use by working data analysts.

Multiwindow displays are a relatively recent development of the computer science community [Morris et al., 1986; Pike, 1983] that are particularly useful as part of a data analysis environment. Windows divide the surface of the display into rectangular regions, each of which operates independently. Most often the windows behave like separate terminals connected to the computer. A user can instigate a long computation in one window and then move to another window and continue to use the machine even as the long computation continues. The multiple windows allow the user to move between various tasks, just as a multitasking operating system allows the computer to work on several jobs at the same time.

The window environment is particularly well suited to data display because each graph can be in a separate window [Stuetzle, 1987]. The collection of windows can be moved around on the screen, much as one would organize graphs on pieces of paper on a desk. But unlike paper graphs on a desk, the shape of windows can be changed, so a graph can be made small for temporary storage in an out-of-the-way place (or made to disappear until recalled) or a graph can have its aspect ratio changed (see Section 2.5) by changing the shape of the window. Furthermore, corresponding points on different graphs can be linked so that operations, such as brushing, performed on points in one window are reflected in other windows.

4. SUMMARY AND DISCUSSION

The amazing computing power now available to data analysts carries with it the potential for new graphical methods—dynamic graphics—that utilize visual input and achieve virtually instantaneous graphical change. High interaction methods represent a new frontier in data analysis and are an important adjunct to conventional static graphics.

Because many early dynamic graphics systems were based on rotation, they are often thought of as tools for exploring multivariate data. However, dynamic graphics methods have the potential to include all areas of data analysis. This is demonstrated by the wide range of methods discussed in Section 2.

The chief reason for the advent of dynamic graphics is the current availability of the hardware and software tools needed for their implementation. Section 3 gave an introduction to these tools, illustrating both the opportunities they afford and the limitations they impose.

One question that frequently arises about dynamic methods is how a data analyst can present results that are based on them. The methods are discovery tools that enable us to uncover interesting structures in data, but once having learned about a particular structure we can almost always design a static display to show it. This was done in this paper for several data sets. In fact, dynamic graphics can be viewed as methods that allow us to move rapidly through a long sequence of static views until the desired or interesting static view is found; having found it we can present it to others.

It is quite easy to invent a dynamic graphical method. It is very difficult to invent one that is a better tool than those already available. This paper presents basic ideas of dynamic methods and computing principles that should allow others to design new and effective dynamic methods for data analysis. The challenge is to use the known techniques as a starting point, to develop new graphical methods and to evaluate those methods as they are used on real data.

Acknowledgments

Jon Bentley, John Chambers, John Hartigan, Daryl Pregibon, Werner Stuetzle and two referees gave us many useful comments on the manuscript.

REFERENCES

Anderson, E. (1935). "The irises of the Gaspé peninsula," *Bulletin of the American Iris Society*, **59**: 2-5.

Andrews, D. F. (1981). "Statistical applications of real-time interactive graphics, *Interpreting Multivariate Data*, V. Barnett, ed., 175-185. Chichester: Wiley.

Becker, R. A. and Cleveland, W. S. (1984). "Brushing a scatterplot matrix: high interaction graphical methods for analyzing multidimensional data," Technical Memorandum. Murray Hill, NJ: AT&T Bell Laboratories.

Becker, R. A. and Cleveland, W. S. (1987). "Brushing scatterplots," *Technometrics*, **29**: 127-142.

Becker, R. A., Cleveland, W. S., and Weil, G. (1988). "The use of brushing and rotation for data analysis," *Dynamic Graphics for Statistics*, W. S. Cleveland and M. E. McGill, eds., 247-275. Pacific Grove, CA: Wadsworth & Brooks/Cole.

Becker, R. A., and McGill, R. (1985). "Dynamic displays of data (video tape)." AT&T Bell Laboratories, Murray Hill, New Jersey: AT&T Bell Laboratories.

Bruntz, S. M., Cleveland, W. S., Kleiner, B., and Warner, J. L. (1974). "The dependence of ambient ozone on solar radiation, wind, temperature, and mixing height," *Symposium on Atmospheric Diffusion and Air Pollution*, 125-128. Boston: American Meteorological Society.

Buja, A., Asimov, D., Hurley, C., and McDonald, J. A. (1988). "Elements of a viewing pipeline for data analysis," *Dynamic Graphics for Statistics*, W. S. Cleveland and M. E. McGill, eds., 277-308. Pacific Grove, CA: Wadsworth & Brooks/Cole.

CORE (1977). "Status report of the graphics standards planning committee of ACM/SIGGRAPH," *Computer Graphics, SIGGRAPH '77 Conference Proceedings*, **11**(3). New York: Association for Computing Machinery.

Chambers, J. M., Cleveland, W. S., Kleiner, B., and Tukey, P. A. (1983). *Graphical Methods for Data Analysis*. Monterey, CA: Wadsworth.

Cleveland, W. S. (1985). *The Elements of Graphing Data*. Pacific Grove, CA: Wadsworth.

Cleveland, W. S. and Devlin, S. J. (June 1988). "Locally-weighted regression: an approach to regression analysis by local fitting," *Journal of the American Statistical Association*, **83**.

Cleveland, W. S. and McGill, R. (1987). "Graphical perception: the visual decoding of quantitative information on displays of data (with discussion)," *Journal of the Royal Statistical Society, Series A*, **150**: 192-229.

Crile, G. and Quiring, D. P. (1940). "A record of the body weight and certain organ and gland weights of 3690 animals," *The Ohio Journal of Science*, **15**: 219-259.

Davies, O. L., ed. (1957). *Statistical Methods in Research and Production*, London: Oliver and Boyd.

Diaconis, P. and Friedman, J. H. (1980). "M and N plots," Technical Report No. 151. Stanford, CA: Stanford University, Department of Statistics.

Donoho, A. W., Donoho, D. L., and Gasko, M. (1986). *MACSPIN Graphical Data Analysis Software*, Austin, Texas: D^2 Software, Inc.

Donoho, D. L., Huber, P. J., Ramos, E., and Thoma, H. M. (1986). "The man-machine-graphics interface for statistical data analysis," *Statistical Image Processing and Graphics*, E. J. Wegman and D. J. DePriest, eds., 203-213. New York: Marcel Dekker.

Fisher, R. A. (1936). "The use of multiple measurements in taxonomic problems," *Annals of Eugenics*, **7**: 179-188.

Fisherkeller, M. A., Friedman, J. H., and Tukey, J. W. (1975). "PRIM-9: an interactive multidimensional data display and analysis system," *Data: Its Use, Organization, and Management*, 140-145. New York: The Association for Computing Machinery.

Fowlkes, E. B. (1971). "User's manual for an on-line interactive system for probability plotting on the DDP-224 computer," Technical Memorandum. Murray Hill, NJ: AT&T Bell Laboratories.

Friedman, J. H., McDonald, J. A., and Stuetzle, W. (1982). "An introduction to real time graphics for analyzing multivariate data," *Proceedings of the Third Annual Conference and Exposition of the National Computer Graphics Association*, **1**: 421-427.

GKS (1985). *Graphical Kernel System (GKS)*, ANSI X3.124-1985. American National Standard Institute, New York. New York: The American Society of Mechanical Engineers.

Gould, S. J. (1979). *Ever Since Darwin: Reflections in Natural History*. New York: Norton.

Heckbert, P. S. (1984). "Techniques for real-time frame buffer animation," *Computer FX'84*, 57-67.

Huber, P. J. (1983). "Statistical graphics: history and overview," *Proceedings of the Fourth Annual Conference and Exposition of the National Computer Graphics Association*, 667-676. Fairfax, VA: National Computer Graphics Association.

Jerison, H. J. (1955). "Brain to body ratios and the evolution of intelligence," *Science*, **121**: 447-449.

Keeling, C. D., Bacastow, R. B., and Whort, T. P. (1982). "Measurement of the concentration of carbon dioxide at Mauna Loa Observatory, Hawaii," *Carbon Dioxide Review: 1982*, W. C. Clark, ed., 377-385. New York: Oxford University Press.

Littlefield, R. J. (1984). "Basic geometric algorithms for graphic input," *Proceedings of the Fifth Annual Conference and Exposition of the National Computer Graphics Association*, 767-776. Fairfax, VA: National Computer Graphics Association.

McDonald, J. A. (1982). "Interactive graphics for data analysis," Technical Report Orion 11. Stanford, CA: Stanford University, Department of Statistics.

McDonald, J. A. and Pedersen, J. (1985a). "Computing environments for data analysis, Part I: Introduction," *SIAM Journal on Scientific and Statistical Computing*, **6**: 1004-1012.

McDonald, J. A. and Pedersen, J. (1985b). "Computing environments for data analysis, Part 2: Hardware," *SIAM Journal on Scientific and Statistical Computing*, **6**: 1013-1021.

Morris, J. H., Satyanarayanan, M., Conner, M. H., Howard, J. H. Rosenthal, D. S. H., and Smith, F. D. (1986). "Andrew: a distributed personal computing environment," *Communications of the ACM*, **29**: 184-201.

Newman, W. M. and Sproull, R. F. (1979). *Principles of Interactive Computer Graphics*. New York: McGraw-Hill.

Newton, C. M. (1978). "Graphics: from alpha to omega in data analysis," *Graphical Representation of Multivariate Data*, P. C. C. Wang, ed., 59-92. New York: Academic Press.

Ottenweller, J. E. Tapp, W. N., and Natelson, B. H. (1986). "Biological aging in the cardiovascular system of syrian hamsters," Department of Neurosciences Technical Report. Newark, NJ: VA Medical Center, New Jersey Medical School.

Pike, R. (1983). "Graphics in overlapping bitmap layers," *ACM Transactions on Graphics*, **2**: 135-160.

Pike, R., Guibas, L., and Ingalls, D. (1984). "Bitmap graphics: SIGGRAPH'84 course notes," Technical Memorandum. Murray Hill, NJ: AT&T Bell Laboratories.

Stuetzle, W. (1987). "Plot windows," *Journal of the American Statistical Association*, **82**: 466-475.

Tufte, E. R. (1983). *The Visual Display of Quantitative Information*. Cheshire, CT: Graphics Press.

Tukey, J. W. (1973). "Some thoughts on alternagraphic displays," Technical Report Number 45, Series 2. Princeton, NJ: Princeton University, Department of Statistics.

Tukey, J. W. (1982). "Another look at the future," *Computer Science and Statistics: Proceedings of the 14th Symposium on the Interface*, 2-8. New York: Springer-Verlag.

Tukey, J. W. (1988). "Control and stash philosophy for two-handed, flexible, and immediate control of graphic display," *The Collected Works of John W. Tukey, Vol. 5 – Graphics*, W. S. Cleveland, ed., 329-382. Pacific Grove, CA: Wadsworth & Brooks/Cole.

Tukey, J. W. and Tukey, P. A. (1985). "Computer graphics and exploratory data analysis: an introduction," *Proceedings of the Sixth Annual Conference and Exposition of the National Computer Graphics Association*, 773-785. Fairfax, VA: National Computer Graphics Association.

Tukey, J. W. and Wilk, M. B. (1965). "Data analysis and statistics: techniques and approaches," *Proceedings of the Symposium on Information Processing in Sight Sensory Systems*, 7-27. Pasadena, CA: California Institute of Technology.

Yule, G. U. (1927). "On the method of investigating periodicity in disturbed series, with special reference to Wolfer's sunspot numbers," *Philosophical Transactions of the Royal Statistical Society, Series A*, **226**: 267-298.

Comment

John W. Tukey

The overall impact of this paper is both substantial and sound. Thus my comments will have to focus on recommended changes in flavoring or on possibilities for the future.

1. THE MUD CAN BE DULL

The Murray Hill tradition in data analysis has long included aspects of "plant your feet firmly on the ground, even if they do sink deep in the mud." The limitation of scatterplot matrices to original coordinates is a case in point. The discussion of (brain weight)/(body weight)$^{2/3}$ in Section 2.1 is another case in point. A scatter of "log brain weight MINUS 2/3 log body weight" against log body weight would be a useful supplement to the scatter of Figure 2, in part because it would offer an expanded vertical scale. In Figure 11, where "abrasion is stated to be the intended response," an additional row and an additional column for "abrasion loss residual" and "tensile strength" would greatly clear up the situation—perhaps leaving brushing the task of finding still subtler behavior.

John W. Tukey is Senior Research Statistician and Donner Professor Science, Emeritus, at Princeton University. His mailing address is 408 Fine Hall, Washington Road, Princeton, New Jersey 08544.

2. HIGHLIGHTING MAY BE INESCAPABLE—BUT IS STILL INADEQUATE

Paper representations of screens with highlighted points are rather weak and wan—and highlighted screens may be somewhat so. Particularly for paper versions, we ought to further enhance the contrast between emphasized and background points. Two easy ways to do this are: a) median +'s or x's for emphasized, with dimensions at least 3 times those of background circles, or b) filled circles for emphasis and little dots for background. This sort of improvement is needed for alternagraphic emphasis as well as for brushed emphasis. (Compare with the last paragraph of Section 2.1 in Becker, Cleveland and Wilks.)

3. DO PANELS MAKE UP A TABLE OR A GRAPH?

To Becker, Cleveland and Wilks, the answer seems to be "clearly, a table!" because they number rows of panels from above down (and put the vertical coordinate first!). For some of us, the answer seems to be "clearly, a graph" so we would number rows of panels from below upward, and put the horizontal coordinate first.

Whichever side you take—if one is to write in text about panel numbers, panel rows and panel columns, in the pictures, they should be labeled clearly enough (e.g. (1, –) and (–, 3) *or* (–, 1) and (3, –)) so it would be really hard for the reader/viewer to miss the point.

If you do adhere to the graphic paradigm, the distinctive diagonal of your scatterplot matrix will run NE-SW and not NW-SE.

4. RECTANGLES, ANYONE?

It would have been helpful if the account of brushing had said—if it is true, as I think—that brushes are rectangular because rectangles can be computed faster. How much faster? Are we near the present boundaries when we include brushing? Or could we afford other brush shapes?

5. COGNOSTICS, ANYONE?

The paragraph in Section 2.4 on the scatterplot matrix assumes that scrolling is our only remedy when p is too large. Another approach would be to use cognostics (e.g., Tukey and Tukey, 1985) to help us rearrange our variables so that the initial view shows the most interesting k of them. Instead of simple scrolling, then, we might hold the first $k - 2$ (or $k - 3$, or $k - 1$) fixed and scroll the other 2 (or 3 or 1).

6. JITTERING OR TEXTURING?

Many of the jitterings suggested by Becker, Cleveland and Wilks are quite effective. But the jittering of the categorical variable in Figures 16 and 17 can be appreciably improved. A form of limited randomized (Tukey, Tukey and Veitch, in preparation) called texturing has appreciable advantages.

7. A GOOD POINT, BUT NOT TO BE OVERWORKED

The second paragraph of Section 2.5 in Becker, Cleveland and Wilks opens with "we often judge the slopes of line segments to determine the rate of change." The rescaling in their Figure 21 shows how we can do fairly well while preserving the original coordinates. If slopes are the main issue, however, using the full rectangle for two panels, one above another, showing year to year *changes* in sunspot number in one and sunspot numbers *themselves* in the other, would give more visual impact?

8. WHICH AXES?

The discussion under "Control" in Section 2.6 is too three-dimensional for adequate insight and visualization. Axes of rotations in k dimensions are $(k-2)$-dimensional manifolds. As a consequence rotations are better described as "from one axis toward another axis" (really "from one coordinate toward another coordinate") with a tacit understanding that $k-2$ orthogonal coordinates remain unchanged.

9. PERSPECTIVE

I was very pleased that "Enhancements" (in Section 2.6) did not argue for the use of calculated perspective, something whose utility I doubt. I am, however, confused about the "rules of perspective" in paragraph 2 of this subsection.

10. LIMITATIONS

The discussion in this subsection of Section 2.6 could well be taken further. We can easily judge near-linearity and direction of curvature for the two front variables. But no three-dimensional scheme, including rotation, lets us judge near-linearity of curves that involve the back variable.

Not enough thought has been given as to how far we can avoid this difficulty—and to how to do this.

It would have helped also if we had been reminded that there are really three clusters in this iris data, and that projection pursuit comes close to resolving the two that naive scatterplot matrices do not.

11. DREIBEINS

The second paragraph of Section 2.6 on "advanced strategies" regrets the loss of "feet on the ground" when we rotate coordinates. It does notice that keeping track of the linear transformation can help us "to infer meaning." It omits, however, the visual presentation of a "dreibein," consisting of labeled projections of all original coordinate axes, emanating from a common point. This does a lot to explain the present coordinate system.

We need to think much harder about how to express this sort of information visually.

More attention needs to be given to the combination of multiple views and rotation. (The last paragraph of Section 2.6 discusses the combination of brushing and rotation—this is good, but the triple combination can be even better.) Two or more views of both point cloud and dreibein may be of considerable value. Doing this may make us want to be able to modify two different rotations (exactly or nearly) simultaneously.

12. SPECIAL ARITHMETIC

Becker, Cleveland and Wilks wisely emphasize the usefulness of integer arithmetic. The overall process of data analysis would probably benefit from the availability of middle decimal point arithmetic, in which much less rescaling would be needed.

13. SCALING FOR DATA DISPLAY

The discussion of "precomputation" assumes scaling in terms of "max MINUS min." Presumably the rescaling should include a translation to keep the location—as well as the separation—of "max" and "min" constant. My own preference would be to fix upper and lower eights of the frequency distribution, rather than max and min.

14. SCATTERPLOT MATRICES

In addition to the need for numbering rows and columns (see Point 3 above), the scatterplot matrices use of numbers in the corners of the "empty" diagonal cells to specify the range covered for each variable could have been mentioned in the text. More important would be a more effective visual display of the scales involved. This could be placed on the margins of the whole matrix or we could put the information in the "empty" cells (in each of a variety of ways).

15. MULTIPLE SCATTERPLOT MATRICES

Becker, Cleveland and Wilks emphasizes scatters of raw variables. Without eliminating this display from use, we could supplement it by such displays of other scatter matrices as: a) simple residuals scatters, with $y_{.x}$ scattered against x where y = rows and x = columns, b) partial residual scatters, with $y_{.REST}$ (where REST is all but y and x) scattered against x and c) full residual scatters, with $y_{.ALL}$ (where ALL is all but y) scattered against x.

16. SUMMARY

The paper of Becker, Cleveland and Wilks has gone a long way from graphic archeology ("you can see it, if you know how to look!") toward graphic impact ("you can't miss it"). But we need to go further (Points 1, 2, 7, 11 and 15). The proposed styles of graphic presentation could have been improved in a few other cases (Points 3, 5, 6, 13 and 14). There are a variety of points where the text could have been clarified (Points 4, 8, 9, 10, 13 and 14). In general, however, one can only praise the paper of Becker, Cleveland and Wilks.

Acknowledgment

This research was supported by the Army Research Office under Contract DAAL03-86K-0073.

Comment
Peter J. Huber

The authors deserve to be congratulated for a competent overview of an area of statistics that is becoming more important (and accessible) every year.

They try hard to give a sober, no-nonsense account of the current state of the art. There is a core of simple-minded, extremely useful techniques—foremost among them methods for identification and labeling. But there is also a halo of experimental techniques, and the cautious statement: "Far more experimentation is needed with these advanced strategies" (Section 2.6) ought to be translated into plain English as: "We could not make sense out of those strategies, but perhaps somebody else will." The lunatic fringe techniques are useful to hatch new ideas, but only few of them will survive.

I believe that newcomers to high interaction graphics—the buzzword "dynamic" is semantically inaccurate, by the way—are still attracted mostly for the wrong reasons, namely by the video game glamor of fancy techniques. Reflection on the intrinsic, practical value of a technique comes only afterward, when the glamor has rubbed off. For example, when we gained direct, hands-on access to decent computer graphics in 1978, Stuetzle and I began to experiment first with the most exotic techniques—like interactively controlled sharpening (see Tukey and Tukey, 1981), combined with kinematic graphics in three dimensions. It took a long while for me to realize that I never would be able to interpret the curious shapes I saw in those sharpened scatterplots.

The following are comments on points where I either disagree with the authors or would put the emphasis differently.

It is important to avoid gimmicks. Built-in side effects can become extremely annoying when a technique is used in a context not anticipated by its designer. For this reason rescaling after a deletion ought not be automatic (Section 2.2), but should require a separate user request.

In Section 2.4, I suspect that the authors' judgment has been colored by accidental features of their implementations (cf. the remarks of Huber, 1987, Section 4). I fail to understand why roping in a region by drawing a line around it should be intrinsically slower than, and inferior to, brushing the interior of that region. Roping is relatively elastic with regard to timing considerations, while the response to brushing can become unacceptably slow in the case of very large scatterplots.

Undeleting is trickier than the authors make it appear (end of Section 2.4). The problem is to *selectively* undelete a few points. It is aggravating if you have to undelete everything and then to start the deletion

---

Peter J. Huber is Professor Statistics, Harvard University, Cambridge, Massachusetts 02138.

process from scratch. A more convenient solution, using alternagraphics, is due to Thoma: alternate with a view showing the deleted points only. If you delete a point in the alternative view, it gets visible in the original view, and vice versa. Incidentally, this is one of many examples where alternagraphic switching is better done by the user hitting a button, rather than by an automatic timer.

In Section 2.6, the authors say that "it is entirely reasonable to implement all three...rotation-control methods...." I would substitute "feasible" for "reasonable." The point to be made here is that each alternative eats up some menu space and potentially confuses the user by mode trapping (you never are in the mode you think you are). I am not aware of *any* data analytic problem where the first or third of the modes offers a substantial advantage over the second. What sometimes is required in addition is a *variant* of the first method, for moving one object (point cloud) as a rigid body relative to some other. In order to do this, one would have to control all six degrees of freedom of rigid body motion simultaneously (for which, incidentally, the mouse is a woefully inadequate interaction device).

As the authors stress, scatterplot matrices are nice because they provide a single integrated view of the data (Section 2.6). But I believe they understate the resolution problems, and that for all but the very smallest values of p an alternagraphic solution is preferable. For example, show only two scatterplots at one time, keeping one fixed, and use the other space to flip through all the plots in one row or column of the scatterplot matrix.

Some shorter comments.

End of Section 3.1. The distributed processor model is not only at a design disadvantage, but it also creates a software maintenance nightmare.

Section 3.3: *Integer Arithmetic.* I believe these considerations are no longer relevant after the advant of coprocessors (8087, 68881, etc.).

Section 3.4. These sad comments on graphics standards, unfortunately, are even true for semistatic graphics (where the only interaction is the all important identification of labeled observations).

Section 3.5. Windows are great if there is enough screen resolution (800 by 1000 or better) and we immediately got hooked on them with our first Apollo in 1982. But just like the proverbial *goto* in programming, extensive use of windows may actually be harmful in data analysis. Data analysis is an experimental science, and a "laboratory journal" metaphor is more appropriate than a messy "desktop," especially since the electronic version can be messed up much more thoroughly than a real one, and in much less time!

Of course, no survey can cover everything in depth. Still, because of their importance, I believe the following topics would have deserved a more thorough treatment: techniques for identifying and isolating clusters (the letter I of the original PRIM-9) and the role and use of colors.

ADDITIONAL REFERENCES

Huber, P. J. (1987). "Experiences with three-dimensional scatterplots," *Journal of the American Statistical Association*, **82**: 448-453.

Tukey, P. A. and Tukey, J. W. (1981). "Graphical display of data in three and higher dimensions," *Interpreting Multivariate Data*, V. Barnett, ed., 189-275. Chichester: Wiley.

Comment

William F. Eddy

The authors are to be thanked for their thorough review of the current state of interactive graphics (although I think I detected a certain bias in favor of methods they and their colleagues have developed). As I started careful reading of this paper I found myself repeatedly asking: What if Option 2 instead of Option 1? It seems obvious that there is considerable work yet to be done in deciding which choices should be made. I will resist the temptation to produce a long list of such questions but rather point in other directions.

1. STATISTICAL ROOTS

Graphical techniques have always been a part of statistics. Nevertheless, I was very struck on reading this paper that graphical statistics is currently in very much the same state that mathematical statistics was about 100 years ago. In fact I went back and reread parts of some of Karl Pearson's long series of papers on the mathematical theory of evolution that was published in the *Philosophical Transactions of the Royal Society of London* between 1894 and 1916. In the very first paper, Pearson gives a graphical method for calculating the first five moments of a probability density function.

Reading Pearson's papers reveals interesting parallels between the state of mathematical statistics then and the state of graphical statistics now:

1. There is a strong emphasis on description rather than analysis.

William F. Eddy is Professor Statistics, Department of Statistics, Carnegie Mellon University, Pittsburgh, Pennsylvania 15213. This report was prepared while the author was a Resident Visitor at Bell Communications Research, Morristown, New Jersey.

2. There is a tendency to modify existing tools rather than develop totally new ones.

3. There is a tremendous joy in new insights developed from the use of the techniques.

4. There is a very limited number of researchers in the area.

The first of these features is common to any very young science; we can only hope that graphical statistics will become more scientific. The second of these features is also indicative of infancy of development in the field; we can only hope that graphical statistics will grow as did mathematical statistics. The third of these features is one of the real attractions of working in a developing field; we can only hope that all statisticians will have the opportunity to use interactive graphical techniques. The fourth of these features means there is a tremendous opportunity to engage in useful research in the area, particularly the kind that can have a lasting influence; we can only hope that many people will be attracted to research in graphical statistics.

I interpret all these similarities in a very hopeful way. The heyday of mathematical statistics (that is, the time in which techniques were developed that actually affected statistical practice) was, say, the first half of this century. Given the accelerated pace of developments now I guess we should look forward to a ten year "Golden Age of Graphical Statistics."

2. DYNAGRAFIX: WHERE DO WE GO FROM HERE?

It is unreasonable to suppose that in the next five or ten years we will have enough computational power for whatever interactive graphical tasks we wish to perform. Our expectations will increase at least as fast as hardware capabilities. This means that at the heart of dynamic graphical techniques there will always be the core problem of how to get the computations done in real time. The evolution of graphic display systems obviously follows the evolution of other computer hardware. First, there were dedicated mainframes, then we had time-shared systems. Now the workstation model described in this paper is becoming ubiquitous. It seems clear that the next step is a distributed approach to interactive graphics. Anyone with a workstation that supports multiple windows and a local area network can use the poor man's distributed approach: open a window on a remote machine for the purposes of computation and display the results on the local workstation. A more serious problem occurs when the remote machine is not able to supply the needed "horsepower". This is all too common in statistics research groups where the remote machine may only be an unused workstation. In this case, it is necessary to use multiple machines and the decomposition alluded to at the end of Section 3.1 of this paper becomes a research dilemma. Decomposition of statistical calculations (not only graphical) for distributed computation is a problem that merits much wider attention.

One of the great failures of statistical research in recent years has been our inability to capture the attention of the larger scientific community. In part this is simply because most science feels it can proceed with little or no statistical input. In part also this is because much statistical research is not relevant to the larger scientific community. And in part also this is because we do not use modern "marketing" techniques for the ideas we have which are generally useful to this community. (For proof by a not-exactly-statistical example, consider the fast Fourier transform with Richard Garwin behind it. See Garwin, 1969.) Dynamic graphical techniques for the analysis of high dimensional data sets provides statistics with the opportunity to transcend this past.

The increasingly widespread use of supercomputers has created a greatly increased need for techniques of data analysis that can rapidly pick through huge quantities of output and find the "interesting" stuff. Dynamic graphical techniques obviously have great potential to assist in this task and help overcome the first two hurdles in the previous paragraph. With respect to the third (marketing), I propose, with due apologies to the late L. Ron Hubbard, that we name these techniques, collectively, DYNAGRAFIX, and we start advertising them as a replacement for AI. A few carefully designed and properly planted success stories for DYNAGRAFIX will either set AI back 50 years or trigger widespread interest in statistics, especially graphical.

Acknowledgment

This research was partially supported by the Office of Naval Research under Contract N00014-84-K-0588.

ADDITIONAL REFERENCES

Garwin, R. L. (1969). "The fast Fourier transform as an example of the difficulty in gaining wide use for a new technique," *IEEE Transactions on Audio and Electroacoustics*, AU-17: 68-72.

Comment: Deja View
Howard Wainer

I was distracted during my reading of this paper by a high-spirited giggle that wafted in over my shoulder. This was surprising because the door to my office was closed and I thought I was alone. As I turned to try to discover the source of this disturbance I caught a glimpse of the high spirit whose presence had been heralded earlier. It was that old schoolmaster H. G. Funkhouser, whose history of graphics, published half a century ago (1937), has provided a jumping-off point for modern chroniclers. In that treatise Funkhouser was forced to describe graphics that, for technical reasons, could not be reproduced. In doing so he frequently (but regretfully) spent the thousand or so words that each picture required. Mark Twain described women's use of profanity as "They know the words, but not the music." Similarly a verbal description of a graphic may contain the facts of the matter, but not the nuance.

The reasons for his posthumous return to my office became clearer. *"Le plus c'est change, le plus c'est le meme chose"* he mumbled. The current authors (Becker, Cleveland and Wilks) faced the same problem; trying to convey both the fact and nuance of dynamic displays within a static medium. That they did not fail is to their credit, but the result is a combination of 19th Century female profanity and the bipedal walking characteristics of Samuel Johnson's dog. They have done a good job of conveying the general idea and many of the details, but the magic is missing. The pale approximation of these exciting new methods that is conveyed on these pages seems to be a prime example of what Ron Thisted has called the *"Instamatic in Yosemite"* syndrome, in which the frustrated describer can only sputter "You had to be there." It may be that we are approaching the limits of what a static medium can effectively communicate in this area.

Over the years I have formed the opinion that the principal goal of our Bell Labs colleagues at scientific meetings is to foment frustration among the rest of us. Time and time again they report methods and procedures that require machinery, budgets and expertise among support personnel that outstrip what the rest of us can practically contemplate. Yet they blithely go on telling us how they spin and mask and brush, while we look at BMD output. I often laugh as a naive assistant professor enthusiastically responds to one of these all too familiar presentations.

"What kind of equipment do I need to be able to do this? asked naive assistant professor.

"We used a BLNFY-87 (pronounced 'Blinfree' = Bell Labs Not For You) in combination with an RB-1," responded Bell Labs colleagues.

Howard Wainer is Principal Research Scientist in the Division of Statistical and Psychometric Research, Educational Testing Service, Princeton, New Jersey 08541.

"Can other equipment be programmed to do this?" naive assistant professor hoped.
"I imagine so, although we haven't tried," tempted Bell Labs colleagues.

What naive assistant professor didn't learn, and what Bell Labs colleagues never mentioned, was the secret ingredient in their work; the RB-1 {one Rick Becker}. Without one of these, the task is made either much more difficult, or impossible.

So much for my envious grousing. Let me get on with it. We should all be grateful to our Bell Labs colleagues for their path-breaking work in this area. They have done a lot of hard thinking to arrive at a set of methods that appear to be enormously helpful. In accomplishing this I assume that a great deal of time was spent in blind alleys pursuing what seemed in prospect like a good idea but, after a lot of work, turned out not to work very well. The polished product we read about here can be used and, perhaps in a less formal communication, we can find out about the less fruitful paths. I assume that anyone interested in pursuing this area will have the good sense to bend an elbow with one of these authors and find out what else they did.

Much to my surprise, I am less envious now than I have been in previous situations. The principal reason for this is that much of what they are describing is available to the rest of us now (or will be shortly) for less than the annual budget of Saudi Arabia.

MacSpin (Donoho, Donoho and Gasko, 1985) costs under $200 and runs on a MacIntosh. It allows one to view scatter plots three dimensions at a time with rotation around any of the three screen axes. Its facility for the identification of data points is, if anything, *better* than what Becker, Cleveland and Wilks describe. In addition to being able to go in either direction ("Where is *'dog'*?" or "What is this point?"), it can also provide three levels of detail about each data point. Automation is easy, and, through its judicious use, so too is alternagraphics. Software development is dynamic, especially in dynamic display, and so I expect that Andrew Donoho will almost surely have implemented even more tools by the time this commentary finds its way into print.

The DataDesk (Velleman and Velleman, 1986) is a statistical analysis program that also contains some rudimentary dynamic capability. It can produce spinning plots; indeed several windows with a three-dimensional plot in each of them (although only one spins at a time).

These two programs, used in concert, provide a powerful tool for the data analyst at a price that we all can manage. Moreover, being commercially available programs with wide usage means that they have been debugged in a way that more narrowly circulated experimental computer packages are not. The appearance of dynamic display capability in MacSpin and later in DataDesk portends well for the future. Who will buy a data analysis program without such accoutrement? Thus, other software developers will be forced to add these capacities to their programs and we will be the richer for it.

Let me conclude by expressing my gratitude to our Bell Labs colleagues in general, and to Becker, Cleveland and Wilks in particular, for their continuing research into data analytic tools that more fully utilize the computing power now available and the human information processing ability imbedded in our visual system. Their imagination and sweat has provided us with the knowledge of a battery of methods that every salt-worthy data analyst would want to have close at hand. Simultaneously, a second set of talented folks are working hard to make these tools available for the rest of us. To both groups I give my heartfelt thanks, and ask that they stop wasting their time reading this and get back to work—I have a data set that I've been looking at, and I think I'm missing something.

ADDITIONAL REFERENCES

Funkhouser, H. G. (1937). "Historical development of the graphical representation of statistical data," *Osiris*, 3: 269-404.

Velleman, P. and Velleman, A. (1986). *The DataDesk Handbook.* Ithica, NY: Data Description, Inc.

Comment

Edward R. Tufte

Even though we navigate daily through a perceptual world of three spatial dimensions and reason occasionally about still higher dimensional arenas with mathematical and statistical ease, the world portrayed by our information displays is caught up in the two-dimensional poverty of endless flatlands of paper and video screen. Escaping this flatland is the major task of envisioning information—for all the interresting worlds (imaginary, human, physical, biological) we seek to understand are inevitably and happily multivariate worlds. Not flatlands.

Such escapes grow more difficult as ties of data to the familiar spatial world weaken and as the number of data dimensions increases. But the history of information displays and statistical graphics—indeed the history of communication devices in general—is nothing but a progress of methods for enhancing the density, richness, efficiency, complexity and dimensionality of communication. Methods for escaping flatland include

---

Edward R. Tufte is Professor of Political Science and Statistics, Lecturer in Law, and Senior Critic in the Program in Graphic Design, Yale University, 3532 Yale Station, New Haven, Connecticut 06520.

layering and separation, micro/macro readings, contours, perspective, narratives, multiplying of images, use of color and dynamic graphics (Tufte, 1983, 1988).

The visual display tasks involved in dynamic graphics for data analysis are very nearly identical with the flatland portrayal of any dynamic physical system. It is, after all, data moving.

For example, when Galileo first looked through his telescope in 1610, he was confronted with displaying the dynamics of sunspots. In "The Starry Messenger" and "Letters on Sunspots," Galileo reported his observations in a large collection of small multiples sequenced on time, recording complex data of moving sunspots on a rotating sun observed from an orbiting and rotating earth (Figure 1).

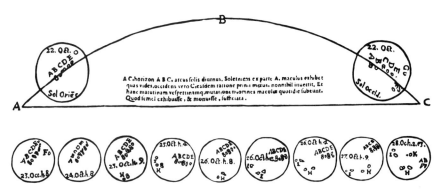

Figure 1. Galileo reports his dynamic observations, sunspots of October 1611. From his "Three Letters on Sunspots," 1613.

Through some 370 years of astronomical research, sunspot records have evolved into data-rich time series. The Maunder butterfly diagram records the distribution of sunspots in latitude only moving over time, sacrificing area for time (Maunder, 1904) (Figure 2).

The modern version partially recovers area in reporting an enormous volume of information (Figure 3).

In these 20th century scientific performances, note the high resolution and high data densities (pushing the resolution of the printing process rather than merely that of the local computer output device) and the skill in the details of production (no jaggies, except those of the data, competent typography). Such resolution, data density and production quality are rarely attained in the statistical research literature on data graphics; indeed, it is a literature routinely filled with low resolution, low density displays and, ironically, with visually clunky graphics. The production values in the literature are amateurish, compared to ordinary workaday professional graphic design.

This same research is generating quite an elaborate jargon, partly no doubt to facilitate technical communication and all that, but it does appear now and again to be an attempt to own a concept by naming it. Giving old ideas new names does not yield new ideas. Of course new words and new usages of old words are needed, preferably reserved for

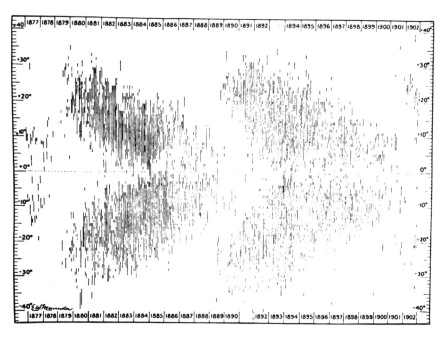

Figure 2. The Maunder butterfly diagram showing the distribution of sunspots in latitude, 1877-1902.

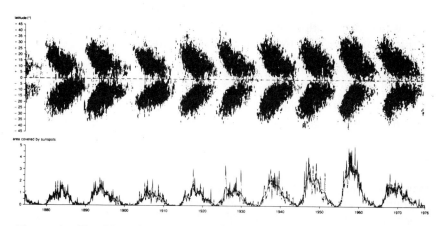

Figure 3. The Maunder butterfly diagram and time series of area of sun covered by sunspots, 1877-1976. From Science Research Council, Royal Greenwich Observatory (Andouze and Israel, 1985).

new ideas of some consequence. Jargonizing the familiar may even impede communication. For example, without ever knowing it (fortunately), Tokyo train passengers have been looking for decades at schedules in the form of what statisticians now call "the back to back stem and leaf display" (Figure 4).

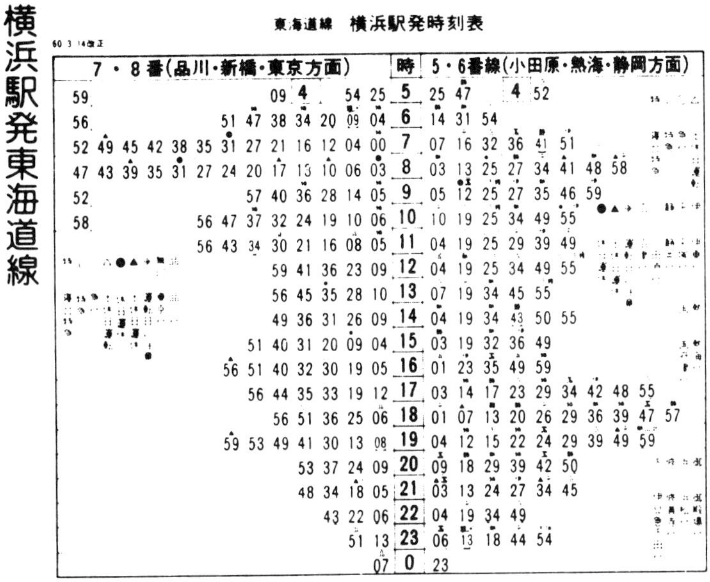

7・8番(品川・新橋・東京方面)	時	5・6番線(小田原・熱海・静岡方面)
59 09 4 54 25	5	25 47 4 52
56 51 47 38 34 20 09 04	6	14 31 54
52 49 45 42 38 35 31 27 21 16 12 04 00	7	07 16 32 36 41 51
47 43 39 35 31 27 24 20 17 13 10 06 03	8	03 13 25 27 34 41 48 58
52 57 40 36 28 14 05	9	05 12 25 27 35 46 59
58 56 47 37 32 24 19 10 06	10	10 19 25 34 49 55
56 43 34 30 21 16 08 05	11	04 19 25 29 39 49
59 41 36 23 09	12	04 19 25 34 49 55
56 45 35 28 10	13	07 19 34 45 55
49 36 31 26 09	14	04 19 34 43 50 55
51 40 31 20 09 04	15	03 19 32 36 49
56 51 40 32 30 19 05	16	01 23 35 49 59
56 44 35 33 19 12	17	03 14 17 23 29 34 42 48 55
56 51 36 25 06	18	01 07 13 20 26 29 36 39 47 57
59 53 49 41 30 13 08	19	04 12 15 22 24 29 39 49 59
53 37 24 09	20	09 18 29 39 42 50
48 34 18 05	21	03 13 24 27 34 45
43 22 06	22	04 19 34 49
51 13	23	06 13 18 44 54
07	0	23

Figure 4. 1985 timetable, Tokaido Line at Yokohama Station, Sagami Tetsudo Company, page 72. Similar designs are used in Tokyo subway station signage to show subway schedules.

Finally, those interested in dynamic graphics should also take a look not only at the skilled paper of Becker, Cleveland and Wilks in hand but also at MacSpin (Donoho, Donoho and Gasko, 1985). MacSpin is a genuine masterpiece, a marvelous interactive program accompanies by a marvelous manual that gracefully explains the straight-forward computer mechanics involved and, more importantly, shows dynamic data analysis in action and chronicles the development of such displays. Alas, even in the MacSpin book, the graphics have the jaggies and murkies, too.

ADDITIONAL REFERENCES

Audouze, J. and Israel, G., eds. (1985). *The Cambridge Atlas of Astronomy.* Cambridge, MA: Cambridge University Press.

Maunder, E. (1904). "Note on the distribution of sun-spots in heliographic latitude, 1874 to 1902," *Royal Astronomical Society Monthly Notices*, **64**: 747-761.

Tufte, E. R. (1988). *Envisioning Information.* Cheshire, CN: Graphics Press.

Rejoinder

Richard A. Becker
William S. Cleveland
Allan R. Wilks

We would like to thank the discussants for their interesting comments. Our responses cover six areas: implementation of the methods of the paper, presentation of the graphs, the underlying software, the computing environment, brushing and rotation.

Since writing the paper we have studied dynamic methods in a color graphics environment—a Silicon Graphics IRIS 2400T workstation. Most of the original paper is based on an implementation of methods in a distributed processing system with an AT&T Teletype 5620 graphics terminal, which is monochrome. Our responses here will reflect more of the experience with the IRIS implementation.

1. IMPLEMENTATION

Experimental vs. **Tested: Field Testing**

Comments by Huber and Eddy made us realize that one deficiency in the paper is an explicit statement about whether the methods in Sections 2.1 to 2.6 are experimental or well tested. We have examined a large number of dynamic methods by field testing, which will be described shortly. With one exception, the methods of Sections 2.1 to 2.6 are those that we tested and judged to be useful tools for data analysis. (The one exception is advanced strategies for rotation control, which we only reported but did not test.) We strongly urge software developers to implement these methods in their software systems. Wainer is quite right—we tried many other ideas that did not work out.

Field testing a method means using it on a variety of data sets including those where data analysis is in progress. At the moment, because the amount of theory about data display is small, extensive field testing is the only way to effectively judge a graphical method. Armchair thinking is not enough. In 1982 Tukey (1981a) wrote the following about the development of a graphical method:

- now we try it [the graphical method] on diversified data—trying to understand when its performance is less than adequate,
- and then we try to understand something of what modifications would help,
- and then we try the modifications,
- and then we repeat the last 3 steps as needed!

There is no substitute for adequate iteration. *Such iteration is the original developer's obligation.*

It is particularly important to try out methods in setting where people are attempting to learn about the world from the data, and where the methodology is a means to an end. One cannot fully assess a method by using just old data sets no longer of interest to anybody and dredged up just to test the method. In the paper, though, we used familiar or easy-to-convey data sets because of space limitations.

NIH Means Not Implemented Here

In his first sentence, Eddy refers to "bias in favor of methods they and their colleagues have developed." "Developed" should be changed to "implemented." Many of the methods of the paper were invented by us. Many were invented elsewhere; we hope the extensive citations and bibliography make this clear. But except for the advanced strategies of Section 2.6, we discussed only methods that we implemented. This is as it must be. We could not write with much insight about methods that we did not field test, and we could not field test a method that we did not implement.

2. PRESENTATION

Excitement

We quite agree with Tukey that "paper versions of screens with highlighted points are rather weak and wan" and with Wainer that the static pages of a journal do not convey the excitement of dynamic graphics. We feel badly about this, but we can do something about it. If the IMS is willing to distribute it, we will provide a videotape that demonstrates the methods. (Perhaps we can call it the DYNAGRAFIX tape.)

ACM SIGGRAPH, the premier computer graphics group, now publishes the *SIGGRAPH Video Review* to provide a means of disseminating a wide variety of computer graphics. Perhaps it is time that statisticians did something along these lines, too.

The Mud

It is true that the scatterplot matrices of our examples do not contain residuals or transformed data, but this was not meant to imply that we do not think graphing them is of substantial value. To miss the value would be to miss some of John Tukey's most creative and important ideas about data display. (See Cleveland (1985) for a graph of (brain weight)–2/3(body weight).) In fact, the importance of graphing such derived variables is precisely what stimulated the first paragraph of Section 3.5. It is important to use these dynamic techniques within the context of a good data analysis system. Rather than burden the dynamic techniques with basic statistical computations, we expect that these will be carried out in a powerful data analysis environment.

Panels of the Scatterplot Matrix

The table vs. graph description of Tukey is nicely put. We doubt, though, whether having the empty panels running along the 45° line, or perpendicular to it, makes an appreciable difference.

Slopes

Tukey makes an excellent point about slopes. But we cannot, of course, avoid slope judgments altogether, and optimizing them when we cannot avoid them makes good sense. We also note that as soon as we plot the slopes of the sunspots we are quite likely to find ourselves studying the slopes of the slopes, so we would apply the 45° principle to such a graph.

Scales

Tick marks and tick mark labels in the empty panels, as Tukey suggests, are clearly needed.

Resolution

Huber has made important points about resolution on the scatterplot matrix and has given a sensible suggestion. In our implementation of brushing on the IRIS, the data analyst can use a pull-down menu to delete a variable shown on the scatterplot matrix or to add one not shown. For data analyses where resolution of the full matrix is a problem, we often spend a large fraction of the analysis time studying subsets of the variables.

Texturing

We look forward to seeing the work of Tukey, Tukey and Veitch.

Dynamic or High Interaction

Huber is right about the term "dynamic." It does not suggest, as it ought to, that the methodology we describe is more than "data moving," as Tufte nicely puts it, and is really data "being moved." Early versions of our paper used "high interaction," a more appropriate word, but as time went along we switched to the more conventional "dynamic" to avoid jargon. Meaning gave way ever so slightly to stype. As distortions of this ilk go, it is not so bad.

Jaggedness Is Not Particularly Important

Tufte's concerns about jaggedness are appropriate and relatively easy to address in static, printed graphics. However, our paper concerns images that appear in real time on a computer display screen. Although we could readily produce higher quality plots (on a phototypesetter, for example), actual screen images are shown in the paper. These images show the jaggedness inherent in most current hardware.

A graphical display in science and technology should be judged mainly in terms of whether it conveys quantitative information and helps us learn about the world. It is unreasonable to produce displays that have a high quality appearance if the costs are too great; there is the cost in dollars to purchase and program extremely high resolution hardware, there is the cost of redeuced information flow in a dynamic method, and there is the cost in analyst time in trying to get the best appearance of a graph. Because jaggedness does not generally interfere with our understanding of the data, it is relatively unimportant in interactive graphics; even a small increase in cost deters us from taking action when the benefit is small.

3. SOFTWARE

Simplicity of Software

Huber mentions that dynamic graphics has "a core of simple-minded, extremely useful techniques." We agree wholeheartedly. One of the difficult tasks in designing software is to find the core ideas and to keep them from being hidden by a clutter of options and features. This is the central idea of software tools as exemplified by the UNIX operating system (Ritchie and Thompson, 1978). Another interesting idea is that of the lifeboat principle in designing software: for each feature that is added, one must be thrown overboard. This leads to simple, yet general, software design.

Software Secrets

Wainer seems to imply that dynamic graphics programming is somehow beyond the realm of ordinary people. We insist it is not. It is a lot of hard work, and at present the effort put into dynamic graphics on one display device does not carry over to other devices. Nevertheless, anyone with sufficient desire and willingness to work should be able to experiment with these techniques.

Arithmetic

We agree with Tukey that special arithmetic can provide advantages over floating point arithmetic. Even with the availability of floating point coprocessors, integer arithmetic is likely to be faster; ultimately integers are used to address the screen on the graphics device.

4. ENVIRONMENTS

Distributed Model

It is interesting to contrast Eddy's statement that "It now seems clear that the next step is a distributed approach to interactive graphics" with Huber's "The distributed processor model is not only at a design disadvantage, but it also creates a software maintenance nightmare." We tend to side with Eddy; software design can be more subtle in the distributed approach, but the payoff is high.

Windows

Huber points out that modern workstations with multiple windows allow a data analyst to become thoroughly confused about what has been covered in an analysis session. Of course, the answer is not to outlaw windows and force the analyst to think in the traditional one-step-at-a-time manner, but to make use of the computer to monitor and record the analyst's actions. Later, the computer can use these audit trails to produce a well organized file describing that was done, as in a laboratory notebook. Much work needs to be done in this area.

5. BRUSHING

Brush Shape

It is true that a rectangular brush makes computations easier. However, we did not pick a rectangular brush for computational convenience.

To condition on one or two variables, the analyst must select rectangular regions whose sides are parallel to the coordinate axes. The brush is rectangular because that makes such selection easy and because conditioning is important and frequently used. When painting in the lasting mode, the brush shape has little effect on ease of use—we have used diamond-shaped brushes and could have implemented other shapes.

Roping

Roping is a natural idea—if given a scatterplot on a piece of paper, most of us would have little trouble using a pencil to draw a curve around a cluster. Roping is an analogue to this. Past implementations of roping did not produce a pencil-like instrument, but rather gave a pin-and-string method in which a polygon was formed by indicating vertices with a pointing device (Newton, 1978).

A roping implementation would have been within our capabilities. However, there are other problems with roping. There is no mechanism for unroping if we should accidentally rope-in more or less than desired. We can envision the need for a rope editor and much added complexity.

Highlighting

The Highlighting issue, as Tukey has pointed out, is an important one. Plotting symbols of different colors provide the best highlighting of all. This, of course, was already known from experiences with static graphical methods (Cleveland and McGill, 1984).

Automatic Deletion

Fowlkes' operation was of the form "show me a normal probability plot with this data point removed from the sample." There would have been no reason to break it into two operations. However, in other settings, such as brushing, we are in agreement with Huber that deletion and redrawing should be separate operations. In fact, this is what we did in our brushing implementation on the 5620.

Undeleting

In Section 2.4 we describe shadow highlight. Although we did not say so explicitly in the paper, for our implementation on the Teletype 5620 the data analyst can also switch from the delete operation to a shadow delete operation; points that were previously deleted are turned on and points that were previously not deleted are turned off.

6. ROTATION

Control

We have to disagree with Huber on rotation control. We found method three (with stereo viewing) to be an extremely effective data analytic method. It provides delicate and accurate control that allows the data analyst to quickly get to a desired orientation of the point cloud. Also, method one is useful when studying how one variable depends on two others; it is very helpful to rotate about the axis of the dependent variable to help us keep the dependence structure firmly fixed in our minds. Of course, we could always use method two to do this by getting the dependent variable axis lined up with one of the screen axes, but this is a nuisance. In fact, having seven rather than three control selections adds very little additional burden to the system.

Perspective

Stereo viewing, as we state in the paper, is an important enhancement. Without stereo, point clouds lose their three-dimensional feeling as soon as rotation stops. In our IRIS implementation, the routines that draw a stereo view also provide perspective and control of the distance to the center of the point cloud. At the moment we agree with the statement of Tukey that viewing from an infinite distance (i.e., no perspective) is sufficient, but this issue needs more study.

Dreibein

We have had no experience with a dreibein and so, have little to offer. We only wonder whether, as described by Tukey, the portrayal of more than three axes should be called a "vielbein."

ADDITIONAL REFERENCES

Cleveland, W. S. and McGill, R. (1984). "The many faces of a scatterplot," *Journal of the American Statistical Association*, **79**: 807-822.

Ritchie, D. M. and Thompson, K. (1978). "The UNIX time-sharing system," *Bell System Technical Journal*, **57**: 1905-1946.

Tukey, J. W. (1987a). "Thoughts on the evolution of dynamic graphics for data modification displays," *The Collected Works of John W. Tukey Volume 5*, W. S. Cleveland, ed. Pacific Grove, CA: Wadsworth & Brooks/Cole.

2

SOME APPROACHES TO INTERACTIVE STATISTICAL GRAPHICS

D. F. Andrews

University of Toronto, Canada

E. B. Fowlkes
P. A. Tukey

Bell Communications Research

ABSTRACT

This paper is a survey of some applications of interactive graphical systems for the analysis of data. The survey, while incomplete, presents a broad range of problems in which interactive graphics are useful. General features desired for interactive graphical systems for statistical applications are given. A variety of hardware and software implementations are discussed with particular emphasis on implementation costs in terms of time, effort, and capital. The cost varies from moderately high to very low.

1. INTRODUCTION

The purpose of this paper is to survey some applications of interactive graphics for the analysis of data. The usefulness of interactive graphics in a broad range of problems and a variety of tools and techniques available and required for these applications are discussed. Some

---

From *Dynamic Graphics for Statistics*, William S. Cleveland and Marylyn E. McGill, eds., copyright © 1988 by Wadsworth, Inc. All rights reserved.

indication is given of the costs in terms of time, effort and machines required to implement various techniques. In many instances this cost is relatively small.

Section 2 presents some applications of interactive statistical computing. This is the framework in which interactive graphical procedures will operate. Section 3 presents some of the uses of interactive statistical graphics both in the analysis of data and in the summarization of the results of analysis. Section 4 contains some general considerations for setting up new graphical facilities. The remaining Sections 5-8 are examples of several types of machines, programs and applications.

2. STATISTICAL USES OF INTERACTIVE COMPUTING

Scientific investigations are iterative; data suggest models or theory which suggest the collection of further data leading to further refined models and so on. Scatter plots of one variable against another can suggest variables which could be included in a model and the form of their influence or effect. Probability plots can suggest distributional anomalies. Residual plots can detect possible non-linearities, changes in variance and other unexpected departures from an hypothesized model. The early stages of a model formulation involve considerable interaction between model and data. The investigator does most of his creative work at this stage. Usually, only after several iterations of this process can a question be formulated which is worth a definitive experiment followed by a formal analysis.

The early stages of model formulation can be speeded considerably with interactive computing. Many approaches may be quickly explored; many transformations or partitionings of the data may be considered; many regressors in many forms may be tried. A recent book by Daniel and Wood [1971] shows in detail the many varied and unpredictable steps of analysis. Their Chapter 5, for example, deals with a single problem involving 21 observations on 4 variables. A great deal of time must certainly have been spent waiting for computer output from the thirteen computer passes which they describe. Much of this time might have been eliminated if an interactive system had been available.

It is inefficient to design a single program to select and investigate a subset of all possible approaches. The number of these is too large and the selection procedure is most often based on the knowledge and judgment of the investigator. However interactive systems provide the investigator with a set of operations which can be applied to the original data or to the results of previous operations. The interactive nature of the system allows the results of one step to guide the next.

If current trends continue, many statistical analyses will be done on computers by people with relatively little formal statistical training or by statisticians with relatively little computational experience. High-level interactive systems can (optionally) prompt a user, supplying him with statistical and or computational information. Thus the inexperienced user can be told the form of input for a particular procedure and be reminded of some of its drawbacks-at the time he is using and interpreting the procedure.

3. WHY INTERACTIVE GRAPHICS?

3.1 Statistical Reasons

The essential feature of interactive statistical computing is to help the analyst to use the results of previous calculations to direct future calculations. If decisions are to be based on data, the data must first be understood. This applies both to decisions made to direct the stages of analysis and to decisions made as a result of this analysis. Only in the most simple situations can the data be summarized by a few numbers. Some questions are most easily answered from pictures. For example: What sort of relationship, if any, holds between two variables y and x? (A piecewise linear relationship might easily be overlooked in a table.) How does a particular batch of numbers depart from a normal distribution? What transformation will improve the situation? How many and which points should be treated separately? What is the shape of some cluster? Is there a pattern in a collection of residuals? Are the data homogeneous or clumped? The importance of graphical methods in answering such questions has been considered by numerous authors-Tukey [1965], Tukey and Wilk [1965], Wilk and Gnanadesikan [1968], and Gnanadesikan [1973] to name a few. If the questions are raised in an interactive analysis, some form of interactive graphics is required. Examples of interactive graphical applications are discussed in Andrews and Tukey [1973a] and Britt, *et al.* [1969].

3.2 Visual Considerations

A further essential aspect of interactive graphics lies with the idea of the pictures themselves. It is very difficult to prescribe *a priori* all the specifications of a good plot: what scale to use, what groupings to use, what data to include. Many of these aspects are best assessed by the eye. Frequently many pictures must be produced before a useful display is generated. With an interactive graphical system this interplay between graph and eye may be speeded up to a reasonable degree.

These considerations are particularly important for producing displays for reports or papers. Recently routines have been developed at Bell Laboratories to produce plots with their associated labels and text, which are of publication quality.

3.3 Educational Reasons

Students of statistics need to know not only where techniques work but also where they fail and why. Students can acquire more experience by attempting many approaches on many examples. To gain this experience without being saddled with a prohibitive programming or calculation effort requires a fast computational facility that is relatively easy to use. Such a need may be met by an interactive statistical computational system. If a graphical facility is added, the results of such analyses may be quickly and easily absorbed. Because of the large numbers of such students to be served, it is likely that a time sharing system on a large computer would be useful for this purpose.

4. CONSIDERATIONS FOR IMPLEMENTATION OF A GRAPHICAL SYSTEM

A facility for interactive graphics necessarily combines certain equipment and software in order to display pictures in a sequence directed by the user.

The type of digital computer that provides the interactive environment is an important factor. Many statisticians have had some experience with time-sharing systems, which have been widely available for about six years. More recently, the proliferation of mini-computers has made the dedicated computer approach accessible to more people.

Floating point arithmetic is normally provided by time-sharing services, but a number of "stand-alone" graphics systems have been built which have only fixed point arithmetic. Since many statistical applications require numerous arithmetic calculations, these systems are typically not very useful to the statistical researcher.

Fortran IV or some comparably high-level language should be available. The prospective user will certainly want to use subroutines and procedures that he has written for other systems or acquired from outside sources, and he will want them to run without extensive rewriting. A number of time-sharing services have offered stripped-down Fortrans which do not provide this flexibility.

The computing system should provide easy access to stored data for statistical computations. It is now practical to consider data bases for iterative graphical applications in the one hundred to one thousand observation range, and it should be possible to input this data in some manner other than typing it in through a keyboard terminal at each session.

A final but extremely important characteristic of the computer is the response time to actions by the user at a keyboard terminal. No system, regardless of the number of powerful and useful operations it provides, is worthwhile if it is too slow to respond to user commands. Those who have had experience with time-sharing services realize the importance of this requirement.

A second area of consideration deals with display devices and terminals. Interactive graphics should not be equated automatically with the availability of typically expensive cathode ray tube displays. Some simple but useful forms of graphs may be implemented using printer plots on keyboard terminals. An example utilizing the "six-line" plots of D. F. Andrews and J. W. Tukey [1973a] will be given later in the paper.

Many installations, however, will use some type of scope as a display device, especially since the cost of such hardware is continually decreasing. The earlier graphical systems had continually-refreshed CRT displays and were equipped with light pens, RAND tablets and other devices for graphical input as well as output. The hardware to support these devices, especially the light pen, makes such systems very expensive. Recent experience at Bell Laboratories suggests that this kind of sophistication is unnecessary for many statistical applications. The simple graphical input functions that a data analyst needs, such as identifying a point or fitting a line by eye, can be handled through keyboard input or some other inexpensive device. Furthermore, the continually refreshed CRT display can be replaced by a storage scope whose screen retains an image after it has been illuminated once. This simplifies the hardware by eliminating the need for picture-refreshing circuitry. The drawback of these devices is that parts of pictures cannot be erased and redrawn without erasing the entire screen, and real-time motion cannot be stimulated. This is a tolerable restriction for many statistical applications.

A person using a cathode ray tube display device will from time to time want permanent prints of some of the pictures he produces. Many computer installations which provide interactive graphics also have microfilm or some other hard copy facility, and in other cases Polaroid snapshots can serve as an acceptable substitute.

A third factor to consider is the need for a basic set of support programs to accompany the graphical facility. In many cases the person who sets up the facility will have to write these programs himself, either because the graphical terminal does not come equipped with support programs, or because the software devised by other people does not suit the statistician's needs. Much of the software that has been developed for general use is designed for complex graphical data structures and has a degree of generality that is not needed by the statistician. Such software typically occupies a large amount of costly core storage and thereby restricts the size of the user's statistical analysis routines.

The statistician setting out to write some basic software for a new graphical terminal will probably want to organize his system in two modules. On a lower level he will need programs that actually construct and transmit graphical commands to the display device. They will probably perform the basic graphical functions of clearing the screen, drawing a line segment between two points, and plotting text starting at a given point. Clearly the coding of this module will depend heavily on the characteristics of both the graphical terminal and the host computing system.

A higher level module can then be written which calls the low-level module to do its plotting. This essentially device-independent set of programs can be coded in a language such as Fortran, and can be more or less sophisticated depending on the needs and tastes of its authors and users. Some design considerations are:

1. It should be simple to use, with an easy-to-remember calling sequence whose arguments include arrays of X and Y coordinates that define separate points or ends of line segments. Other arguments or other entry points can provide information for the title of the plot.

2. It should automatically organize the available plotting surface, leaving borders for titles, labels, etc.

3. The module should provide automatic scaling from the user's data space to the fixed coordinate system of the low-level module, based on the data being plotted.

4. Several reference values with optional grid lines should be printed automatically along the edges of the plot, and should be chosen as multiples of some convenient unit.

5. The package should allow for the plotting of several sets of data on the same plot, with distinct plotting characters for each set.

6. The package should provide a facility for plotting single points, labeled points, or connected line segments.

7. Parameters relating to the scale, the title, and the organization of the graph on the plotting surface should all have sensible default values, and should be easily accessible for modification.

8. The package should provide a simple way to control any special graphical features that the terminal may have, such as a movable cursor.

9. The package should occupy a small amount of core storage so that it will run effectively in the time-sharing environment.

5. TELETYPEWRITER PLOTS

It was pointed out above that interactive graphics should not be equated with the availability of scopes and the associated hardware. One of the least expensive graphical facilities consists of a teletypewriter or similar terminal which has access to a time-sharing system. The cost of such a set-up is about $100 a month for rental of the terminal, plus the costs for computing and telephone connection.

Teletypewriter-like terminals are proliferating. Many universities, industries, businesses and government establishments are providing time sharing services for such terminals. They may be used for interactive statistical computing and various interactive graphical capabilities can be implemented.

Scatter plots can be produced by teletypewriter terminals in a straightforward way. However, the relatively low resolution due to fixed line and letter spacings usually forces the user to generate a relatively large plot. The slow rate of the transmission line and of the terminal currently restrict the output to about 30 characters a second, which applies to blanks as well as to printing characters, and causes an ordinary full page plot to take several minutes. This delay is too great for most interactive applications where many plots are required. Some editing and formatting of the plot can help. For example, Figure 1 is a kind of plot

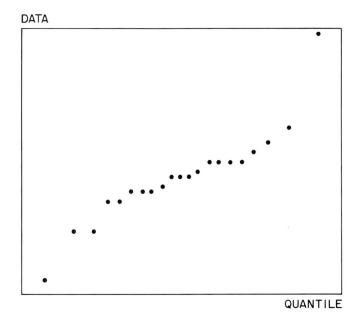

Figure 1. Standard Probability Plot. Requires about 100 vertical positions.

that would require a large number of vertical positions, but by tilting it as in Figure 2 we can produce essentially the same plot with many fewer vertical positions. This takes better advantage of the available resolution. Another technique is to code in additional resolution with different characters as in Figure 3. Such plots requires some getting-used-to, but the effort may be worth the savings in time.

An example of what can be done is provided by a procedure developed by D. F. Andrews and J. W. Tukey primarily for displaying residuals [1973a] which uses only 6 lines, labeled –TWO, –ONE, –, +, +ONE, and +TWO. These plots are particularly useful to statisticians if

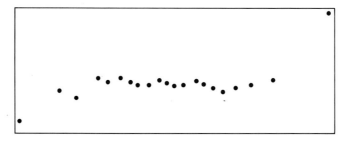

Figure 2. Tilted Probability Plot. Requires only 10 positions. This modification produces increased resolution and plotting speed.

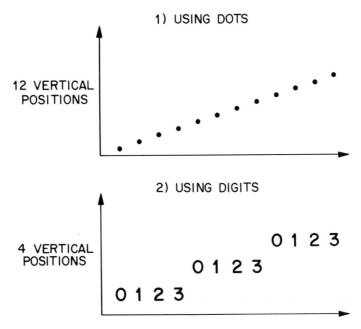

Figure 3. Plots of a Straight Line. See Section 5 for further explanation of digits.

the data have been rescaled to be comparable with a Gaussian distribution with unit variance. Figure 4 is an example of a 6-line plot of residuals. Such a plot takes about 15 seconds to print. The following explanation of how the 6-line plot represents floating point numbers is adapted from Andrews and Tukey [1973a]. The 6-line plot represents numbers in a mixed digital system using digits to the base 4. For numbers between −3 and +3, numbers are simply truncated after one quaternary place. The encircled character in Figure 4 represents a residual with value 1.26 (decimal). This 1.26 (decimal) becomes 1.1002... (quaternary) and then 1.1 (truncated quaternary) which is represented as +ONE 1, where '+ONE' indicates the line on which the character '1' is printed. Numbers between −4.24 and −3 or +3 and +4.25 are treated similarly, using the improper quaternary digits 4,5,6,7 and 8. Thus 3.3 (quaternary) would be represented as '+TWO 7' and −4.0 (quaternary) as '−TWO 8'. Values above 4.25 are represented as '+TWO 9'; values below −4.25 as '−TWO 9'. Representative values on both ordinates and abscissae for three groups of points are printed below the plot. Each group is composed of roughly one sixth of the points. The groups are formed by ranking the X's, and taking one group in the center and one group each to the left and right of center. Gaps of approximately one sixth of the points are left between the groups. The representative values are midmeans (means of the middle half of the sorted data) of the X and Y values for the groups.

The six-line plotting programs include several fitting options as well as facilities to rescale, relocate, remove a fitted straight line, remove a fitted quadratic, and expand or contract the plot along a vertical axis.

Probability plots, traditionally displayed as almost linear configurations of points running diagonally across the page are extremely wasteful of resolution and plotting time on these terminals. If a straight line is fitted to such points in a suitable way the residuals resulting from such a fit will have much less variation and may be displayed with greater precision on a device with the same resolution. They may be usefully plotted by the "six-line" program for studying distributional shape. The coefficients of the fitted straight line summarize location and spread.

```
+TWO                               9                       0         0
+ ONE   ①           O
   +    2                  O       1
   -                       3       2                                 1
- ONE                      3
- TWO                                      9
        REPRESENTATIVE VALUES FOR Y   0.73    -0.46    0.73
        CORRESPONDING VALUES FOR X   50.00    58.00   72.50
```

Figure 4. A Plot of Residuals vs. an Independent Variable. Note four large residuals on top and bottom lines.

The plotting program consists of a set of small Fortran subroutines that are linked together. These may be easily incorporated into other programs. More discussion of the plotting procedure appears in Andrews and Tukey [1973a], and the computer programs are discussed in Andrews and Tukey [1973b].

6. INEXPENSIVE CRT DISPLAYS

If the statistical investigator working on a time-sharing system can afford to invest between $5,000 and $12,000 in a graphical display facility, and has a few weeks of time to spend writing some graphical support programs, he can augment his teletype or printer/keyboard with a cathode ray tube display that will provide additional resolution. At least two such devices are available. One of these, the Tektronix 4010 terminal, is a single compact unit with a keyboard and a storage scope, and costs just under $5,000. The second which will be described in more detail, is the Graphic 101 terminal. It was designed and assembled at Bell Laboratories and costs between $10,000 and $12,000.

The Graphic 101 terminal as shown schematically in Figure 5 consists of an ASR33 Teletype (which can be replaced by a Texas Instruments keyboard/printer), a Tektronix 611 storage scope, a data phone, and a PDP-8/E mini-computer. The PDP-8/E is equipped with one additional special circuit board which was designed at Bell Laboratories and is built by the computer manufacturer.

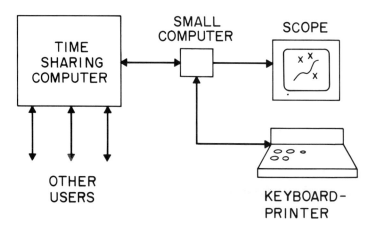

Figure 5. Schematic Diagram of Graphic 101 System.

The terminal communicates with a time-shared computer in ASCII code over a single voice-grade phone line. In one mode of operation it behaves just like a teletypewriter, sending and receiving text. In the other mode, ASCII characters received from the computer are interpreted as graphical commands and cause line segments and characters to be plotted at appropriate places on the screen. The mini-computer is not available to the user for computations; it is programmed to act only as the interface between the various components.

Notice that this kind of system, when running in graphical mode, is strictly for output. It has no light pen or other graphical input device. This simplifies its design and makes it relatively inexpensive. The major difference between the Tektronix 4010 and the Graphic 101 systems is that the computer in the latter system allows it to accept a richer set of graphical commands. This means that a picture may be encoded more compactly so that it can be transmitted and drawn more quickly.

Transmission speed is still a drawback here. Complex displays can take a minute or two, but simple ones can be drawn in a few seconds.

A basic time-sharing software package for the Graphic 101 system which runs under the Time Sharing System on the Honeywell 6070 was developed for the statistical community at Bell Laboratories. It was written to meet the specifications and modular design constraints discussed earlier, and required several weeks of skilled programming effort spread over a year. The package is written almost entirely in Fortran and requires about 2.5K words or 10K bytes of core memory which is about 10% of the total memory available to the time-sharing user at this installation. (Fortran overhead requires, in addition, about 20% of the total.)

A time-series forecasting system which utilizes this basic package has been developed by D. M. Dunn and is discussed in an unpublished Bell Laboratories memorandum. This system was designed in connection with the development of procedures for forecasting demand within the Bell System. The system offers commands which enable the user to read and write time series on permanent disc storage, to alter names and titles of series, to create new series by applying various transformations, to invoke several different forecasting algorithms, to calculate residuals and confidence limits, and to plot series in various combinations. A command is available to send selected pictures to a microfilm printer for hard copy.

Figures 6 and 7 show some plots produced by this system. Figure 6 is an original data set which consists of the total number of residence telephone lines in service in the Flint, Michigan, area for each month between 1954 and 1970. This plot took about 45 seconds to be transmitted to the Graphic 101 terminal. Next, the system was instructed to create a new time series by computing a monthly forecast from the original series. The algorithm used was an exponentially weighted moving average. The second plot, Figure 7, shows the forecast superimposed on the original

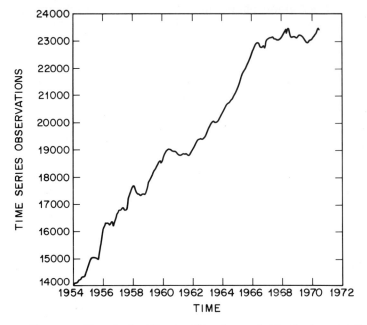

Figure 6. Time Series Plot for Flint Sunset Main Stations. Adjusted for cutovers: solid line.

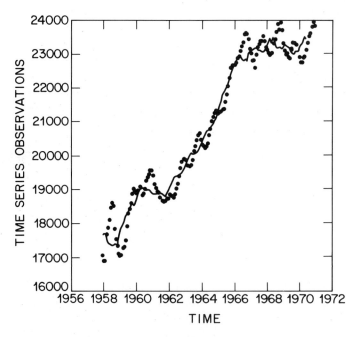

Figure 7. Flint Sunset Main Stations. Actual values = stations adjusted for cutovers; solid line = forecast values.

series. This somewhat inadequate forecast did not account for the trend in the data clearly evident in Figure 6. This suggested the removal of the trend by construction of a series of lagged differences of the original data.

7. MODERATELY EXPENSIVE CRT DISPLAYS

If the statistical investigator wants to interact directly with the pictures of a CRT display and would like the speed and response that is afforded by a dedicated computer, a medium sized computer like the Honeywell DDP-224 may provide the answer.

Such an installation is available at Bell Laboratories in Murray Hill. It provides an environment well suited for the design and implementation of an on-line interactive graphical program. The DDP-224 has 32K memory of 24 bit words, and hardware floating point arithmetic so that the large number of arithmetic calculations required to alter some Q-Q plots in real time, for example, can be done efficiently. Figure 8 shows a schematic diagram of the DDP-224 system with the wide variety of peripheral equipment that the 224 can accommodate, including graphical display devices such as a CRT and a color television, a tape drive, disc units, card reader, etc. Also, it has a Fortran IV compiler, and graphical displays can be generated by Fortran callable routines which plot points,

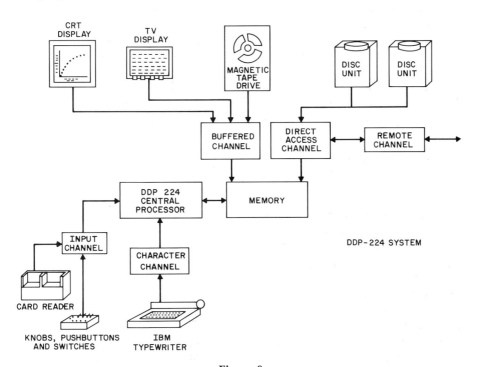

Figure 8.

vectors, and characters. The graphical data structure is extremely simple. The cost of this equipment is approximately $100K.

A probability plotting system was designed for the DDP-224 by E. B. Fowlkes and is discussed in an unpublished Bell Laboratories memorandum. The system was designed with certain definite features of operation in mind, features which are important for any system of this general type. First, the system must be usable by people with little or no programming training. This was achieved in part by having the system prompt the user on how to proceed at each step via typed messages. Also, errors made by the user do not cause the program to terminate. Instead, typed messages help the user to recover from error conditions.

The probability plotting system is structured into a set of 14 commands most of which operate on the probability plot which is currently displayed on the CRT. Applying a command to a probability plot generates a new probability plot which may serve as the operand for yet a new command. Commands are entered through the typewriter. Pictures may be saved on removable disc packs and hard copy of probability plots is available.

Calculations and flow of the program are controlled by the knobs, pushbuttons, and switches in a manner that will be illustrated by the following example.

Example

The data are taken from a 2^5 factorial experiment reported by Cuthbert Daniel [1959]. This experiment was designed to study the effects of five composition variables and one process variable on the tensile strength of a certain type of steel. An analysis of the data from this experiment was made by constructing a half normal probability plot of the 63 ordered contrasts. Figure 9 shows the half normal plot as it appeared on the CRT display. It shows the ordered contrasts plotted against the percentage points from the reference distribution (half normal in this example).

The lack of linearity in the configuration can be explained in part by the large gap between the top two points. The *OMIT* command in the system may be used to delete the last point which corresponds to a contrast that is possibly measuring a real effect. Two knobs move a small box about on the screen. When the box is positioned over a point to be omitted and a button pressed, the program searches through the coordinates to find the point inside the box.

The point is omitted, and the Q-Q plot redrawn as in Figure 10 in less than 1 second. Notice that omission requires that all percentage points (viz. the abscissa positions on the plot) be recalculated.

MODERATELY EXPENSIVE CRT DISPLAYS 87

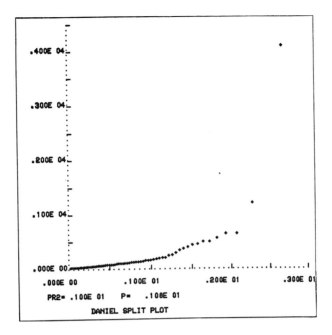

Figure 9. Daniel Split Plot.

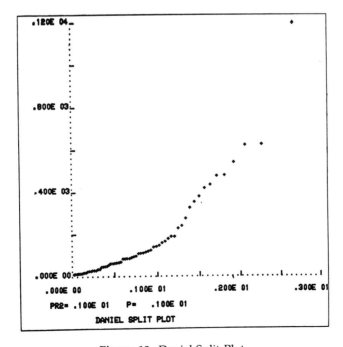

Figure 10. Daniel Split Plot.

Further analysis of the experimental procedure used to collect these data revealed a difficulty which contributed to the non-linearity of the resulting plot. Thirty-two heats of steel were first made, one for each of the 32 combinations of the 5 composition variables. Each level of the process variable, denoted F, was then applied to half of each heat. Thus there was plot splitting on factor F.

The *SPLIT* command is used to investigate the effect of the plot splitting. This command operates on the identifiers of the contrasts. A new probability plot is generated from the contrasts whose identifiers either do or do not match a certain pattern specified by the user. The probability plot in Figure 11 shows contrasts whose identifiers do contain the letter F. The configuration is now reasonably linear.

The set of contrasts remaining resembles a random half normal sample with constant variance. The slope of a straight line fitted to these contrasts provides a scale parameter estimate. In order to be conservative in the estimation, the *SUBPICTURE* command may be used to form an enlarged picture of a subset of the smallest contrasts. A character "P" is positioned with 2 knobs first to the lower left-hand corner of the subpicture, then to the upper right-hand corner. When a button is pressed, the subpicture is framed as in Figure 12. Pushing the button again enlarges the subpicture to fill the screen.

The *SLOPE* command may now be used to fit a straight line to the probability plot. The end of the line is positioned as before. After

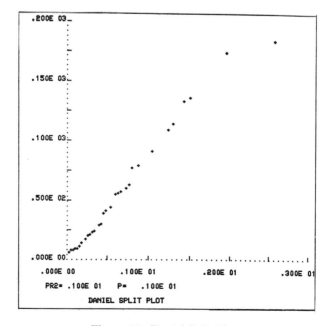

Figure 11. Daniel Split Plot.

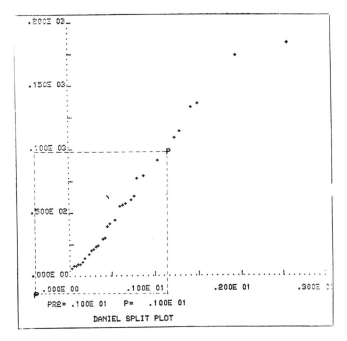

Figure 12. Daniel Split Plot.

anchoring the end by pushing a button, the line is drawn using polar coordinates, with one knob to control the radius and another the angle. At the end the calculated slope is displayed in the margin of the plot.

The system has been used in various applications including the determination of the distribution of lengths of conversations for WATS calls and the determination of the distribution of breaking points for certain plaques made from polyethylene plastics and subjected to stress tests. Each of these problems was characterized by a large number of test conditions with corresponding sample data and several candidate probability models. The system helped to speed up the process of choosing appropriate models.

REFERENCES

Andrews, D. F. and Tukey, J. W. (1973a). "Teletypewriter plots for data analysis can be fast: 6-line plots, including probability plots," *Journal of the Royal Statistical Society, Series C*, **22**: 192-202.

Andrews, D. F. and Tukey, J. W. (1973b). "Algorithm AS61 6-line plots," *Journal of the Royal Statistical Society, Series C*, **22**: 265-269.

Britt, P. M., Dixon, W. J., and Jennrich, R. I. (1969). "Time-sharing and interactive statistics," *Bulletin of the International Statistical Institute, Proceedings of the 37th Session*, **43**(1): 191-207. Voorburg, The Netherlands: International Statistical Institute.

Daniel, C. (1959). "Use of half-normal plots in interpreting two-level experiments," *Technometrics*, **1**: 311-341.

Daniel, C. and Wood, F. S. (1971). *Fitting Equations to Data.* New York: Wiley.

Gnanadesikan, R. (1973). "Graphical methods for informal inference in multivariate data analysis," *Bulletin of the International Statistical Institute, Proceedings of the 39th Session,* **45**(4): 195-206. Voorburg, The Netherlands: International Statistical Institute.

Pardee, S., Rosenfeld, P., and Dowd, P. G. (1971). "G101 - A remote time-share terminal with graphical output capabilities," *IEEE Transactions on Computers,* **C-20**: 878-881.

Tukey, J. W. (1965). "The future of processes of data analysis," *Proceedings of the Tenth Conference on the Design of Experiments in Army Research Development and Testing,* 691-729. Durham, NC: U.S. Army Research Office.

Tukey, J. W. and Wilk, M. B. (1965). "Data analysis and statistics: techniques and approaches," *Proceedings of the Symposium on Information Processing in Sight Sensory Systems,* 7-27. Pasadena, CA: California Institute of Technology.

Wilk, M. B. and Gnanadesikan, R. (1968). "Probability plotting methods for the analysis of data," *Biometrika,* **55**: 1-17.

3

PRIM-9: AN INTERACTIVE MULTIDIMENSIONAL DATA DISPLAY AND ANALYSIS SYSTEM

Mary Anne Fisherkeller
Jerome H. Friedman
Stanford Linear Accelerator Center, Stanford, CA

John W. Tukey
Princeton University, Princeton, NJ

ABSTRACT

(Submitted to A.E.C. Scientific Computer Information Exchange Meeting, May 2-3, 1974.)

PRIM-9 is an interactive data display and analysis system for the examination and dissection of multidimensional data. It allows the user to manipulate and view point sets in up to nine dimensions. This is accomplished by providing all 36 two-dimensional projections along the original axes at the push of a button, along with the ability to rotate the

---

Supported by the U.S. Atomic Energy Commission under Contract AT(043)515.
Prepared in part in connection with research at Princeton University supported by the U.S. Atomic Energy Commission.
SLAC-PUB-1408, republished by permission of the SLAC Publications Office, Stanford, CA.
This paper appears in *The Collected Works of John W. Tukey, Volume V—Graphics*, W. S. Cleveland, ed. Pacific Grove, CA: Wadsworth & Brooks/Cole Advanced Books & Software, 1988.
From *Dynamic Graphics for Statistics*, William S. Cleveland and Marylyn E. McGill, eds., copyright © 1988 by Wadsworth, Inc. All rights reserved.

data to any desired orientation. These rotations are performed in real time and in a continuous manner under operator control. From the parallax effect, arising from the dynamic aspect of this continuous rotation, one perceives a third dimension (depth into the screen).

PRIM-9 gives the operator the ability to perform manual projection pursuit. That is, by rotation and view change he can look at his data from all possible angles in the multidimensional space and try to find those that provide interesting structure. The system also allows interactive masking and isolation. The user can conveniently mask on any or all of the current variables, thus isolating interesting structures found along the way. These interesting structures can then be further analyzed alone, or may be subtracted from the total sample to simplify the search for still other structures. In addition to this strategy, the user may invoke an automatic projection pursuit algorithm. Starting at any projection (view), this numerical algorithm will search (in much the same manner as a human operator) for those projections that provide interesting structure. The system also incorporates the ability to save either temporarily, or permanently, any interesting view that is found. The operator can return to these views at any later time or reproduce them on a hard copy device.

1. INTRODUCTION

PRIM-9 is an interactive computer graphics program for Picturing, Rotation, Isolation, and Masking — in up to 9 dimensions. It is implemented on the Graphics Interpretation Facility of the Stanford Linear Accelerator Center, Stanford University. This facility consists of an Information Display's IDIIOM refresh CRT and a Varian 620/i minicomputer, linked to an IBM 360/91.

PRIM-9 is a result of a continuing program of research into techniques for applying computer graphics to exploratory data analysis. A general introduction to its properties and uses is documented in a 25-minute sound motion picture entitled "PRIM-9" [1973], produced by the Computation Research Group of the Stanford Linear Accelerator Center, Stanford University. This note details its properties with emphasis on the human engineering aspects of its implementation and on the various data analysis problems to which it can be applied.

PRIM-9 has been developed toward ends of very different breadth and distance: first, to gain insight into what can be learned by looking at the numerical aspects of data in more than two aspects at a time and, second, to implement a tool for pictorial examination and dissection of multidimensional data. The development of PRIM-9 has grown through many stages, and many of the early techniques that were implemented and then later discarded may turn out to be central to other data display systems. The resulting system as it is currently implemented is especially straightforward in concept. Its emphasis is on picturing and rotation on the one hand and masking and isolation on the other. Picturing means an ability to look at the data from several different directions in the multidimensional space. Rotation means, as a minimum, the ability to turn the data so that it can be viewed from any direction that is chosen. Picturing and Rotation are essential abilities, valuable alone, but their usefulness is greatly enhanced when combined with Masking and Isolation. Masking is the ability to select suitable subregions of the multidimensional space for consideration. That is, only those data points that lie in the subregion are displayed. Isolation is the ability to select any subsample of the data points for consideration. That is, only those points in the selected subsample are displayed. It is important to note that masking is tied to coordinates. If we rotate the data points, different points will enter and leave the masked region. Isolation, on the other hand, is tied to the data points. Under all operations of the system (except further isolation), the isolated sample remains the same. In the absence of rotation, masking and isolation are equivalent; however, in the presence of rotation the distinction is essential.

By interactively applying picturing, rotation, isolation and masking to his data, the user can, in particular, perform projection pursuit. That is, he can look for those views that display to him interesting structure. He can isolate any structures so found and study them separately and/or remove them from the remaining sample, simplifying the search for still further structures. In this way, he may gain considerable insight into the multidimensional properties of his data. As an aid to this process, the present version of PRIM-9 also provides an automatic projection pursuit algorithm [Friedman and Tukey, 1974]. This algorithm assigns to each view a numerical index that has been found to closely correspond to the degree of data structuring in the projection. When invoked, the algorithm will search for those views of the multidimensional data that maximize this projection index. At any point in the interactive session the user may invoke the automatic projection pursuit algorithm. Starting at the current view, the algorithm will find the view corresponding to the first maximum of the projection index, uphill from the starting view. Manual projection pursuit can then continue from this new (hopefully more structured) view. The algorithm can also be invoked at the beginning of the session, starting with the various original or principal axes of the data. The resulting views can then provide useful starting points for further investigation.

The next section gives a brief description of the hardware and software configuration of the Graphic Interpretation Facility of the Stanford Linear Accelerator Center, on which PRIM-9 was developed. This is followed by several sections describing the implementation of the various features of the PRIM-9 system. The note then concludes with a section discussing various techniques for applying these features to some multidimensional data analysis problems.

2. THE GRAPHIC INTERPRETATION FACILITY

The Graphic Interpretation Facility of the Stanford Linear Accelerator Center (SLAC) is described in detail elsewhere [Beach, 1969; Beach, Fisherkeller, and Robinson, 1971]. A brief description emphasizing those properties relevant to the implementation of PRIM-9 is included here.

The primary computing resource at SLAC during the development of PRIM-9 was an IBM SYSTEM/360 Model 91, with 2048k bytes of storage, operating under OS/MVT, with HASP.* The two basic components of the Graphic Interpretation Facility are a Varian Data Machines 620/i computer (Varian Data) with 8k 16-bit words of storage, and an IDIIOM Display Console (IDIIOM) with a 21-inch CRT made by Information Displays, Inc. When the display is operating, the 620/i memory contains a program for the 620/i to execute and a program (display file) for the IDIIOM to execute. These two programs run concurrently, with the IDIIOM stealing cycles from the 620/i. The instructions (orders) in the display file may display characters, points, or straight line segments, perform unconditional or subroutine jumps, count and index, or interrupt the 620/i. Characters, points, and lines are positioned on a 1024 by 1024 raster unit grid on the face of the CRT. The 620/i instruction set includes, in addition to the usual set for a standard mini-computer, instructions to start and stop the IDIIOM in its execution of the display file, and instructions to read and reset registers associated with the display operation.

Interaction at the console is by means of a solid state alphameric keyboard, a light pen, and a function keyboard with 32 buttons. Under program control, portions of the display file may be designated as light pen sensitive or insensitive. When the light pen is pointed at a sensitive item on the CRT, the 620/i will be interrupted. The function buttons generate interrupts on both the closing and the opening of the switch. This means that software can produce either (1) single impulse for each depression of the button (cycling) or (2) a repeated impulse controlled by software timing, occurring so long as the button remains down. This last facility, especially when combined with automatic reversal (see below), is of great importance in allowing effective control. Plastic overlays may be placed on the function keyboard to identify the purpose of each button.

The 620/i and SYSTEM/360 are connected through an IBM 2701 Parallel Data Adapter Unit. Data may be transmitted in either direction through this link. The 620/i can interrupt the SYSTEM/360 and determine whether the SYSTEM/360 is trying to read or write; however, the SYSTEM/360 is not able to interrupt the 620/i.

Although the Facility can operate in a stand-alone fashion, most of our work (including the implementation of PRIM-9) has been done using

---

* The current resource includes in addition to the S/360-91, twin IBM SYSTEM 370/168's, operating under VS/2R1.6 with ASP.

it in conjunction with the SYSTEM/360. The SYSTEM/360 provides fast computation and mass storage, while the 620/i maintains the display.

A package of PL/1 procedures, the IDIIOM Scope Package, has been provided for writing highly interactive programs on the IBM 360/91 without burdening the writer with all of the details of programming the 620/i and IDIIOM. The user of this package can, by means of procedure calls, control the display console in addition to having all of the facilities of PL/1 available to him. Procedures are supplied to construct graphic display elements which will display information on the IDIIOM CRT, and transmit elements as well as interrupts between the SYSTEM/360 and 620/i.

3. IMPLEMENTATION

3.1 Scaling and Coordinates

The data is stored (in the 360/91) in initial coordinates either as scaled before entry or as rescaled to fit into the display region. It remains in this form. Rotations to current coordinates are performed by a transformation matrix. An ongoing step of rotation involves (A) updating this transformation matrix (multiplication by a simple rotation matrix) and (B) passing the data through the modified transformation matrix and transferring the relevant coordinates to the 620/i. Thus in the PRIM-9 system, it is important to distinguish between the *initial* coordinates and the *current* coordinates. The *initial* coordinates are defined to be an orthogonal set of fixed axes in the multivariate space. The input data to PRIM-9 consists of a number of ordered sets of numbers. Each of these numbers is assigned to the initial axis corresponding to its position in the ordered set. These numbers then become the values of the initial coordinates and each ordered set of numbers becomes a point in the Eucleadian space spanned by the initial axes. The *current* coordinate axes are another orthogonal set spanning the same space that is connected to the original set by a similarity transformation. That is, each of the current coordinates is a particular linear combination of the initial coordinates. So long as only rotations are used (see 3.6 for exceptions), these linear combinations are restricted to those that do not change the interpoint distances in the multidimensional space.

Rotation has the effect of continuously changing the linear transformation between the current and initial coordinates. At the beginning of the session, the initial and current coordinates coincide, but as soon as a rotation is performed the current coordinates become distinct from the initial ones.

3.2 Picturing

For picturing, PRIM-9 provides the choice of any two of the *current* coordinates as horizontal and vertical axes. Two buttons cycle through the choices. One button changes the coordinate displayed in the vertical direction, while the other button changes the coordinate shown along the horizontal direction. In this way, it is easy to go through the $\binom{NDIM}{2}$ (NDIM is the total number of coordinates) possible displays, as well as getting from one particular display to another. Choosing a display involves selecting two integers i and j, where i is the y (vertical) coordinate and j is the x (horizontal) one. Pushing the first button increments i by one and the second, similarly, increments j, both modulo NDIM. In order to speed up the selection process, all unnecessary combinations have been eliminated by requiring that i be greater than j. That is, i cycles from 2 to NDIM while j cycles from 1 to $i - 1$.**

3.3 Rotation

Being able to see projections on all coordinate pairs can be very useful. But it is not enough. To be able to get reasonably to any two-dimensional projection means either a way to call for the projection that we want, or a way to move about in the multidimensional space. Since we usually do not know just what we want, and when we do we will find it difficult to learn to call for it in a general way, we need a way to move about. Continuous controlled rotation is a natural way to move about (change projection) in a multidimensional space.

In the implementation of controlled rotation, one naturally thinks of turning a knob. However, the configuration of the Graphic Interpretation Facility does not support a knob. Thus, we are forced to implement rotation through pushing buttons. The naive approach to the control of a single rotation by buttons involves two buttons, with these responses:

— to one button: rotation "to the right" at a constant angular rate so long as the button is depressed;

— to the other button: rotation "to the left" at the same constant angular rate so long as the button is depressed.

This type of control has two serious flaws:

— if the rotation rate is slow enough for fine adjustment, the delay time for large rotations is undesirable — so undesirable as to be nearly impractical;

** If we were programming this action again, we would use a constant-step version of the increasing-step-and-reversal control described under rotation.

— if used naively — two buttons per rotation — it is too easy to use up too many buttons. (PRIM-9, in its present version, makes as many as $\binom{9}{2} = 36$ different rotations available.) Both of these disadvantages can be overcome, the first by an "increasing-step-and-reversal" control, the second by "time-sharing" the rotation drive.

The type of rotation control used in PRIM-9 has two features:

(1) rotation reversal — rotation in one sense so long as the single button is held down — when the same button is released and then again depressed, rotation in the opposite sense so long as the button is held down;

(2) rotation by increasing (accelerating) steps. These steps are presently taken as 1, 1, 2, 4, 8, 16, 32, ... times a small (unit) angle. (Note the repetition of 1.) The largest possible step is limited by a speed limit settable at 1, 2, 4, 8, 16, 32, 64, or 128.

This combination of rotation reversal and acceleration gives the operator fast and easy approach to a desired data orientation. The rotation starts out slowly but quickly accelerates to the speed limit so long as the button remains depressed. When the operator sees an interesting data orientation on the CRT screen, he releases the button. Due to human reaction time, both in perception and in releasing the button, the rotation usually overshoots the desired data orientation. Depressing the same button again causes the rotation to proceed in the opposite direction, starting out at the slowest speed, and again accelerating. When the desired orientation again comes into view on the screen the button is released. Now, due to the slower speed (the acceleration usually has not reached the speed limit), the overshoot is much less (in the opposite direction). Again, depressing the button causes rotation reversal at the slowest speed allowing the operator to home in on the desired data orientation. In the usual case, at most two reversals are necessary. This strategy allows rotation in both directions, minimizes the delay time for large rotations, allows a slow rotation for fine adjustment, and requires only one button to drive the rotation.

In order to specify a rotation, one needs not only to specify the sign and magnitude of the rotation angle but also the rotation axis. A general rotation axis in a multidimensional space (for example, in terms of its direction cosines) is complicated to understand and time-consuming to specify. In PRIM-9, directly available rotations are confined to those associated with pairs of the *current* coordinates. This makes both control and understanding relatively easy. A rotation axis is specified by two integers i and j. These integers specify the current coordinate axes that participate in the rotation. That is, coordinates i and j rotate while the other NDIM $-$ 2 coordinates axes, orthogonal to i and j, remain fixed. It is possible to get from any one two-dimensional projection to any other, with relative ease, by combining these selected rotations in the correct amount and sequence (this is the multidimensional analog of the Euler angle specification of a rotation in three dimensions). The two integers (i and j) that specify the coordinates that participate in the rotation are selected in exactly the same manner, as discussed in the previous section for choosing the current projection axes. The selection is controlled by two buttons that cycle through the $\binom{\text{NDIM}}{2}$ possibilities. One button cycles through $2 \leq i \leq$ NDIM in increments of one while the other, similarly, cycles through $1 \leq j \leq i - 1$.

If the axes that define the rotation coincide with the current projection axes, then the data points will simply move in circular orbits about their relative mean in the projection. If neither axis corresponds to a current projection axis, then the display will remain unchanged because the rotation is orthogonal to the current projection axes. Both rotation axes are "invisible" to the screen. Useful rotations occur when one of the axes that participates in the rotation corresponds to a current projection axis, while the other is one of the NDIM − 2 axes orthogonal to the current display plane. In this case, the operator sees the projected points change their positions on the screen as the invisible coordinate is rotated against the visible one. When an interesting pattern emerges, as a result of the rotation, the operator can home in on it as described above. The invisible coordinate can then either be rotated against the other visible one or the operator can change to another invisible coordinate. He can then rotate these new coordinates to try to sharpen the structure even further. Continuing in this manner, the operator may manually iterate to a data orientation that provides an informative view of his multidimensional point cloud.

Easily recognizable continuous rotation provides an additional advantage. Its dynamic effects let one see an additional dimension, not instantaneously, but yet at the same time — in the best sense of those words. This is due to the relative motions of the points associated with parallax. The points that are close in the invisible rotation coordinate move across the screen more rapidly than those that are farther away. This parallax effect gives the illusion of a third dimension (depth into the screen) and this aspect seems to be a very useful complement to the two aspects (horizontal and vertical) provided directly by the screen.

To help in the interpretation of a particular projection, a display of the initial coordinates axes, as projected onto the current projection plane, is easily switched (with a pushbutton) in and out of view. This display allows the user to see graphically the linear combinations of the original coordinates that comprise the two axes of the current projection plane. Another button (neutral) returns the display to its initial state. That is, the current coordinates are reset to the initial ones.

3.4 Masking

Masking is the ability to select any subregion of the multidimensional space, and have only those data points that lie in the subregion displayed on the screen. Masking is used in connection with isolation and also has some interesting uses of its own (these are discussed in the last section). It is important to note that masking is tied to the coordinates. Under rotation, the data points will enter and leave the masked region.

The flexibility of PRIM-9's masking, like the flexibility of its rotation, is limited, so that it is easy to control and understand. PRIM-9 allows simple masking on any or all of the current coordinates. That is, those points for which $X_i < F_i$ or those for which $B_i < X_i$, or both — for a single i or several i's — are caused *not* to appear on the screen. Here X_i ($1 \leq i \leq \text{NDIM}$) are the current coordinates and F_i, B_i are forward and backward bounds on each of these current coordinates.

Masking is controlled by five buttons. One button toggles the masking on and off. A second button cycles (one step per press) through the integer, i, ($1 \leq i \leq \text{NDIM}$) which identifies that current coordinate to which the mask is to be applied or altered. The third identifies which edge F (front), B (back) or J (joint) is to be driven. These three options are cycled through by successive pushes of the button. The fourth button drives the selected mask in a continuous motion using the same "increasing-step-and-reversal" control technique described above for rotation. The fifth button allows the rapid selective removal of the mask for a particular coordinate (identified using the second button) by resetting both the front and back mask edges to their outside positions.

This is in contrast to the effect of the first button which, when toggled off, removes the masks on all coordinates simultaneously. The J (joint) drive moves the front and back masks together in the same direction and speed, maintaining a fixed separation between them. This allows driving an unmasked zone of chosen width back and forth on a particular coordinate.

If the masking coordinate is one of the current projection axes, then driving the mask away from its outside position and toward the center of the screen will cause the points to disappear along an advancing line, accelerating to the speed limit so long as the drive button is held down. When the button is released and again depressed, the masking line will reverse, starting at the slow speed. The masked points will reappear as the mask boundary retreats. As for the case of rotation, this control allows the operator to arrive at and home in on a desired mask position easily and quickly.

A masked coordinate need not be a current projection axis. For example, one can mask on a current coordinate orthogonal to the projection plane. As the mask moves from its outside position on this "invisible" coordinate, points will disappear from various regions of the screen giving insight into the relationship contained in the data between the invisible coordinate and the two visible ones. In particular, the joint drive allows driving an unmasked zone of chosen width back-and-forth on the invisible coordinate. This will cause points to appear and disappear as the mask passes through the various values along the invisible coordinate. More sophisticated techniques using moving masks are discussed in the last section.

3.5 Isolation

Isolation is the ability to select an arbitrary subset of the data sample at any point in the analysis, and perform upon this subset or its complement (the full sample with the subset removed), all operations that one can perform on the full sample. Experience with PRIM-9 has shown that isolation is a most essential adjunct to picturing and rotation. It extends the system from being a purely linear device to a piecewise linear device, greatly increasing its effectiveness and power. Some of its more standard applications are discussed in the last section.

As implemented in PRIM-9, isolation begins with masking. The data points to be isolated are defined by constructing a mask (in terms of the current coordinates) that just contains the points to be isolated. As the mask is applied, the points that are masked out disappear from the screen, making it relatively easy to interactively construct a mask that just includes the desired subset of points to be isolated. It might seem, at first thought, that because of the limited flexibility of PRIM-9's masking, it would be difficult to construct mask boundaries that include arbitrary point subsets. This turns out not to be the case. This is due mainly to the fact that the current coordinates, on which the mask is defined, can be interactively rotated to an arbitrary orientation with respect to the initial coordinates, and that the isolation can be applied repeatedly in defining the subset. In this way, the user can construct a piecewise linear approximation to any boundary surface in the multidimensional space to sufficient accuracy to just include the points to be isolated.

When all of the undesired points have been masked out, so that only the subsample to be isolated appears on the screen, the user presses a button to invoke the isolation. This causes a menu of options to appear. Figure 1 is a photograph of this menu. The user selects the appropriate option by touching a light pen to those places on the screen that define the option.

The numbers at the right reference the sixteen isolates that can be simultaneously defined. Isolate "0 = ALL" always references the total sample and "15 = RESIDUAL" is reversed for special use with the "residual after" option. This leaves fourteen isolates that can be arbitrarily defined and stored by the user.

Touching the light pen to one of the seven options to the left causes a question mark to appear either to the right of the option or at a blank space within it. Touching the pen to one of the sixteen numbers causes the selected number to replace the question mark. When all of the question marks have been replaced by numbers and the command is correct, then touching "APPROVED" initiates the action. The commands are:

FILL n: Save currently masked subset at n.

RECALL n: Recall isolate saved at n and replace current subset with it.

SUBSELECT ON $n1$ FILL $n2$: Save at $n2$ the intersection of the current subset with the isolate stored at $n1$.

INTERSECT $n1$ AND $n2$ FILL $n3$: Save the intersection of isolates $n1$ and $n2$, at $n3$.

UNION $n1$ AND $n2$ FILL $n3$: Save the union of isolates $n1$ and $n2$, at $n3$.

NOT $n1$ BUT $n2$ FILL $n3$: Save the intersection of the complement of isolate $n1$ with isolate $n2$, at $n3$.

RESIDUAL AFTER n ALSO: Replace isolate 15 by the intersection of the complement of isolate n with the current isolate 15.

If only a number is light-penned, "RECALL" is the default command. As each subset is isolated, the remainder may be saved with the last command. An asterisk appears to the left of those numbers that represent isolates that are currently in use. When an isolate has been saved or recalled, it becomes the current subset. If instead of a light pen hit, the button invoking the isolation is simply pushed a second time, the display will alternate between the current subset and the entire data sample. The current isolate number always appears at the bottom of the screen.

3.6 Scale and Location Transformations

The location may be displaced in either the positive or negative direction and/or the scale can be expanded or contracted on any current coordinate axis. This is accomplished by specifying a "key" integer, i, ($1 \leq i \leq$ NDIM) representing the coordinate to be transformed. A single button cycles through the possible values of i. Another button displaces the origin of the selected coordinate while a third button scales the coordinate. These latter two buttons drive the displacement or scale change in a continuous motion, using the "increasing-step-and-reversal" control technique described earlier.

3.7 Saving Views (Projections)

Quite often during a session, the user finds a projected view of his data that is sufficiently interesting to warrant saving it so that he may return to it later in the session, at another session, or perhaps record it permanently on a hard copy device.

PRIM-9 provides for both temporary (within a session) and permanent (between sessions) view saves. The permanent saves can also be transferred to a hard copy device at the user's discretion. This facility allows the user to continue with an analysis after he has found an interesting view. If further analysis should not improve the structuring, or perhaps even worsens it, the user can simply recall the saved view and begin again on a different track.

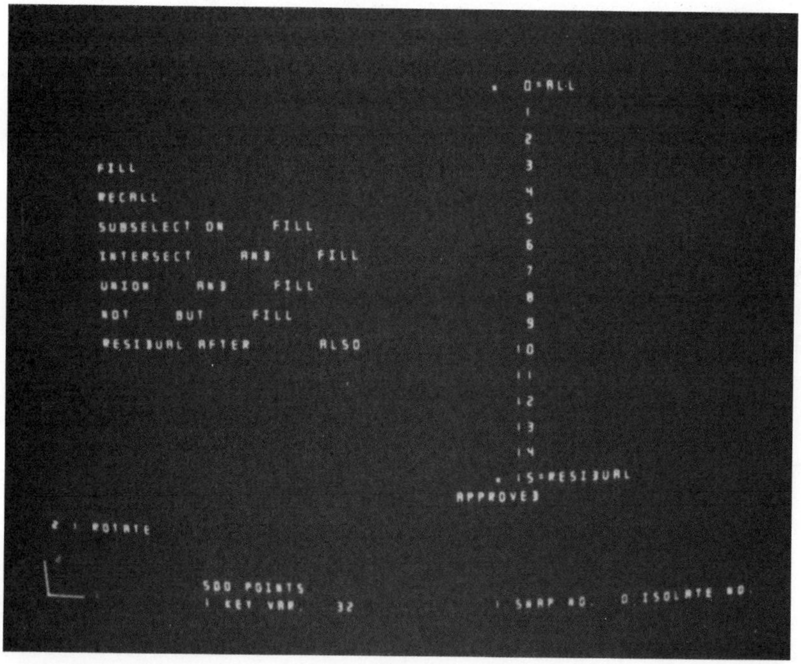

Figure 1

Saving a view (projection) consists of storing the transformation matrix connecting the initial coordinates to the current coordinates along with the mask bounds at the time of the save. Up to six different views may be saved on a temporary basis and another eighteen may be saved in a permanent disk data set. The temporarily saved views are simply identified by their numbers, i ($1 \le i \le 6$), while the permanently saved views on the disk are given identifying names typed by the user on the keyboard. If no such name is typed, then the system assigns a default identifier, which is the date and time of the save. To retrieve one of these views, the user depresses a button, which presents a menu on the screen listing the names of all of his saved views. He selects a view by touching the light pen to the appropriate name.

3.8 Automatic Projection Pursuit

The PRIM-9 system provides the user with a sort of "automatic pilot" for rotation. That is, at the user's request, the system will invoke a numerical algorithm that automatically searches for data orientations (projection directions) that display interesting structure. This algorithm is detailed elsewhere[2] and only its general properties, as they relate to the implementation on the PRIM-9 system, are discussed here.

The automatic projection pursuit algorithm assigns to each projection a numerical index, $I(\hat{k}, \hat{l})$, that corresponds to the degree of data structuring present in the projection. Here $\hat{k}$ and $\hat{l}$ are the two orthogonal unit vectors representing the particular two-dimensional projection of the NDIM-dimensional data. The more structure present in the projection, the larger $I(\hat{k}, \hat{l})$ becomes. The essence of the algorithm is to find those projection directions ($\hat{k}$ and $\hat{l}$) that maximize $I(\hat{k}, \hat{l})$, subject to the constraint $\hat{k} \cdot \hat{l} = 0$. This projection index, $I(\hat{k}, \hat{l})$, is constructed to be a smooth function of its arguments so that sophisticated numerical maximization algorithms can be employed, minimizing the CPU cycles required.

The automatic projection pursuit algorithm can be invoked in two ways from the PRIM-9 system. At the simplest level, the user simply depresses a button. Starting with the current projection (appearing on the screen), the algorithm finds (and displays on the screen), the projection corresponding to the first local maximum of the projection index, $I(\hat{k}, \hat{l})$, uphill from it. In this search the horizontal coordinate, $\hat{k}$, is held fixed while the vertical coordinate is varied in the (NDIM − 1)-dimensional subspace orthogonal to $\hat{k}$. Depressing the button a second time causes the search to continue, but this time the vertical coordinate is held fixed at the previous solution value, $\hat{l} = \hat{l}^*$, while the horizontal coordinate, $\hat{k}$, is varied in the (NDIM − 1)-dimensional subspace orthogonal to $\hat{l}^*$, seeking a further maximum of $I(\hat{k}, \hat{l}^*)$. Pressing the button a third time causes a further maximization, this time holding the horizontal coordinate $\hat{k}$ fixed at its solution value, $\hat{k}=\hat{k}^*$, and again varying the vertical one, but this time in the subspace orthogonal to the new horizontal coordinate $\hat{k}^*$. This procedure of alternately holding one coordinate fixed while varying the other in the subspace orthogonal to the first, can be continued (by repeated pushes of the button) until either an interesting view appears or until it converges (subsequent searches produce no change in the projection). At each stage in this iterative process, the results are displayed on the screen as the current projection.

The second mode of invoking automatic projection pursuit allows starting the search at projections other than the current one displayed on the screen. This is accomplished by depressing a different button. This causes a menu to appear on the screen listing the options available. These options allow the choice of any two of the initial coordinates or principal axes of the current data set (isolate) as starting axes for the automatic projection pursuit. This menu also allows the interactive modification of some of the parameters of the automatic algorithm as well as simply displaying the data along the various principal axes without the corresponding search. The user selects from among these options by touching the light pen to the appropriate places on the screen where the options appear. After starting the search at two of these alternate axes, the user continues it by repeatedly pressing the first button as described above.

4. DISCUSSION

PRIM-9 has been available for production data analysis for a relatively short time so that our experience with it is necessarily limited. We have probably not yet discovered many of its most interesting and useful applications. It has, so far, been employed in the exploratory analysis of multivariate high energy particle physics data, and in both supervised and unsupervised multivariate discrimination analysis in pattern recognition problems.

In the exploratory data analysis application, the system essentially functions as a cluster detection and separation device. By repeated application of view change and rotation (both manual and automatic), the user tries to find those data orientations that reveal to him interesting structure or clustering. Using the automatic projection pursuit algorithm (started at the later principal axes of the data) to find interesting starting projections, the user then manually iterates the rotation to try to find structure or to sharpen any structure found. This iteration process usually proceeds as follows. A visible coordinate is rotated against one of the invisible ones until the clustering is as sharp as possible. This invisible coordinate is then rotated against the other visible one toward the same end. Another invisible coordinate is then chosen to be rotated against the two visible ones, hopefully increasing the cluster formation and separation. After all of the NDIM − 2 invisible coordinates have been rotated against the current visible ones, the whole process can be repeated (since in the process of these rotations the current coordinates have been considerably changed) so long as progress is being made. At any point, the automatic projection pursuit algorithm can be invoked (this usually happens when progress being made by manual rotations is slow).

If a view is found that separates the data into two or more clusters, this view can be used (via masking and isolation) to isolate the clusters into separate data subsamples. These subsamples can then be analyzed individually to see if there are different projections that reveal still further clustering within each of the subsamples. Even if no further clustering is found in an isolate, rotating it in a controlled manner in the multidimensional space can still give considerable insight as to its multivariate properties. The parallax effect of the continuous real time rotation is very useful in providing the third-dimensional effect of depth into the screen.

Moving masks can also be useful in gaining insight into the multidimensional characteristics of data point sets. The simplest of these (as mentioned above) is the sliding of an unmasked zone of chosen width back and forth on the various coordinates invisible to the screen, one at a time. Another more sophisticated approach might be called the "concealed generalized episcotister." Its operation would be roughly as follows. Let each coordinate run between −1 and +1, let a, b, c and d be small fractions, and let the data for this example be six-dimensional.

picture = coordinates 1 and 2

masks = F_3 at $-a$, B_4 at $+b$, F_5 at $-c$, B_6 at $+d$

rotation = play with (4,3), (5,3), (6,3), (5,4), (6,4) (6,5)

The unmasked hypervolume is a hexadecant in the four coordinates 3, 4, 5 and 6. Rotating these coordinates among themselves essentially rotates this hyper-conical (linearly, not spherically hyper-conical) region around among the (four-dimensional projections of the) points, thus generalizing the rotation of a conventional (sector-disk) episcotister. This approach seems often rewarding in gaining the first clues as to major structure in a point cloud.

If, as another example, we suspect that the "core" (in one or more of the invisible coordinates) of a visible concentration of points is curved, the natural mode of inquiry is to (in order):

(1) rotate until the concentration seems as strong as possible (this will tend to eliminate "tilts" of the core);

(2) mask on one invisible coordinate, coming in from each side until perhaps one-sixth to one-fourth of all points are blotted out from each side;

(3) operate mask-countermask rapidly;

(4) rotate among the invisible coordinates at whim, repeating number (3) all the while.

Appearance of a displacement oscillating with the mask-countermask frequency is an indication of existence, direction, and amount of curvature.

In the pattern recognition applications, the system functions as both a linear and piecewise linear device in supervised and unsupervised discrimination analysis.

As an example of its use as a linear device in a supervised application, consider the problem of finding the best linear discriminant direction in the multivariate space for separating two known data classes. Here, the points corresponding to each class are correctly identified from some external source and the problem is to find the best direction separating the two classes when the data are projected onto that direction. The analysis proceeds as follows. A coordinate is added which contains the information identifying the class of each data point. This extra coordinate is pictured (vertically) against each of the data coordinates (horizontally) in turn, starting with the *second* coordinate.

While viewing each such projection, the data is rotated among all the NDIM coordinates with the objective of achieving maximal *overlap* of the two classes in each horizontally pictured coordinate. This process is repeated iteratively until it is impossible to increase the overlap in each of the NDIM − 1 projections (2 through NDIM). At that point the direction of the current first coordinate represents the optimal discrimination axis, and picturing the extra (class identifying) coordinate against this first one, directly displays the discrimination achieved.

It is the presence of isolation in conjunction with rotation that gives the system its piecewise linear capability. This is easily demonstrated by considering the simple example of two-dimensional data set, consisting of two classes whose boundaries are outlined in Figure 2. Clearly, there is no single one-dimensional discriminant direction that can completely separate the two classes (A and $B = B_1 + B_2 + B_3$).

Applying rotation (both manual and automatic), the user might find a projection that achieves a good partial separation (such as P_1 in Figure 2). Using this projection, a mask is constructed at M_1 and the B_1 sample is isolated from its complement $A + B_2 + B_3$. These two isolates are then rotated separately, searching for further structure. In the case of subsample B_1, this will yield a null result. For the subsample complement to B_1, however, the user may iterate to projection P_2 which does exhibit further structure. Using this projection to apply a mask at M_2, the B_2 sample is isolated from its complement $A + B_3$. Continuing with this procedure, the user can further iterate to projection P_3. Applying a mask at M_3 in this projection completely isolates Class A from the remainder of Class B (B_3). Thus, the application of the piecewise linear mask M_1, M_3, M_2 completely separates the two data classes. For the supervised case, where it is known in advance that there are only two classes, this completes the solution. For the unsupervised case, one can separately analyze the B_1, B_2 and B_3 isolates as well as their union ($B_1 + B_2 + B_3$), using rotation and masking to determine whether they represent separate classes or whether they are connected, thus representing a single class.

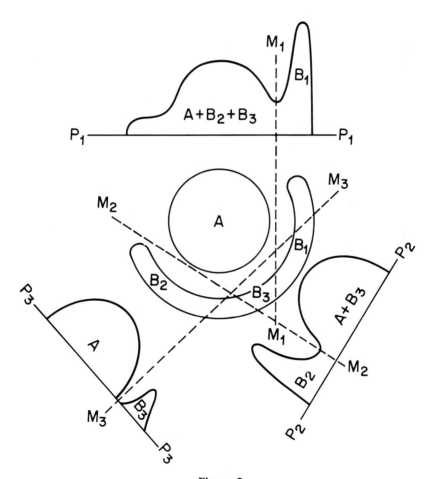

Figure 2

For this simple illustration, the dimensionality of the data was two while the projection pursuit was in one dimension. In PRIM-9, the data dimensionality can be as large as nine while the projection pursuit is in two dimensions. However, the basic principles are the same. One applies rotation and view change, trying to find projections with structure. When structures are found, they are isolated and the isolates are each individually studied, seeking further clustering. If found, these are further isolated and so on. When no more structuring can be found among the isolates, then they are analyzed separately and together to try to understand their multivariate properties.

5. CONCLUSIONS

PRIM-9 has been in use for a short time examining and dissecting multivariate data. Its direct value for this purpose will have to be learned by experience. We have learned from its development that pictorial systems to be effective must, as did PRIM-9, go through many stages of trial and error learning. We now understand that the details of control can make or break such a system. We now recognize that these details of control must be adapted to what is available and to the people who are to use the system. If we had ten knobs and six switches instead of 32 buttons and a light pen, we would have realized the same four essentials in a quite different way. We now recognize the great value of the dynamic aspects of the display, especially easily recognizable rotation. Two aspects, horizontal and vertical, are always before us. We now have a strong feeling that the third aspect which supports these two best is this dynamic aspect of rotation, more useful than stereoscopy, color, flicker, or distinctive characters. We have learned that this sort of pictorial facility can be useful in two quite different ways: first, directly in the interactive analysis of multivariate data, and second, as a source of ideas and approaches for the development of computer algorithms for multivariate analysis. The automatic projection pursuit algorithm, for example, was developed by observing the systematics of the interaction between the users and the PRIM-9 system. These systematics were encoded into a computer algorithm which has been very successful in seeking optimum projections.

Finally, and above all, we have learned that the four essentials of Picturing, Rotation, Isolation, and Masking need to work together and that from them much can be learned.

REFERENCES

Beach, R. C. (1969). "The SLAC scope package for the IDIIOM — a collection of PL/1 procedure which may be used to control the IDIIOM display console," Report No. CGTM No. 80. Stanford, CA: Stanford Linear Accelerator Center, Computation Research Group.

Beach, R. C., Fisherkeller, M. A., and Robinson, G. A. (1971). "An on-line system for interactive programming and computer generated animation," Technical Report, SLAC-PUB-939. Stanford, CA: Stanford Linear Accelerator Center.

Friedman, J. H. and Tukey, J. W. (1974). "A Projection Pursuit Algorithm for Exploratory Data Analysis," *IEEE Transactions on Computers*, **C-23**: 881-890.

IDIIOM Technical Description. Information Displays, Inc., 333 North Bedford Road, Mount Kisco, NY 10549.

"PRIM-9." (1973). Produced by Stanford Linear Accelerator Center, Sanford, CA. (S. Steppel, ed.) Bin 88 Productions. (Film)

Varian Data 620/i Computer Manual. Bulletin No. 605-A, Varian Data Machines, 2722 Michelson Drive, Irvine, CA.

4

KINEMATIC DISPLAY OF MULTIVARIATE DATA

D. L. Donoho
P. J. Huber
E. Ramos
H. M. Thoma
Harvard University

ABSTRACT

Graphics devices are now available for kinematic display of three-dimensional scenes. As statisticians, this technology interests us because of its possible applications to multivariate data analysis. Kinematic graphics seem well suited both for informal exploratory purposes, and for diagnosing problems with formal statistical procedures.

To explore the potential of this technology, we have developed an interactive data analysis package and embedded within it extensive kinematic display capabilities. With this system, the interplay between informal "looking at data" and formal steps like "fitting of models" can be rapid and direct. Moreover, because the system is flexible and extensible, we can use it to analyze data in settings as diverse as cluster analysis and multiple time series analysis.

Reprinted from (1982) Proceedings of the Third Annual Conference and Exposition of the National Computer Graphics Association, **1**: 393-398, by permission of the National Computer Graphics Association.

From *Dynamic Graphics for Statistics*, William S. Cleveland and Marylyn E. McGill, eds., copyright © 1988 by Wadsworth, Inc. All rights reserved.

A videotape illustrating the use of our system and some possible applications will be shown at the conference.

1. BACKGROUND

A *kinematic display* device shows a "movie" of a rotating three-dimensional scene; that is, it shows a sequence of incrementally rotated views of the scene, updated sufficiently rapidly to produce the effect of three-dimensional perception in the human observer.

Kinematic graphics may be obtained, for simple scenes, using a modest home computer equipped with TV monitor. For serious applications, several kinematic display systems are now commercially available: these have high resolution video display units (VDU's), built-in arithmetic processors to handle rotation and change of viewpoint, a variety of user input devices (joysticks, data tablets, dials, switches, and tracker balls) and extensive software libraries for high-level graphics development. Using these specialized systems, one can deal with quite complex 3-d scenes: the more powerful systems can give flicker-free displays of scenes consisting of several thousand objects undergoing smooth motion. Such devices have clear applications in disciplines where the manipulation of 3-d images is of central importance, architecture, mechanical design, flight simulation, and macromolecular chemistry being obvious examples.

As statisticians, we are interested in the use of such devices in the analysis of statistical data. While it is clear that graphics and graphical devices have an important role to play in the *presentation* of statistical data, the reader may not see how they can be used in the *analysis* of statistical data. Indeed, statisticians have developed an apparatus for making inferences about uncertain hypotheses based on statistical data, an apparatus which is algebraic and formal rather than graphical. In one of the most common data-analytic procedures, linear regression analysis (the fitting of a linear equation to data by least-squares) the algebraic approach is plain: one speaks in terms of (independent and dependent) *variables*, (significant) *coefficients*, (best) *equations*, and one assesses the adequacy of a data description by examining ratios of sums of squares.

This formal approach to data analysis has a number of recognized shortcomings. By proceeding formally, one may overlook the obvious and choose an entirely wrong type of model. Or the model may be basically correct, but a particular technique used for fitting the model (e.g., least-squares) might be invalidated by some overlooked problems in the data (e.g., outliers). These problems are addressed by two recent approaches to data analysis:

1. The development of informal techniques for *exploratory* data analysis [Tukey, 1977; Velleman and Hoaglin, 1981]; and

2. The implementation of *diagnostic* techniques for identifying failures in standard analysis techniques [Gnanadesikan, 1977; Belsley, Kuh, and Welsch 1980].

Both approaches rely heavily on graphical display and on the eye's superior pattern recognition ability to avoid the "blind spots" associated with narrow, formal approaches. An elegant, elementary discussion of the importance of graphics for good statistical analysis is given by Anscombe [1973].

Traditional ink and paper graphics, such as scatter plots, are basically limited to handling two or fewer variables at a time (although see Tukey and Tukey [1981] or Gnanadesikan [1977] for ways of squeezing several variables onto a traditional display). On the other hand, much data analysis is *multi*-variate; in the typical setting, one has an array of "observations" — e.g., data on 100 people — consisting of d "measurements" (variables) — e.g., each person's height, age, weight, sex, ... — and one is interested in relations between measurements and among the observations. There is a standard set of formal techniques for answering questions about such data: for predicting weight as a function of height, age, and sex there is regression analysis; for finding groups of individuals with similar characteristics there is cluster analysis; for identifying factors on which men and women differ most there is discriminant analysis, and so on. These formal techniques generalize straightforwardly to highly multivariate situations; whereas as exploratory or diagnostic tools based on ink and paper graphics will not generalize so easily.

The role for kinematic devices in multivariate settings stems from a simple observation: a set of d-variate measurements may be viewed as a *cloud of points* in d-dimensional space (this is familiar already from the case $d=2$ where we make scatter plots). Now with kinematic devices, one may inspect any 3-dimensional projection of such a d-dimensional cloud; by cycling through choices of the 3 dimensions, one may in a short time look at the data from a variety of viewpoints and get a feeling for the shape of the entire cloud. The computer might help here by automatically searching for projections showing interesting patterns, according to some definition of "interesting" [Friedman and Tukey, 1974].

A number of exploratory purposes might be served by the study of such 3-d projections. Inspection might reveal, for example, that the data contain outliers, or clumps, or corners, or an asymmetry of some sort. One would then consider cleaning up, reexpressing or otherwise preconditioning the data before formal analysis. Inspection might also indicate that a formal analysis would be pointless, either because there is no pattern in the data, or because the pattern is so clear.

The study of 3-d projections of the data could also be useful for diagnostic purposes. Note that many formal data analysis procedures may be viewed as measuring shape properties of the d-dimensional cloud of data points. Linear regression fits a hyperplane to the cloud; principal components fits a hyperellipse; discriminant analysis fits a pair of hyperellipses; cluster analysis fits a tree; ... Thus with kinematic displays, the data analyst might visually compare the shape imputed to the cloud by the formal procedure with the actual shape of the cloud, as viewed in 3-d projections. One might particularly want the computer to find and display that projection in which the data and fit agreed least well.

As Frank Anscombe has said, "Good data analysis is never routine." Kinematic graphics can challenge the user of formal techniques by exposing features of the data that are not captured by the "usual" analysis; this is an important and even exciting possibility.

2. OUR SYSTEM

Over the last two and a half years we have been developing software for applying kinematic displays to statistical data analysis.

2.1 Philosophy

We have been interested in the contributions such displays might make, not to the analysis of one particular type of data, but to the analysis of continuous multivariate data in general. We have had in mind an image of data analysis as extraction of information from a point cloud, but as it turns out, the system we have built may profitably be used with non point-cloud data, such as time series.

The basic design goals we set for our system were as follows:

1. That the system be *interactive*, the dialog between user and machine taking place in a tight loop. Ideally, this requires a closed, complete environment within which data analysis could take place without having to leave the system for e.g., writing programs or manipulating data files.

2. That the system be *flexible*, able to handle most types of continuous data, do most types of data analysis, and able to massage (reshape, cut, paste, transform) data into the form required by one or another analysis procedure.

3. That the system be *easy to use*, so that simple problems could be solved directly by persons with a minimal degree of training.

4. That the system be *extensible*, so that sophisticated users, who have learned from experience that certain combinations of basic operations recur throughout their interaction, be able to instruct the system to perform such operations automatically.

For the display software itself, there were the following additional desiderata:

5. The system should be *expert* at presentation of 3-d point clouds, or 3-d projections of higher-dimensional point clouds. It should allow the user to view the cloud in any scale and from any desired viewpoint.

6. The system should be *self-documenting*, able to tell the user, on request, what data he is looking at, from what point of view, and able to identify specific data points or variables.

7. The system should make it possible to *supplement* the usual point cloud display by attaching labels to points or to groups of points, and by superimposing line figures on the point scatter.

The motivation behind these goals should be clear. Note that, to some extent, the goals are in conflict; there is certainly tension between 1,3 and 2,4. Moreover, some are more crucial than others: 4 and 7, for example, are important mainly for sophisticated data analysis; but for this they are indispensable.

2.2 Current System

We have implemented a data analysis system on a VAX 11/780 computer owned by researchers in Chemistry at Harvard. The graphics device used is the Evans and Sutherland Multi Picture System (MPS). The MPS has a high resolution, black and white, calligraphic VDU. The transformations needed for kinematic display are all performed by the matrix arithmetic processor which is built into the MPS; thus the graphics software needs little host CPU support, and the reaction time of the graphics hardware is good even when the VAX is heavily used. User interaction with the graphics software is handled by a data tablet and data pen.

Our system is implemented as one large FORTRAN 77 program. It consists of software for data analysis — ISP (Interactive Statistical Processor) — and for kinematic display — PRIM-H (Projection, Rotation, Identification, and Masking – Harvard; named after the original system of this kind, PRIM-9, designed by Tukey, Friedman, and Fisherkeller).

ISP creates an environment for data analysis in which the objects to be manipulated are arrays of floating point numbers. ISP can manipulate these arrays arithmetically using a notation which looks like FORTRAN but which is like APL in scope. ISP can do matrix algebra, as well as reshape, cut, and paste arrays. It accepts a command language closely paralleling that of the VAX/VMS operating system under which it runs, and recognizes more than 70 commands, including commands for the core procedures of exploratory and classical data analysis. Moreover, ISP is programmable: macro commands may use structured facilities for looping, flow of control, and recursive function calls.

The display subsystem, PRIM-H, is accessed via the "primh" command to ISP. In the simplest usage one simply types

primh data

where *data* is the name of the data array to be inspected. PRIM-H then initializes the MPS and a menu appears on the screen. Using the data pen, the user assigns variables to the x, y, and z display axes, and PRIM-H displays the point cloud defined by this choice of projection axes. By now pointing to appropriate items on the menu, the user may rotate and rescale the point cloud, thus viewing it from any desired position. PRIM-H will display, on command, the observation labels of any data points the user points to with the data pen; it will also connect with line segments any pairs of data points the user selects using the pen. The interaction stays in a tight loop here: the user keeps his eyes on the screen; communicates with the PRIM-H by pointing to areas of the screen; and in particular does no typing.

Figure 1 shows an image of the MPS screen; PRIM-H is here being used to look at the Iris data made famous by R. A. Fisher [see Kendall, 1975]. These data consist of 4 measurements on 50 plants from each of

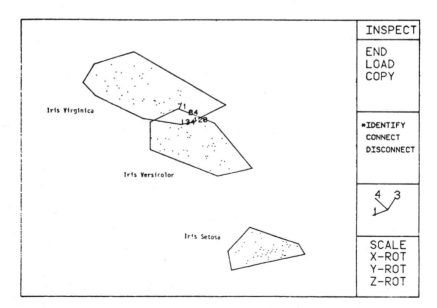

Figure 1. Measurements on 3 species of Iris. Polygons mark species boundaries. Observation numbers mark cases in overlap of two polygons.

three species of Iris. Here, the three variables *sepal length*, *petal length*, and *petal width*, (1, 3, and 4) have been assigned to the x, y, and z axes of the display system; the viewpoint has been rotated so that the three species *Setosa*, *Virginica*, and *Versicolor* are well separated. The user has indicated the boundaries of the species subclouds by drawing the convex hull of each species in this view. Note that the overlap of the species is very slight — only 4 points out of 150 points fall in more than 1 of the hulls. PRIM-H has here been used to do a discriminant analysis "by hand".

For more sophisticated applications, the user may prepare additional arguments to PRIM-H directing that the point cloud display be supplemented with markers for certain points, or with line drawings. If the result is a type of display he plans to use more than once, the user may design an ISP macro command that will set up the display automatically. An illustration of the importance of supplementation is given in Figures 2 and 3. The point cloud displayed in Figure 2 consists of 3 metabolic measurements made on 145 persons (both normal and diabetic) taking a glucose tolerance test [Reaven and Miller, 1979]. The cloud has been described as boomerang-shaped. The reasons for this shape becomes apparent in Figure 3, where the display has been supplemented by markers labelling each case as normal, chemical diabetic, or overt diabetic. The two diabetic groups make up the arms of the bommerang; the normal patients the bend. The figure has also been supplemented with curves dividing these three groups; the groups are almost perfectly separated in this view.

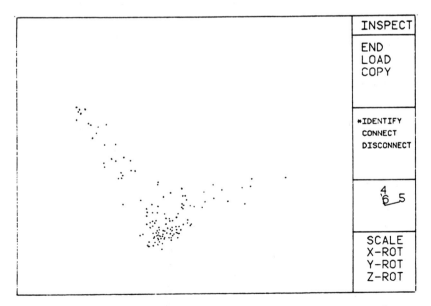

Figure 2. Data from Reaven and Miller (1979). Note "boomerang" shape.

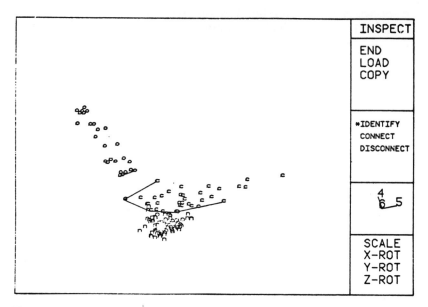

Figure 3. As in Figure 2. Markers: c = chemical diabetic, o = overt diabetic, n = normal.

2.3 Reactions

We rate our system as satisfactory in achieving most of the design goals laid out above. It can be applied to a wide variety of data analysis tasks, thereby allowing us to explore the application of kinematic displays to multivariate data analysis in general. Moreover, the speed and power of the VAX-MPS combination make data analysis using our system pleasant, because interaction is fast and smooth.

We have found our system sufficiently flexible that it may be used even in the analysis of data which do not fit the point cloud model. Figure 4 shows a display of several time-series using ISP-PRIM-H. Each series is an 84 hour segment of half-hourly readings on the activity of squirrel monkey which has been deprived of external time cues (e.g., light/dark). Adjacent series start 28 hours apart. The display was set up by an ISP macro command which breaks a time series into such segments and displays it on the MPS. The user has here added by data pen some curves connecting the daily peaks in activity; the shift in peaks indicates that the monkey's activity has a period of about 25 hours.

3. DIRECTIONS FOR FURTHER WORK

With a basic system in place, we are now using it intensively and plan to make a reasoned presentation of its substantive merits to an

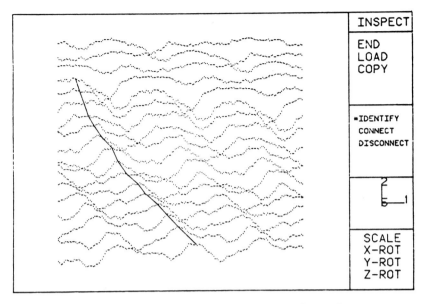

Figure 4. Smoothed activity index of squirrel monkey. Curve connects peaks which occur about every 25 hours.

audience of statisticians. The use of such a system poses interesting theoretical questions: Is exploring data by looking at projections "safe" — if you look at enough different projections of structureless data, will you find structure just by chance? If it is safe, is it "effective"? — in what sense can the information in a d-dimensional point cloud be extracted from a few of its 3-dimensional projections? The method, properly applied, appears to be both safe and effective, even allowing for the fact that we do not know the statistical properties of the eye as pattern detector.

We plan further software development in three areas.

1. We are developing a library of ISP macro commands which use PRIM-H to diagnose problems with standard multivariate data analysis procedures.

2. We are refining the design of both PRIM-H and ISP to take into account inadequacies in the original implementations (e.g., certain features of PRIM-H, such as relative brightness of data and menu, are not under interactive control but should be; and character data are somewhat awkward to handle within ISP).

3. We plan, ultimately, to develop a portable, "downsized" version of our system which will run on small computers with inexpensive color raster displays.

Many of the interesting possibilities raised by kinematic displays have just become open to us, and remain to be explored. However, our experience has established an elementary but very important feature of the approach: in the words of a medical researcher who has used our system, "it allows you to check that you haven't overlooked something obvious in a 'standard' analysis of your data."

Acknowledgements

This research has been supported by NSF contract MCS-79-08685 and by ONR contract N0014-79-C-0512. The data for Figure 4 were supplied by Dr. Phillippa Gander, Physiology, Harvard Medical School.

REFERENCES

Anscombe, F. J. (1973). "Graphs in statistical analysis," *American Statistician*, **27**: 17-21.

Belsley, D. A., Kuh, E., and Welsch, R. E. (1980). *Regression Diagnostics*. New York: Wiley.

Friedman, J. H. and Tukey, J. W. (1974). "A Projection Pursuit Algorithm for Exploratory Data Analysis," *IEEE Transactions on Computers*, **C-23**: 881-890.

Gnanadesikan, R. (1977). *Methods for Statistical Data Analysis of Multivariate Observations*, New York: Wiley.

Kendall, M. G. (1975). *Multivariate Analysis*. London: Chapman and Hall.

Reaven, G. M. and Miller, R. G. (1979). "An attempt to define the nature of chemical diabetes using a multidimensional analysis," *Diabetologia*, **16**: 17-24.

Tukey, J. W. (1977). *Exploratory Data Analysis*. Reading, MA: Addison-Wesley.

Tukey, J. W. and Tukey, P. A. (1981). "Graphical display of data sets in 3 or more dimensions," *Interpreting Multivariate Data*, V. Barnett, ed., 189-275. Chichester: Wiley.

Velleman, P. F. and Hoaglin, D. C. (1981). *Applications, Basics, and Computing of Exploratory Data Analysis*, Boston, MA: Duxbury.

5

AN INTRODUCTION TO REAL TIME GRAPHICAL TECHNIQUES FOR ANALYZING MULTIVARIATE DATA

Jerome H. Friedman
Stanford University

John Alan McDonald
Werner Stuetzle
University of Washington
Stanford University

ABSTRACT

Orion I is a graphics system used to study applications of computer graphics — especially interactive motion graphics — in statistics. Orion I is the newest of a family of "PRIM" systems, whose most striking common feature is the use of real-time motion graphics to display three-dimensional scatter plots. Orion I differs from earlier PRIM systems through the use of modern and relatively inexpensive raster graphics and microprocessor technology. It also delivers more computing power to its user; Orion I can perform more sophisticated real-time computations than were possible on previous such systems. We demonstrate some of Orion I's capabilities in our film: "Exploring data with Orion I".

This research was supported in part by the Department of Energy under contracts DE-AC03-76SF00515 and DE-AT03-81-ER10843, the Office of Naval Research under contract ONR N00014-81-K-0340, and the U.S. Army Research Office under contract DAAG29-82-K-0056.

Reprinted from (1982) *Proceedings of the Third Annual Conference and Exposition of the National Computer Graphics Association*, **1**:421-437, by permission of the National Computer Graphics Association.

From *Dynamic Graphics for Statistics*, William S. Cleveland and Marylyn E. McGill, eds., copyright © 1988 by Wadsworth, Inc. All rights reserved.

1. INTRODUCTION

This paper accompanies a film that demonstrates some programs written for the Orion I workstation. Orion I is an experimental computer graphics system, built in the Computation Research Group at the Stanford Linear Accelerator Center (SLAC) in 1980-81. It is used to develop applications of interactive graphics to data analysis.

We intend this paper for an audience familiar with computer graphics, but not necessarily with statistics. We hope that it provides some perspective on the place of interactive graphics in statistics and an introduction to several areas of current research.

2. GOALS

2.1 Description and Inference

The goal of statistics is to analyze data. A *data set* consists of, say, n observations. For each observation, we measure p variables. It will be a useful idealization to assume that the values each variable takes on are reasonably represented by real, or floating point, numbers. Then each observation can be thought of as a vector in a p-dimensional real vector space. A data set is then a set of vectors. This type of data is typical of the field of statistics known as 'multivariate analysis'.

How we analyze a given data set is context dependent. Sometimes we are given data that arises from an experiment designed to answer a particular question. In this case, we may have a great deal of prior knowledge about our data, and have confidence that we know what to expect in it. This is a situation in which *statistical inference* may be appropriate. An inference is a generalization from a given data set to some larger population (real or hypothetical) from which the data set is presumed to be a sample.

Often, however, we are given data to analyze about which we have very little prior knowledge. The goal then is to "just look at the data and see what is going on". Statisticians call this *description* or *exploratory data analysis* [Tukey, 1977; Mosteller and Tukey, 1977]. We aim to describe features of the particular data set at hand and not to draw (formal) conclusions about larger populations. There are two parts to description: *exploration*, to discover interesting features of our data, and *summary*, to report what we have discovered.

In order to make inferences, we need a probability model of the process by which our data set is sampled from the larger population.

Our inferences will only be reasonable if the probability model is a reasonable approximation of the way the data is generated. In order to construct a reasonable probability model, we must have a great deal of prior knowledge about the data. A common and serious flaw in standard statistical practice is the use of inappropriate models.

In classical multivariate analysis, one typically assumes that observations are independent samples from a multivariate Gaussian distribution. The Gaussian distribution models only ellipsoidal shaped point clouds; it is inappropriate when the data set seems to come from a distribution that has more a complicated shape. In our experience, most real data sets, especially those with many (more than three) variables, have some features that are clearly not Gaussian. The Gaussian distribution, and other very similar models that characterize classical multivariate analysis, have received a great deal of attention from theoretical statisticians. This interest is not because they are reasonable models of the generation of any data. Rather, the Gaussian distribution is favored because it is easy to analyze mathematically.

Description is a more primitive and fundamental problem than inference. If inference is to be done successfully, then we must use good models. In order to get enough prior knowledge about a problem to construct a reasonable model, we must, at some time, look at data.

2.2 The Value of Interactive Graphics

In statistics, graphical methods are used primarily for description. Our research, which emphasizes interaction and real-time motion, concentrates on the exploratory part of description. We want to develop methods that allow us to discover, understand, and summarize the multivariate "structure" of our data set. Since, in general, we have little reliable prior knowledge about our data, we are especially interested in methods that allow us to discover *unanticipated kinds* of structure in data. We also want to be able to respond to unexpected circumstances in an intelligent way.

This is where interactive graphics has much to contribute. First of all, computer graphics can quickly expose different views of data to human perception, so that the data analyst can detect many kinds of patterns in the data. The analyst can interpret apparent patterns using his knowledge of the context in which the data arose. Also, the analyst can select directions for further exploration using results so far obtained.

Thus, interactive graphical methods combine human talents for perception of patterns and judgment using the full context of a problem with a machine's ability to do rapid and accurate computation.

2.3 An Example of Structure

We have intentionally left "structure" undefined. What we mean by "structure" is any apparent pattern or interesting feature in a graphical (or numerical) description of data. Structure is a subjective perception. Quantitative measures of specific aspects of structure can be very useful, but we need to avoid too restrictive definitions.

In developing new graphical methods, it is useful to have some examples of structure in mind. We discuss next a simple kind of structure that is illustrated in our film: "Exploring data with Orion I".

A typical, and simple, thing that we would like to know about a set of data is the following: Does the data set naturally separate into groups or clusters?

In the film we look at a data set presented by Harrison and Rubinfeld [1978; see also Belsley, Kuh, and Welsch, 1980]. They measured 14 variables for each of 506 census tracts in the Boston Standard Metropolitan Statistical Area. They were interested in examining the dependence of housing price (represented by the median value of owner-occupied houses) on air pollution (represented by nitrogen oxide concentration). The remaining 12 variables measured other quantities thought to influence housing price, such as crime rate, average number of rooms per house, etc.

Harrison and Rubinfeld tried to determine the dependence of housing price on pollution by forming a prediction rule (a linear regression) for median housing value as a function of nitrogen oxide concentration and the other 12 variables. The effect of pollution alone on housing price was presumed to be reflected in the partial dependence of the prediction rule on nitrogen oxide.

In the exploration movie, we do not consider explicitly the dependence of housing price on pollution. We concentrate instead on clustering.

Before fitting any prediction rule, it is natural to consider whether it is appropriate to fit one rule for all the data. If the data set separated in a natural way into distinct and internally homogeneous groups (or clusters) then we would want to consider fitting different rules for each group.

Statisticians have developed many algorithms for clustering data. These algorithms usually rely on a notion of distance in the data space to partition a data set into isolated clumps. Observations are considered similar if they are close together and dissimilar if they are far apart. A cluster is a group of points that are close to each other and far from any other points.

With Orion I, we can use other criteria besides separation to partition a data set. In particular, we can use subjective perception of patterns to define natural groupings. A data set may naturally divide into two groups, which follow clearly distinct patterns. Yet the difference between the groups may not be easily summarized by any natural measure of distance.

In the film, we show how the Harrison-Rubinfeld data can be divided naturally into several groups. The major division turns out to be between urban and suburban-rural census tracts. However, the urban and suburban tracts do not form isolated clumps. Instead, in certain views, they lie in intersecting, perpendicular planes. Because the planes intersect, the groups are not isolated in any distance measure. But the separation in the two groups is obvious when seen.

3. METHODS

The really challenging problem in multivariate data analysis is to discover and understand structure involving more than three variables at once.

Suppose our data set involves only a single variable. Then we can see most of what we want with a histogram.

We look at two-dimensional data with a conventional scatter plot. In conventional scatter plots, observations are represented by points, which are plotted at horizontal and vertical positions corresponding to the values of two variables.

To look at three-dimensional data, we draw a three-dimensional version of the scatter plot. Real-time motion graphics makes it possible to draw pictures that appear three-dimensional. We subject the data to repeated small rotations and display the projection of the rotated data as points on the two-dimensional screen. If we can compute and display rotations fast enough (10 per second) then we get an illusion of continuous motion. Apparent parallax in the motion of the points provides a convincing and accurate perception of the shape of the point cloud in the three-dimensional space.

In a conventional scatter plot, it is easy to read off approximate values of the two variables for any particular point. In moving, three-dimensional scatter plots estimating the values of the three variables for any particular point is difficult, if not impossible. However, perceiving the shape of the point cloud is easy. Because we are interested in detecting patterns, we are much more interested in overall shape than in individual points.

There is no completely satisfactory method that lets us look at more than three variables at a time. Three basic approaches to many (more than three) dimensional structures are being studied with Orion I. They are:

1. *Higher dimensional views:* we try to represent as many variables as possible in a single picture.

2. *Projection Pursuit:* we try to find a low dimensional picture that captures the structure in the many dimensional data space.

3. *Multiple Views:* we look at several low dimensional views simultaneously; by making connections between the low dimensional views we hope to see higher dimensional structure.

3.1 Higher Dimensional Views

One way to see high dimensional structure is to try to invent pictures that show as many dimensions at a time as possible.

One of the simplest ways to add dimensions to a picture is through color. We start with a three-dimensional scatter plot. We can add a fourth variable to the picture by giving each point in the scatter plot a color that depends on the value of a fourth variable. With an appropriately chosen color spectrum, we can easily see simple or gross dependence of the fourth variable on position in the three-dimensional space. Our ability to perceive distinctions in color does not compare to our ability to perceive position in space; we should expect to miss subtle or complicated relationships between a color variable and three position variables. Color works best for a variable that takes on only a small number of discrete values.

There are many more tricks that let us add dimension to a picture. For example, we can represent each observation in the scatter plot by a circle, rather than by a simple point. The radius of the circle can depend on the value of a fifth variable.

In a simple scatter plot, observations are represented by featureless points. We add dimension to the picture by replacing points with objects that have features such as color, size, and shape. These features can be used to represent variables in addition to those represented by a point's position in the scatter plot. These "featureful" objects, sometime called *glyphs*, are well known [Gnanadesikan, 1977].

With each new dimension, the glyphs become more complicated and the picture becomes more difficult to interpret. We need more experience to determine which are good ways of adding dimension to glyphs and to understand the limitations of each method.

3.2 Projection Pursuit

The basic problem with adding dimension to a single view is that the picture quickly becomes impossible to understand. An alternative is to restrict our picture to a few (3 or less) dimensions and then try to find low dimensional pictures that capture interesting aspects of the multivariate structure in our data. This is the basic idea of projection pursuit [Fisherkeller *et al.*, 1976 and Friedman *et al.*, 1981].

In general, we could consider any mapping from the many dimensional data space to a low dimensional picture. The projection pursuit methods that have been developed so far restrict the mappings considered to orthogonal projections of the data onto 1, 2, or 3 dimensional subspaces of the data space. More general versions of projection pursuit would include methods similar to multidimensional scaling [Gnanadesikan, 1977].

The original projection pursuit algorithm [Friedman and Tukey, 1974] used a numerical optimizer to search for 1 or 2 dimensional projections that maximized a "clottedness" index, which was intended as a measure of interesting structure.

Automatic projection pursuit methods have been developed for more well defined problems: nonparametric regression, non-parametric classification, and non-parametric density estimation. In these problems, we build up a model that summarizes the apparent dependence in our data set of a response on some predictors. Interesting views are those in which the summary best fits or explains the response.

We are, of course, not restricted to a single low dimensional view. Models of a response will usually be constructed from several views.

Interactive versions of the automatic projection pursuit methods are being developed for Orion I. The system allows a user to manually imitate the Rosenbrock search strategy [1960] used by the numerical optimizer in the automatic versions. However, a human being can search to optimize subjective criteria, using perception and judgment, instead of having to rely on a single, numerical measure of what constitutes an interesting view.

3.3 Multiple Views

This approach is inspired by the *M and N Plots* of Diaconis and Friedman [1980]. A *two and two plot* is one kind of M and N plot; two and two plots are used to display four-dimensional data. To make a two and two plot, we draw two-dimensional scatter plots side by side; the scatter plots show different pairs of variables. We then correct corresponding points in the two scatter plots by lines. To get a picture that is not confusing, we will often not draw all the lines. Diaconis and Friedman give an algorithm for deciding which lines to draw.

Our idea is a modification of the above. Instead of connecting corresponding points by lines we draw corresponding points in the same color.

Briefly, this is how the program works. On the screen there are two scatter plots, side by side, showing four variables. There is also a cursor on the screen in one of the two scatter plots. The scatter plot that the cursor is in is the *active scatter plot*. We can move the cursor by moving the trackerball. Points near the cursor in the active scatter plot are red. Points at an intermediate distance from the cursor are purple. Points far from the cursor are blue. Points in the nonactive scatter plot are given the same color as the corresponding point in the active scatter plot. The colors are continuously updated as the cursor is moved. We can also move the cursor from one scatter plot to the other, changing which scatter plot is active.

Using color instead of line segments to connect points has a disadvantage; it is not as precise at showing us the connection between a pair of points representing the same observation. However, we are not usually very interested in single observations. More often, we want to see how a region in one scatter plot maps into the other scatter plot; the combination of local coloring and the movable cursor is a good way of seeing regional relationships in the two scatter plots.

Another advantage of color over line segments is that it is possible to look at more than two scatter plots at once. Connecting corresponding points with line segments in more than two scatter plots at a time would produce a hopelessly confusing picture. With color, on the other hand, it is no more difficult to look at three or more scatter plots at once than it is to look at two.

4. MACHINERY

4.1 PRIM-9

Orion I is the youngest descendant of a graphics system called *PRIM-9*, which was built at SLAC in 1974 [Fisherkeller, Friedman, and Tukey, 1976]. PRIM-9 was used to explore up to 9 dimensional data. It used real time motion to display three dimensional scatter plots. Through a combination of *projection* and *rotation*, a user of PRIM-9 could view an arbitrary three-dimensional subspace of the 9 dimensional data. *Isolation* and *masking* were used to divide a data set into subsets.

The computing for PRIM-9 was done on a large mainframe (IBM 360/91) and used a significant part of the mainframe's capacity. A Varian minicomputer was kept busy transferring data to the IDIIOM vector drawing display. The whole system, including the 360/91, cost millions of dollars. The part devoted exclusively to graphics cost several hundreds of thousands of dollars.

4.2 Other PRIMs

Successors to PRIM-9 were built at the Swiss Federal Institute of Technology in 1978 *(PRIM-S)* and at Harvard in 1979-80 *(PRIM-H)* (3).

PRIM-H improves on PRIM-9 in both hardware and software. It is based on a VAX 11/780 "midi-" computer and an Evans and Sutherland Picture System 2 (a vector drawing display). It incorporates a flexible statistical package (ISP). However, the system has a price tag still over several hundreds of thousands of dollars. The VAX is shared with perhaps two dozen other users. Computation for rotations is done by hardware in the Evans and Sutherland. No demanding computation can be done on the VAX in real-time.

4.3 Orion I

There are two ways in which the Orion I hardware is a substantial improvement over previous PRIM systems: price and computing power. The total cost for hardware in Orion I is less than $30,000. The computing power is equivalent to that of a large mainframe computer (say one half of an IBM 370/168) and is devoted to a single user. The hardware is described in detail by Friedman and Stuetzle [1983].

The basic requirement for real time motion graphics is the ability to compute and draw new pictures fast enough to give the illusion of continuous motion. Five pictures per second is a barely acceptable rate. Ten to twenty time per second gives smoother motion and more natural response for interaction with a user.

Orion I and the earlier PRIMs use real time motion to display three-dimensional scatter plots. We can view three-dimensional objects on a two-dimensional display by continuously rotating the objects in the three-dimensional space and displaying the moving projection of the object onto the screen. In a scatter plot, the object we want to look at is a cloud of points. A typical point cloud will contain from 100 to 1000 points. So our hardware must be able to execute the viewing transformation, which is basically a multiplication by a 3×3 rotation matrix, on up to 1000 3-vectors ten times per second. The system must also be able to erase and draw 1000 points ten times per second.

The important parts of the hardware are:

1. a SUN microcomputer, based on the Motorola MC68000 microprocessor. The SUN microcomputer is a MULTIBUS board developed by the Stanford Computer Science Department for the Stanford University Network. It is the *master processor* of the Orion I system. It controls the action of the other parts of the system and handles the interaction with the user. The SUN is programmed mostly in Pascal; a few critical routines for picture drawing are in 68000 assembly language.

2. a Lexidata 3400 *raster graphics frame buffer*, which stores and displays the current picture. The frame buffer has a resolution of 1280×1024. There are 8 bits of memory for each pixel, which determine the color through a look up table. The Lexidata 3400 contains a microprocessor that is used for drawing vectors, circles, and characters. Color raster graphics devices like the Lexidata are cheaper and more flexible than the black and white line drawing displays used in earlier Prim systems.

3. an *arithmetic processor* called a 168/E. The 168/E is basically an IBM 370/168 cpu without channels and interrupt capabilities. It was developed by SLAC engineers for the processing of particle physics data and has about half the speed of the true 370/168. Because it has no input/output facilities, it is strictly a slave processor. The 168/E servers us as a flexible floating point or array processor. It is programmed in FORTRAN.

A programmer of Orion I works on:

4. the *host*, an IBM 3081 mainframe. The host is used only for software development and long term data storage. Programs are edited and cross-compiled or cross-assembled on the 3081. Data sets are also prepared on the 3081. Programs and data are downloaded to the SUN board and the 168/E through a high speed serial interface that connects the MULTIBUS and the 3081 (by emulating an IBM 3277 terminal). Strictly speaking, the 3081 is not considered a part of the Orion I workstation; it does not have an active role in any of the interactive graphics on the Orion I.

A user of Orion I deals most of the time with:

5. a 19 inch *color monitor* that displays the current picture.

6. a *trackerball*, which sends two coordinates to the SUN computer. In Orion I, the trackerball provides the angles of a rotation; the apparent motion of a point cloud on the screen mimics the motion of the trackerball under a user's hand. The trackerball has six switches, which are used for discrete input to programs.

The part of the system that currently limits what can be done in real time is the Lexidata frame buffer. Its speed of erasing and redrawing pictures does not keep up with the computation speed of the SUN board and the 168/E. In fact, the 168/E can execute our most demanding real time computation, a sophisticated smoothing algorithm, at much more than ten times a second. This smoothing algorithm [Friedman and Stuetzle, 1981] is much more demanding than simple rotation.

It is worth noting that rotations can be done in real time by the SUN board alone. Thus a system could be built with all the capabilities of earlier Prim systems using only a SUN board or some other 68000 based microcomputer and without a 168/E. Such systems are now (Feb. 1982) commercially available for about $25,000 for basic hardware and about $40,000 for a complete, stand alone system with hard disk, UNIX-like operating system, resident high level language compilers, etc.

REFERENCES

Belsley, D. A., Kuh, E., and Welsch, R. E. (1980). *Regression Diagnostics*. New York: Wiley.

Diaconis, P. and Friedman, J. H. (1980). "M and N plots," Technical Report No. 151. Stanford, CA: Stanford University, Department of Statistics.

Donoho, D., Huber, P. J., and Thoma, H. (1981). "The use of kinematic displays to represent high dimensional data," *Computer Science and Statistics: Proceedings of the 13th Symposium on the Interface*, 274-278. New York: Springer-Verlag.

Fisherkeller, M. A., Friedman, J. H., and Tukey, J. W. (1976). "PRIM-9: An interactive multidimensional data display and analysis system." *Proceedings of the 4th International Congress for Stereology*. Gaithersburg, MD: National Bureau of Standards Special Publication 431.

Friedman, J. H. and Stuetzle, W. (1981). "Projection pursuit regression," *Journal of the American Statistical Association* **76**: 817-823.

Friedman, J. H. and Stuetzle, W. (1983). "Hardware for kinematic statistical graphics", *Computer Science and Statistics: Proceedings of the 14th Symposium on the Interface*, 163-169. Amsterdam: North-Holland.

Friedman, J. H. and Tukey, J. W. (1974). "A Projection Pursuit Algorithm for Exploratory Data Analysis," *IEEE Transactions on Computers*, **C-23**: 881-890.

Gnanadesikan, R. (1977). *Methods for Statistical Data Analysis of Multivariate Observations*, New York: Wiley.

Harrison, D. and Rubinfeld, D. L. (1978). "Hedonic prices and the demand for clean air," *Journal of Environmental Economics and Management*, **5**: 81-102.

Mosteller, F. and Tukey, J. W. (1977). *Data Analysis and Regression*, Reading, MA: Addison-Wesley.

Rosenbrock, H. H. (1960). "An automatic method for finding the greatest or least value of a function," *Computer Journal*, **3**: 175-184.

Tukey, J. W. (1977). *Exploratory Data Analysis*. Reading, MA: Addison-Wesley.

6

CONTROL AND STASH PHILOSOPHY FOR TWO-HANDED, FLEXIBLE, AND IMMEDIATE CONTROL OF A GRAPHIC DISPLAY

John W. Tukey
Princeton University, Princeton, NJ

ABSTRACT

A philosophy of two-handed control and screen-prompting for data-view modification systems is proposed.

Emphasis is on simplicity and easy learnability for the user. This implies more than one level of tentativeness — of "first try, and then accept or decline."

Five buttons, each of which has a continuing consistent interpretation, supported by simple screen prompts, and motion of a tracker ball, seem able to provide as flexible control as may be desired.

The philosophy, illustrated by examples proposed for PRIM-81, emphasizes (1) simple controls, adequately prompted; (2) continuing meaning for controls; (3) prompts displayed compactly on the same screen as the main display; (4) no need to move hands or take eyes off the screen.

Prepared in connection with research at Princeton University, supported by the DOE, and research at Bell Laboratories.

This paper appears in *The Collected Works of John W. Tukey, Volume V—Graphics*, W. S. Cleveland, ed. Pacific Grove, CA: Wadsworth & Brooks/Cole Advanced Books & Software, 1988.

From *Dynamic Graphics for Statistics*, William S. Cleveland and Marylyn E. McGill, eds., copyright © 1988 by Wadsworth, Inc. All rights reserved.

Later parts discuss the record-keeping aspects of a successive-modification display system, particularly in their impact upon the user through required control and status displays. A record-keeping philosophy (stash philosophy) is outlined for a data-view-modification system in which (display or) potential display moves along a — probably branched — trail with certain points being marked (as *nodes*). The marked points are those displays (or potential displays) to which it seems most likely that the user will return — either to look again, or to start a new branch of the trail. Again we illustrate general ideas with (incomplete) applications to the proposed control of PRIM-81.

We then consider the problem of relating an apparently helpful coordinate system to the initial coordinate system from which the display began, and offer an apparently feasible answer, making use of projection-pursuit regression.

1. FOREWORD

Some of the key ideas set forth here developed during active discussions with Jerome H. Friedman, Mary Anne Fisherkeller and Werner Stuetzle of the Stanford Linear Accelerator Center (SLAC). These ideas were based, in substantial part, on experience with PRIM-9 at SLAC (JWT, JHF, MAF) and PRIM-ETH at the ETH (WS). The present account attempts to codify and clarify the position to which many stages of (presumed) improvement have brought us.

2. THE GENERAL TARGET

We discuss here the two-handed, screen-prompted control of a dynamic point-cloud display system, with special reference to the present plans for "PRIM-81." This system:

1) is based in part on the Picturing-Rotation-Isolation-Masking capabilities developed for PRIM-9 [Fisherkeller, Friedman, and Tukey, 1974];
2) embodies further capabilities, particularly those involving non-linear transformations of the coordinates of the points;
3) utilizes modern hardware (including microprocessors) to make such facilities usefully dynamic; and
4) emphasizes both logical simplification and effective screen-prompting to reduce the operator's need either for complex memorization or looking away from the screen.

It is not the purpose of this account to discuss the deeper internal structure of the proposed system (including relationships between host and dedicated computers, etc.). Parts E, F, and G, however, discuss the logical and control structure needed for trail-leaving, recapture, and grafting.

It is hoped to focus on a control which involves two hands in relatively fixed positions and a screen display that is as nearly self-prompting as possible.

The formulation we will discuss assumes;

1) one hand on a tracker ball capable of being rolled in any combination of two directions;
2) the other hand on an array of five buttons, one for each finger (including the thumb);
3) a one-column prompting menu, on the screen adjacent to the data display field, so that eyes do not need to leave the monitor.

Many of the principles exemplified here would be applicable in a fairly wide variety of other graphical control systems, but we shall confine our example to this one.

Parts E, F and G of the present account discuss the control environment and stash philosophy for a data-view-modification system in which (display or) potential display moves along a — probably branched — trail with certain points being marked (as *nodes*). The marked points are those displays (or potential displays) to which it seems most likely that the user will return — either to look again, or to start a new branch of the trail. Again we shall illustrate general ideas with (incomplete) applications to the proposed control of PRIM-81.

Finally, in Part H, we discuss synthesizing descriptions of the relationships between two sets of coordinates, when these coordinates are only given at the data points.

A. GENERAL CONSIDERATIONS

3. TREES, MENUS, NAMES, AND CUES

A *menu* is a list of *items*. Each *item* starts with a *name*, which we plan to hold to 6 characters in the interest of saving both space and impact for better use by the display of the data itself. The name is usually followed by one or more *cues*, which, for similar reasons, we hold to 2 characters. All examples in this account will use *at most* 1 cue, but more would be included if really needed. Thus the prompting menus we describe here need involve, with two spaces, only 1 + 6 + 1 + 2 = 10 character columns.

Three kinds of cues will be used here:

a) reminders of direction of menu change (IF item is taken):

   UP (up to menu above)
   DN (down to menu below)
   WO (to the worker tree)
   PE (to the peeker tree)
   TR (to the trace tree; see parts E and F below)

b) cues that the tracker-ball is enabled in a particular way:

   ·T (rotation about 1 axis effective) [· is a blank]
   TT (rotation about 2 axes effective)

c) a number of one or two digits showing a choice of a coordinate, node, subnode, number of iterations, etc. (Actual display: numerical. Exhibits below: schematic mnemonics, as follows:)

   ## (a coordinate number)
   NN (a node number; mainly in parts E and F)
   SS (a subnode number; mainly in parts E and F)
   KK (number of times a process is implemented)

The *menus* are thought of as basically arranged in *trees*, branching downwards from a top (or root) menu in each tree, where connections are possibilities of transfer. There need be no natural limitation on the number of steps downward along any one branch.

We feel free to interconnect one tree to another, when it may be well to remember at which menu we left a particular tree, or to cross-link menus in a single tree, or to introduce links that skip up or down two or more levels. The tree structure is introduced for conceptual and manipulative convenience, not as a restriction.

It is to be hoped that supplementary information, including that which might otherwise be called up by a "HELP button," will be presented, automatically, on a separate screen. This keeps distractions away from the main screen. Thus variable numbers can be adequate cues on the working screen, since their translations, if useful, could be on the auxiliary screen (but see Section 10).

4. TENTATIVE AND FINAL ACTION

We are trying to control what is done to our display in the light of what we see. This means that "trial" must be the watchword. We need to try things without committing ourselves to the actions that we try. So that the normal pattern is

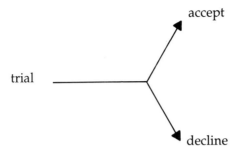

where "decline" means return to the state where that particular trial began.

This should — and does — occur at a variety of levels:

1) within an item, where we try a change, and then either accept (TAke) or decline (SKip) that change;
2) within a menu, where we reach some combination of changes, and then either accept (RETURN) or decline (NULL-R) all the actions taken in that occupation of that menu;
3) between trees, where we may transfer to the peek tree to look around, anticipating a return to the status in the work tree from which we came (but where we may want to have exceptional facilities, e.g. PSTASH in Section 12 below Exhibit 3, that let us retain what we have done in the peek tree).

Throughout, the sequence ought to be:

1) try;
2) consider, guided by what can be seen;
3) accept or decline.

B. SOME DETAILED CONSIDERATIONS

5. WHOA-BACK CONTROL

In PRIM-9, where all control was by buttons (supplemented, later, by a light-pen), rotations (and other continuous motions) were driven by what we might call a "whoa-back" control operated by double-interrupt buttons that made one signal on depression and another on release. (In PRIM-81, such continuous motions will be controlled by using the tracker ball.)

Whoa-back control steps forward, at regular time intervals, during the time from button depression to release. On the next depression of that button, however, the direction of motion reverses. And there is a similar reversal at every succeeding use. In controlling rotations, etc. it was found helpful to use an accelerated stepping scheme where successive steps are of relative sizes 1, 1, 2, 4, 8 , ... , with doubling up to some adjustable step-limit. (The two 1's are important, one 1 is not enough.)

In PRIM-9, lists (e.g., coordinate numbers) were stepped through as closed cycles with one step ahead per button press (and no backwards motion!). The deficiencies of this form of control were painfully obvious, however. It is intended that such list control in PRIM '81 should be on a whoa-back basis. (What the desirable step-sequence should be will have to be found out by trial.)

Of the five buttons controlled by the button hand, two, conveniently called "SCan" (for the index finger) and "PlcK" (for the second finger) are assigned to provide different forms of "whoa-back" control. (We will discuss their general and specific functions later.)

For hardware implementations, it is important to notice, as J. H. Friedman has stressed, that since we can easily afford to have a computer poll the buttons at frequent intervals (tens of milliseconds, perhaps) we do not require buttons with additional interrupt capabilities.

6. MENU DISPLAY CONSIDERATIONS

What we want from our screen prompting is:

1) information as to where we are (which tree, which menu, which item);
2) information as to what we can do where we are;
3) information on where we could move.

Logically, this seems most naturally described by

0) a menu, regarded as a cycle of items (potential actions);
1) an arrow pointing at one item, or other form of highlighting;
2) supplementary information — cues — following that one item, and perhaps, more detailed information on the "help" screen;
3) a list of the names of the other items in the menu, with or without the corresponding cues.

(Notice that (1) here tends to match (1) above, etc.)

Pictorially, we could use a list in fixed position for a cyclic menu, moving an arrow up and down to show where we are. Rather than trusting an arrow, however, we should plan to increase our emphasis by adopting two or more of:

(a) a distinctive mark, perhaps a square;
(b) a distinctive intensity, high "at the mark";
(c) a distinctive color;
(d) details (cues) following only the name "at the mark";
(e) reverse video.

Exhibit 1 shows a possible realization of a particular menu, ROTATE, to be discussed below, both as it might appear on the screen and as we would represent it in later exhibits.

If we keep the names in a fixed position we have a sensible *absolute* implementation of a cyclic menu.

Exhibit 1

An instance of ROTATE

As on the screen					As represented in later exhibits	
Format 1			Format 2			
WO			WO			
ROTATE			ROTATE			
H-PICK	11		H-PICK		H-PICK	##
V-PICK	21		V-PICK		V-PICK	##
B-PICK	3		B-PICK		B-PICK	##
• HB-VB	TT	•	HB--VB	TT	HB-VB	TT
-HV-	T		-HV-		-HV-	T
NULL-R	UP		NULL-R		NULL-R	UP
RETURN	UP		RETURN		RETURN	UP
NODE-M			NODE-M		NODE-M	

Notes:

• This is the mark.

1) The functions of the various items will be discussed below, in connection with Exhibit 2.
2) A compromise between format 1 and format 2, where any numerical cues appear for items, but other cues only appear at the mark, may be desirable.

But there *seems* to be much to say for a *relative* implementation, in which the marked line always appears (vertically) in the middle of the list, so that the names move up and down when the cyclic menu is scanned. (We have already introduced "SCan" as the name of the button used to drive such scanning.) It will take empirical trial to learn whether relative or absolute menus will provide easier control. (Perhaps a mixture will be still better.)

7. BUTTON ACTIONS

We now set out a simple 5-button logic for the button hand. With (0) for thumb, the proposed assignments are:

0) TAke, a button with the general meaning of "accept," "do it," etc.
1) SCan, a button to move up or down the cyclic menu (using whoa-back control, as just described).
2) PIck, also using whoa-back control, a button to drive one up or down some other auxiliary list (perhaps of coordinate or node numbers). (In other implementations, perhaps, right or left along a string of cues.)
3) PEek, a button to shift between trees (or, more generally, to "peek" at the display when the screen is otherwise occupied).
4) SKip, a button to decline the present attempt and move on, no action having been taken during that occupation of that line.

Besides these 5 buttons, we have the tracker ball, which may or may not be active. If active, it can be in one of two states:

5) "TT," where rotations about two axes control separate quantities.
6) "·T," where rotation about one axis controls a single quantity (and the other rotation has no effect).

Since it is the intention that TAke, SKip, and PEek are all effective at any instant, there is no need to provide cues for these buttons.

In any occupation of an item, use of any button other than SCan, or use of the tracker ball, if enabled, locks us into that line until either TAke or SKip is used, accepting or declining the change(s) we have made. It may be helpful to prompt this by removing the intensity or color change from the name, but not the cue (in other embodiments the cues) once we are locked into the line. (Since TAke and SKip are always enabled, we can always unlock ourselves easily.)

We plan for at least 3 trees—work, peek, and trace. In more complex versions, we may need still others, but not just because we have more things to do. A separate tree is needed when we want to "peek," interrupting our main activity to get guidance, but ordinarily returning to the exact place we left, leaving behind all that was done while peeking.

Thus, when we are in the trace tree, trying to find the nice picture we once had, or trying to collect information about how we got from picture A to picture B, we will often want to "peek" at the display. And when we are in the work tree—possibly in the middle of some action—we will often want to "peek" at the present state of the data in some other view.

Various schemes for implementing the PEek button are consistent with this. Actual trial will be required to learn which is more effective.

We might:

1) Once in the trace tree, have the PEek step us back and forth to the work tree, so long as no work tree action is attempted; and once firmly in the work tree (action attempted) have PEek step us back and forth to the peek tree. A specific return (not using PEek) from (the top of) the work tree to the trace tree would also be needed.
2) Have PEek step us 1, 2, 3, 1, 2, 3, . . . , through trace tree, work tree, and peek tree. (Simplest to program, not good from user's point of view.)
3) As 1, except that the specific return is replaced by having a part of the display flash TRACE for a second, every 6 seconds, during which second (delayed by reaction time) use of PEek transfers control to the trace tree.

If color display is used, menus from different trees should appear in different colors, so that the user is adequately warned about which tree is occupied. (This is not to be taken as interfering with a special color for the item at the mark.)

Similar control schemes with fewer than 5 buttons are quite feasible, but we have avoided them in order to simplify and clarify the functions of each button.

8. CONTROL LOGIC

SCan brings us to the right item, PIck takes us to the right place in any auxiliary list. We may then want to go on within the item to further picking—or to ball-rolling—but we may also be through with the item, and want to pass on to another.

TAke indicates that we are through with picking, and wish to accept what we have. SKip, which says "forget what we were just doing," should *almost* always take us on to a new name (though we should not insist on this).

In most circumstances we should jump to the same next item, whether we use TAke or SKip. This item need not be—and often will not be—the "next" item in the order of the cyclic menu.

One extreme case arises for items where TAke shifts us to another menu, either lower or higher in the menu tree, something quite different from the usual action of SKip.

We need to store a record of what we have done to our data, and how we have viewed it. We shall return to this question below (Parts E and F, Section 18ff).

9. ROTATE AS AN EXAMPLE

We now describe a simple menu, providing rotation of the depicted point cloud.

In n-space, rotations are poorly described by "axes" (which are $n-2$-dimensional linear subspaces). Since motion is confined to two-dimensional planes, all parallel to one another (and making up an $(n-2)$-parameter hyperstack) it is convenient to describe a rotation by naming two directions in such a plane (and hence in all such) and saying that the rotation is from one toward the other, by a given amount. If we are willing to combine simply described rotations in order to achieve less simply described ones, it suffices to use, as elementary rotations, only those from one current coordinate toward another current coordinate. (Each rotation, of course, changes how the current coordinates are expressed in terms of initial coordinates. Thus we are soon far from initial coordinates.)

If we are to see directly, in our view fixed by two current coordinates, something of the effects of a rotation, that rotation needs to affect one or both of the currently viewed coordinates. Our preparation for rotation, then, involves choosing

1) the initial horizontal (H) and vertical (V) coordinates, and
2) a single distinguished back (B) coordinate from the list of all current coordinates (of which there may be, say, 8 to 20 at any one time).

Our rotations then redefine the H, V, and B current coordinates. If we want to do other rotations, we have to return to the HVB-picking phase, make new choices, and continue.

ROTATE appears as a name on a higher menu, and TAke, when ROTATE is at the mark in that menu, transfers to the ROTATE menu whose names and cues are as in Exhibit 2. Normal entry (as shown by =>) will be at H-PICK, although some uses might well (as shown by (=>)) enter at HB—VB, assuming coordinate picking has already been accomplished.

If we enter at H-PICK, the item at the mark would be something like

H-PICK 07

where "07" represents the coordinate number (for a *current* coordinate) now in use as the horizontal axis. (We may want to suppress leading zeros.) Manipulation of PIck will replace this "07" with whichever of the available coordinates (perhaps 8-20 in number) is desired. Pressing TAke (*or* SKip, if we wish to stay with the original "07" coordinate) will then jump us to

V-PICK 13

which shows 13 as the present vertical coordinate number. After similar manipulation, we jump first to

B-PICK 04

to pick the third, selected "back" coordinate, and then to

HB-VB TT.

In this latter situation, rolling the tracker ball right or left rotates in the plane of the horizontal and back axes, while rolling the tracker ball up or down rotates in the plane of the vertical and back axes.

When we tire of making further changes in such rotations, either TAke (or SKip, if we wish to forget what we have just been doing with HB-VB) will jump us to

-HV-·T

where right-left on the tracker ball will produce rotation in the horizontal-vertical plane. (TAke or SKip now take us back to HB-VB, since this seems the most likely next name.)

Exhibit 2

Pattern of menu ROTATE

	Name & Cue		(SK jump)	(TA jump)	(Function of TA)
=>	H-PICK	##	(V-PICK)	(V-PICK)	(picks H)
	V-PICK	##	(B-PICK)	(B-PICK)	(picks V)
	B-PICK	##	(HB-VB)	(HB-VB)	(picks B)
(=>)	HB-VB	TT	(B-PICK)	(-HV-)	(rotates H→B and V→B)
	-HV-	T	(HB-VB)	(HB-VB)	(rotates H→V)
	NULL-R	UP	(RETURN)	(up)	(annuls this rotation)
	RETURN	UP	(NULL-R)	(up)	(accepts this rotation)
	NODE-R	UP	(RETURN)	(up)	(marks node and accepts)

Notes:

1) Only "names and cues" displayed on screen.

2) ## = translates to current coordinate number.

3) H is the horizontal axis, V the vertical, and B a selected back axis.

4) H→B means rotates from H toward B, etc.

In any of these lines, before making a change with PIck, TT, or ·T, the use of SCan can bring us to any line we choose. (After use of PIck, TT, or ·T, we cannot SCan until we jump—until we use SKip or TAke.)

If we go to

NULL-R UP

the use of TAke eliminates all the rotations we have made since entering ROTATE, and moves us to the next menu *up* the menu tree (to the menu, and item, from which we entered ROTATE). (This assumes that the *data table* in effect at entry to ROTATE was stored explicitly or implicitly and is available to NULL-R.) (The use of SKip jumps us to RETURN.)

If we go to

RETURN UP

the use of TAke leaves the rotation we have been trying out in effect and jumps us one step up the tree, again to the menu and item from which we came, while that of SKip returns us to NULL-R.

* all linear transformations *

If we wish to have available all, possibly inhomogeneous, linear transformations,—and not just orthogonal linear transformations of determinant +1 (all rotations)—we need to add two more elementary operations to our current-coordinate to current-coordinate rotations. These are probably most conveniently added to ROTATE and taken either in the form

Name & cue	(jumps)	(functions of TA)
SHOVUP.T	(SWELUP)	displaces current picture upward (increases all V coordinates by a + or − constant)
SWELUP .T	(HB-VB)	expands/contracts V by a constant

or in the alternative form

Name & cue	(jump)	(functions of tracker ball)
SHOSWE TT	(HB-VB)	shift up/down: expand/contract vertically

Choice between these two will require trial.

* additional rotations *

This choice prevents us from carrying out certain procedures, such as that which PRIM-9 specialists will recognize as the "generalized episcotister," in which rotations among concealed variables interact with masking (not with isolation) to affect the current view. These could be provided either by adding a separate menu, or adding

    C-PICK (picks a C-axis)

and

    -BC- (rotates from B toward C)

to the menu as given. (It would seem quite undesirable to combine -HV- and -BC- into HV-BC, since accidental use of BC should be discouraged.) These two additional items, if added, should appear below NODE-M and jump to each other.

10. GEOMETRIC NATURE OF PRIM ROTATIONS

An important aspect of PRIM-9, PRIM-81 and their (future) descendants is an attachment to current coordinates—and a corresponding detachment from initial coordinates. PRIM-modification systems emphasize the "geometric object" provided by the data—and de-emphasize how that data was initially given. They look for "good pictures" and leave to a follow-up process learning how these good-picture coordinates are related to the initial coordinates.

Thus if we rotate first among current (initial) coordinates 1, 2, and 3 and then among current coordinates 3, 4, and 5, what we can say is that

- new current coordinates 1 and 2 involve initial coordinates 1, 2, and 3;
- new current coordinates 3, 4, and 5 each involve initial coordinates 1, 2, 3, 4, and 5;
- initial coordinates 1 and 2 are involved in current coordinates 3, 4, and 5 in proportional ways (their coefficients differ by common multipliers).

Thus some attention to the details of the relation of current coordinates to initial coordinates may be helpful.

In PRIM-9, this information was occasionally stored in matrix form, and frequently shown graphically in terms of an n-Bein, a figure showing labelled lines or arrows along the original coordinate axes, as seen in terms of the two current coordinates momentarily used for H and V.

Such a display may well be useful in PRIM-81, but its generation cannot be made by vector drawing, since twists and other nonlinear actions will convert parts of axes—and arms of n-Beins—into curves in the H-V plane. Much thought, and considerable experimentation, is likely to be needed before such possibilities are clear.

If we are confining ourselves to linear modifications, so that current coordinates are given by a matrix times initial coordinates, we may well want to assist our understanding in other ways. For 4-body and 5-body final state data, for example, where several (or many) coordinate systems are natural, we might want to provide stepwise regressions of (unique) current coordinates on (redundant) natural coordinates. (This is likely to be material for the auxiliary screen.)

Even if nonlinear modifications are being used, stepwise regressions confined to the data points might give helpful guidance. (See also Part G.)

C. THE PEEK TREE

11. GENERAL STRUCTURE

While certain aspects of the peek tree structure are related to questions of what is to be saved, and how, we can usefully cover many aspects here.

The nature of "peeking" is that we anticipate returning to the action (or actions) we have just been engaged in. What if we are in the middle of such an action? Unless we automatically undo the actions taken during peeking, we may find it tedious—perhaps difficult—to return to exactly where we were. A good peeking system, then, ought usually to return us to just that position in just that action in just that menu in just that tree from which we left.

An overnaive view would be that we must be careful to take only reversible actions while peeking, since we must be able to undo them. A little more thought shows us that it suffices to remember carefully where we were, and to jump back to that state when we stop peeking. Memory can be much more trustworthy than applying inverses.

12. OCCASIONAL STASHING

Moreover, we ought to be prepared for the rare occurrence that what we have done while peeking is valuable in its own right. In such

Exhibit 3

Possible structure of PEEKER, the top (= root) of the peek tree

Name & Cue	(REsit)	(SK jump)	(TA jump)	(function of submenu called)
=> 3-ROTA DN	(G-LOOK)	(G-MAKE)	down	(rotates among 3 coordinates)
G-MAKE DN	(G-MAKE)	(G-BOOL)	down	(makes groups, in various ways)
G-BOOL DN	(G-BOOL)	(G-LOOK)	down	(Boolean combinations of groups)
G-LOOK DN	(*-ROTA)	(N-PEEK)	down	(sets control of groups displayed)
=> A-ROTA DN	(G-LOOK)	(N-PEEK)	down	(rotates among all directly available coordinates)
P-WORK DN	(*-ROTA)	(*-ROTA)	down	(optional, see text)

Notes:

1) "REsit" describes position on *re*turn from submenu indicated.
2) If H, V, B were already set, and in use in the nonlinear item from which we entered, entry is at 3-ROTA, else at A-ROTA; *(in *-ROTA) is set to 3 or A accordingly.
3) Details for G-MAKE, G-BOOL and G-LOOK are not given anywhere in this account; equivalent menus must also appear in the work tree.
4) "A" means whatever number of coordinates are directly available.

special cases, we will want to keep a record of what we have done, in a useful form. In this account, we shall only specify a naive label (PSTASH) for an action to do this, leaving the necessary detailed actions for exposition elsewhere.

13. PEEKER STRUCTURE

We may want to go so far as to allow ourselves any chain of actions in the peek tree that could be carried out if the peek tree contained a complete copy of the work tree. We shall cover all reasonable possibilities by describing a peek tree that contains the minimum actions — rotations and groups — and an unspecified menu P-WORK which gives access to none, some, or all of the other facilities of WORKER.

Exhibit 3 shows the resulting partial structure for PEEKER.

We also need to provide for the rare occurrence that we need to save what has been done in PEEKER. This calls for at least 2 more items, namely

   NODE-R .. (P-WORK) (P-WORK) (not applic)

and

   PSTASH WO (P-WORK) (back to WO) (not applic)

the first of which sets nodes as we peek, in preparation for a possible use of the second. PSTASH stores a trail of what was done in PEEKER and does the necessary bookkeeping.

If we are prepared to copy the part of the data in the front (dedicated) memory on demand, we can implement, either by itself or as an extension of PEEKER, in place of the more restricted PEEKER, such an N-PEEK (menu or submenu) which has full freedom to do anything in the middle of anything else, since "inversion" merely means giving up the copy, however malformed, and going back to the original.

The most user-convenient structure would be completely recurrent in the sense of allowing N-PEEK to make any use of WORKER (see the next section) on the copy of current coordinates that is to be destroyed at the end of "peek."

D. MORE SPECIFICS FOR WORKER

14. SNAKES AND SNAKE-MAKING

The simplest input to guide a data-modifying action is a constant. We have discussed (Section 5) entering integers, with PIck, and continuous values, with the tracker ball.

The next simplest input is a smooth function of one variable, which we shall call here a *snake*. Since we want all our actions to be recordable and reproducible, we cannot allow free-hand snakes. So we need to (1) choose a few-parameter family of functions and (2) choose a way to parametrize (choose coordinates for) this family which offers convenient control.

Initially we make the choices

1) family = all polynomials of ≤ a given degree,
2) coordinates = coefficients of Chebyshëv polynomials

Better families should be available — polynomials are not really nice — but we will probably have to wait for experience with polynomial snakes before we make better choices.

We start with 6-parameter families. This means a general 5th-degree polynomial, or a 6th-degree polynomial through the origin. Planning for both seems desirable.

Since we are working on a finite interval, which we may as well take as $-1 \leq x \leq +1$, the nice equal-ripple characteristics of Chebyshëv polynomials (on that interval) makes them a natural first choice. Moreover, they are easily calculated by a variety of simple recursions.

It seems natural to begin with the lowest coefficients, which affect the broadest aspects of the function, and then work our way upward. In each specific application, we will be watching the effect of changing our function on some specific display, trying to choose these coefficients to make the picture "nicer." Different applications will have different meanings for "nicer." (If we are twisting, for example, we may be looking at points near a twisted strip, trying to untwist that strip as well as we can.)

* snake making *

We plan to use these polynomials as functions of an axis coordinate that is *fixed for the time being*. This offers us the possibility of either (i) computing the $T_i(x)$ once for all, and storing them temporarily, or (ii) computing them dynamically.

The 5-SNAKE portion of a snake-making menu is naturally displayed as

Name & cue		(jump)	(function of TAke)
LOSNAK	TT	(MESNAK)	(accepts α_0, α_1)
MESNAK	TT	(HISNAK)	(accepts α_2, α_3)
HISNAK	TT	(LOSNAK)	(accepts α_4, α_5)
SEESNA		(LOSNAK)	(displays $x, s(x)$)

where the 5-SNAKE value is (note, $T_0(x) \equiv 1$):

$$S_5(x) = \alpha_0 + \alpha_1 T_1(x) + \alpha_2 T_2(x) + \ldots + \alpha_5 T_5(x)$$

and the tracker ball (as "TT") controls α_0, α_1 in LOSNAK; α_2, α_3 in MESNAK; and α_4, α_5 in HISNAK.

If we do not need the constant term, then we can use the 6-SNAKE value

$$S_6(x) = \alpha_1 T_1(x) + \alpha_2 T_2(x) + \ldots + \alpha_6 T_6(x)$$

where the tracker ball now controls α_1, α_2 in LOSNAK; α_3, α_4 in MESNAK; and α_5, α_6 in HISNAK.

* a possibility *

The usefulness of replacing polynomials by splines whose knots are organized and controlled as a binary decomposition is an open question.

15. TWISTS

Rotations are determined by two directions and an amount. The two directions determine a stack of parallel planes (a one-dimensional stack for three-dimensional data; an $(n-2)$-dimensional stack for n-dimensional data). The amount says how much we rotate in each of these parallel planes, the same amount in each.

The simplest generalization of a rigid rotation keeps the stack of planes, but rotates by different amounts in different planes. If the amount is to be a simple function of what plane we are in, it will depend on a simple coordinate for that plane. The simplest coordinates are linear functions of position. Since these are coordinates for (*not* in) the planes, we can always find an equivalent coordinate whose "axis" is at right angles to the planes.

We also have to specify the origins for these plane rotations, and presumably will want to use the natural centers for the two coordinates used to define the planes. In that case, we can consider the coordinate axis at right angles to these as the axis of all the rotations.

In a suitably rotated coordinate system, then, the point with coordinates

$$(\text{axis, from, to,} t_4, t_5, \ldots, t_n)$$

(where "axis," "from," and "to" are the numerical values of 3 coordinates) belongs to a 3-space of all points

$$(-, -, -, t_4, t_5, \ldots, t_n)$$

and to a plane of all points

$$(\text{axis}, -, -, t_4, t_5, \ldots, t_n) .$$

In each 3-space we rotate the planes about the common axis, made up of points

$$(\text{axis}, 0, 0, t_4, t_5, \ldots, t_n)$$

making different rotations for different values of "axis." Clearly this ought to be called a twist.

All these 3-spaces are parallel to one another. All the individual twists are the same, going into one another under the rigid displacements that take any one 3-space into any other.

* twist-making *

Let now A, F, and T with values $x_A, x_F,$ and x_T be the "axis," "from," and "to" coordinates. The basic nonlinear portion of the twist is then

$$(x_A, x_F, x_T) \rightarrow (x_A, x_F \cos\theta + x_T \sin\theta, -x_F \sin\theta + x_T \cos\theta)$$

where, plausibly,

$$\theta = S_6(x_A),$$

a 6-parameter snake. Except for calculating many $\cos\theta$'s and $\sin\theta$'s, there is no difficulty in finding and applying either the 6-TWIST just described or its 5-TWIST analog.

It seems likely that dividing $0 \leq \theta(radian) \leq \pi/2$ into 16 equal intervals will allow adequately precise rapid calculation by

$$\left. \begin{array}{l} \cos\theta \doteq A_i + B_i\theta \\ \sin\theta \doteq C_i + D_i\theta \end{array} \right\} \quad i = 1 \ldots, 16$$

since the interval lengths will be $\pi/32 < .1$ and the curvature correction will be less than (half-length)²/2 from zero or (half-length)²/4 = (length/4)² from a midchoice (of the constant term), giving a maximum error of .0006. Thus computing time for reasonable sample sizes should not cause great difficulty.

* twist-control *

A reasonable menu for twist control is set forth in Exhibit 4. Notice that acceptance of the twist requires storage of both the linear consequences of (AF-AT) as well as the nonlinear consequences of the twist itself.

* other kinds of twists *

If we were preparing to untwist spiral nebulae, we would want

$$\theta = \text{a function of } ((\text{"from"})^2 + (\text{"toward"})^2)^{1/2}$$

instead of

$$\theta = \text{a function of "axis"}$$

It is easy to invent other generalizations, but harder to see where they would be useful.

* analogs for SHOVUP and SWELUP *

We have introduced TWIST as a natural generalization of ROTATE. We earlier suggested expanding ROTATE to include items like SHOVUP and SWELUP. What about their generalizations?

It seems relatively clear that if we include twists, we should include these other generalizations, perhaps in the form of Exhibit 5, which might — or might not — be incorporated in the twist menu.

Exhibit 4

A reasonable twist menu

	Name & Cue		(jump)	(function of TA)
=>	A-PICK	##	(F-PICK)	(picks A = axis)
	F-PICK	##	(T-PICK)	(picks F = from)
	T-PICK	##	(V-PERM)	(picks T = toward)
	V-PERM	ZZ	(self)	(cycles through three 2-views)
	AF-AT	TT	(ENDPIC)	(rotates from A toward F and T)
	ENDPIC		(LOTWIS)	(irretrievable move down)
	LOTWIS	TT	(METWIS)	(adjusts first 2 α's)
	METWIS	TT	(HITWIS)	(adjusts next 2 α's)
	HITWIS	TT	(LOTWIS)	(adjusts last 2 α's)
	V-PERM	ZZ	(self)	(cycles through three 2-views)
	SEESNA	QQ	(note 4)	(displays snake vs. axis)
	NODE-M		(RETURN)	(sets node, does *not* move up)
	NULL-T	UP	(above)	(declines twist, moves up)
	RETURN	UP	(above)	(accepts twist, moves up)

Notes:

1) The first 6 lines are replaced by "axis = ##," "from = ##," and "toward = ##" once a transition (irretrievable) across the solid line is made. (The lower portions might well be blanked out until this occurs.)
2) ZZ is AF, FT, or TA according to which 2 of A, F, T are presently displayed as H and V.
3) QQ is SN when snake is *not* displayed, WO else.
4) Jump is to self if snake is *not* displayed, else to LOTWIS.

16. RE-EXPRESSION

We turn next to another sort of menu, one devoted to the elementary re-expressions of, most probably, initial coordinates — but conceivably, current coordinates. (While our SHOVE-SWELL operations could be used for this purpose, polynomials are likely to prove unpleasant, as we shortly note.)

We need to provide both flexibility and parsimony in our re-expressions. The greatest need is for monotone re-expressions of single variables. Monotonicity over the span of the data can be essential, but more monotonicity — monotonicity over the half-line $(0, \infty)$ — or even

Exhibit 5

SHOVE and SWELL control
(name of menu not yet fixed)

Name & cue		(jump)	(functions of TA)
V-PICK*			accepts V coordinate
A-PICK*			accepts A coordinate
H-PICK			accepts H coordinate; default H = A.
LOSHOV	TT**	(MESHOV)	accepts β_0, β_1
MESHOV	TT**	(HISHOV)	accepts β_2, β_3
HISHOV	TT**	(LOSHOV)	accepts β_4, β_5
LOSWEL	TT**	(MESWEL)	accepts γ_0, γ_1
MESWEL	TT**	(HISWEL)	accepts γ_2, γ_3
HISWEL	TT**	( )	accepts γ_4, γ_5
NULL-R	UP	(up)	declines actions; goes up
RETURN	UP	(up)	accepts actions; goes up
INTACC	(V-PICK)		accepts actions; goes to V-PICK (for an additional change)

Notes:

*) This item dims once any action is taken for LOSHOV, MESHOV, HISHOV, LOSWEL, MESWEL, or HISWEL; it cannot be reached again until NULL-R, RETURN, or INTACC has been pushed.

**) When at these items, the HV-display shows as the V-coordinate,

S_5^* (A)(V-coordinate at menu entry + S_5^* (Acoordinate))(A_N)

where

$$S_5^*(x) = \beta_0 + \beta_1 T_1(x) + \ldots + \beta_5 T_5(x)$$

$$S_5^{**} = 1 + \gamma_0 + \gamma_1 T_1(x) + \ldots + \gamma_5 T_5(x).$$

over the full line $(-\infty, \infty)$ — is often very well worthwhile. This means that we will not dare to use polynomials (snakes á la Section 14 are out, for this use).

The instances where a graphical approach to re-expression seems most likely to be helpful involve a situation where the re-expression of u is guided by some other variable v, and the desire is to make the dependence of v on $\psi(u)$, (where $\psi(\cdot)$ is the re-expression) *as nearly linear* as possible.

To do this effectively, it will ordinarily be desirable to "smooth" v before comparing it, graphically, with $\psi(u)$. The kind of smoothing that one wants is not at all to be point-for-point. We can quite reasonably have a smoothed picture containing fewer points that we started with.

A useful smoothing component that we propose to use is conveniently labelled V3REF and consists of the following subcomponents

Exhibit 6

Structure of RE-EXP
(used to straighten relation of R to its guide G by simple re-expression; displays R or its re-exp vs. G throughout except for VLEVEL)

Name & cue			(REsit)	(SK jump)	(TA jump)	(functions of TA)
	NULLRO	TA	-	(R-PICK)	(R-PICK)	annuls rotation
=>	R-PICK	## TA	-	(G-PICK)	(G-PICK)	picks re-expee
	G-PICK	## TA	-	(K-SMOO)	(K-SMOO)	picks guidance
	K-SMOO	k TA	-	(GUIDED)	(GUIDED)	smooths k times
						(k holds till changed)
=>	GUIDED	DN	(V-DATA)	(V-DATA)	down	[to submenu]
	STARTO	.T		(LOG-R)	(LOG-R)	(sets start per T)
	LOG-R			(K-SMOO)	(K-SMOO)	(alternates logs with antilogs*)
	V-TREX	TT	-	(V-TREM)	(NULLRX)	(shows re-expee vs. guide)
	V-TREM	TT	-	(V-TREX)	(NULLRX)	(shows re-expee vs. guide)
	-GUIDE	.T	-	(NULLRX)	(NULLRX)	(shifts R by guide, new
						display is R-G, expanded)
	NULLRX	UP	-	(RETURN)	(up)	(annuls entry)
	RETURN	UP	-	(NULLRX)	(up)	(accepts entry)
	NODE-R	UP	-	(RETURN)	(up)	(sets node and accepts)

Notes:
* if already logged, TA antilogs, and vice versa.
1) => means "jumps in to sit at last used of these"
2) After RETURN from GUIDED; V-TREX starts re-exp at the trex corresponding to last VUERAW (in GUIDED); V-TREM starts it at the trem corresponding to the last VUELOG (in GUIDED).
3) Entry at R-PICK from above, entry at GUIDED from REGRESS, return accordingly.
4) SKip of SCan over GUIDED for fully manual operation.
5) re-expee = re-expression of coordinate being re-expressed.
6) Nodes are discussed in Part E, below.

(we assume sorting on u before starting and grouping into steps of no more than 1000th the range)

V) dealing with any vertical ties — any sets of (u,v)'s with common u are replaced by single points $(u, \text{med } v)$;

3R) iterated medians by 3's (EDA Chapter 7);

E) the end-value rule (EDA);

F) dealing with consecutive flat ties — any strings of (u,v), as already smoothed, with common v are replaced by single points (med u,v).

At either step V or step F, or both, we may reduce the number of points, perhaps drastically.

Repeating the process can produce further smoothing, since the F step of the previous repetition offers new opportunities for 3R to work on in the present repetition.

Since a) the numbers of points go down, and b) once we reach monotonicity the process stops, it seems reasonable to suppose that we will have (transient) memory-space for all our steps of V3REF smoothing.

With this introduction we can propose the basic structure of RE-EXP shown in Exhibit 6. (Not all details, including the use of trexes (trivially re-expressed exponentials) and trems (trivially re-expressed monomials), have been discussed.)

The submenu GUIDED could then be structured as in Exhibit 7. (See Tukey, 1982 for some background.)

Exhibit 7

Structure of GUIDED
(used to find functional form and rough coefficients for
simple re-expression matching R to its guide, G)

Name & cue	(SK jump)	(TA jump)	(function of TA)
=> LODIDI TA	(KSMELT)	(KSMELT)	log divided diffs = lodids
KSMELT k TA	(VUERAW)	(VUERAW)	smelts k times (1) (k prompts PIck)
VUERAW TT TA	(VUELOG)	(NULLGU)	controls eye-fitted line
			lodids vs. raw re-expee
VUELOG TT TA	(VUERAW)	(NULLGU)	controls eye-fitted line
			lodids vs. log re-expee
NULLGU TA UP	(RETURN)	(up)	annuls visit to this menu
RETURN TA UP	(NULLGU)	(up)	accepts *last* fitted line
NODE-M TA UP	(RETURN)	(up)	sets node and accepts

Notes:

1) See Tukey 1982 for details.
2) Nodes are discussed in Part E, below.

17. SOME OTHER CAPABILITIES

Some of the other capabilities planned for PRIM-81 include:

- a vertex-by-vertex polygon generator, which can be used both for grouping (analogous to isolation in PRIM-9, setting flags on data points) and masking (setting "masks" in terms of *current* coordinates);
- Boolean facilities for both grouping and masking;
- control/display procedures to exercise and learn more about binary decomposition algorithms;
- facilities, in the host computer, for carrying out projection pursuit and several of its generalizations;
- a blending menu, to combine "current" coordinates from two or more states of the display (two or more nodes).

The relation of PRIM-81, with its "geometric object" emphasis, to similarly controlled systems directed toward regression and related procedures still remains rather unclear. (But ponder Part G.)

E. WHAT MAKES UP A TRACE?

18. BASICS

We are considering a system under human control which works with

1) a data table (actual or simulated); and
2) displays of various aspects of that table

modifying either or both as directed. There are two retrieval needs to be met by storing a trail of the modifications and selections made starting from a given point:

1) we may need to retrieve an intermediate state (data table, and program status — and hence a display);
2) we may want to record — or study — the concatenation of steps that took us from one state to another.

If we assume, for an instant, that the notion of "steps" is already well defined, we could adopt an unrealistic extreme where we store both the state after every step — at every subnode — and also store the details of the entire chain of steps leading to that subnode. This would make unnecessary demands on storage, without easing access. We only need to do much less, to store the whole state at apparently important points —

nodes — (with perhaps some additional nodes to break up long chains of steps between nodes). And we only need to store step information once for each link from one subnode to the next. (Each node is, of course, also a subnode.)

Thus we need:

- node stashes, each of which contains the appropriate data table and a status entry recording the tree, menu, item, and parameter values in use at the node;
- link stashes, each of which describes the longest chain of concatenated steps that can be compactly described, thus telling us how to get to the subnode that ends the link from the subnode that begins it. (Status information for the subnode should be included.)

For linear modifications, like rotations, where the concatenation of several modifications is another linear modification, we can include a chain of such steps in a single link without any difficulty. The matrix of the concatenated linear modifications is just the product of the individual matrices. This overall matrix is all that we need store (for that link) besides the final status.

Nonlinear modifications of the sorts we are likely to consider using do not seem to have any natural way to concatenate a chain of even two such modifications. So we expect each individual nonlinear modification action to be a link on its own, with subnodes (or nodes) both immediately preceding and immediately following.

19. SETTING SUBNODES AND NODES

When we make a transition between any two of

- a linear modification;
- a nonlinear modification; or
- a different nonlinear modification (which may be of the same kind, but with different parameters, e.g., a twist around a different axis)

the system *must* set a subnode automatically.

The simplest way to ensure this is to:

a) set a potential subnode when a nonlinear action is tentatively undertaken, unless a subnode has already been set in that position — and make that subnode actual whenever the nonlinear action is TAken;
b) set a subnode following the nonlinear step, whenever it is TAken;
c) and store the necessary information as a link stash for the link joining these two subnodes.

* nodes *

Setting nodes, except at the beginning and end of branches and sessions, could be a wholly human procedure, based on a recognition that the current state, as reflected by the current data table and one or more displays (which may not be identified), is

- highly suggestive, or
- relatively satisfactory, or
- needs to be followed up in two or more ways.

But we may also want to count subnodes, and automatically promote a subnode to a node when it is the 10th, 12th, 15th and 20th subnode. While this would ensure easier access to all subnodes, it may not be needed.

20. RECALL OF STATUS

If we want

- to look again at a promising display, or
- to go back and start a branch of actions,

we should be able to pick out the node we wish. (If we want a subnode, we want the most closely preceding node as a first step.) So we need facilities to

- run up and down the tree of nodes, and to
- step forward one subnode at a time.

It seems simplest, from the point of view of control, to number, not just the individual node, but that node together with a chain of (one or more) links emanating from it, so that an individual node may have two or more numbers, but a numbered node (unless the end of branch) has a unique next numbered node.

21. SEGMENTS

In some suggested implementations of data modification display systems, there may well be two or more layers of memory with only some of the coordinates in an outer layer at one time. The additions to stash and display planning to take care of this are postponed to a later point (Part G, Section 26ff).

F. TRACE DISPLAY AND CONTROL

22. DISPLAY

We give a reasonable form of trace display in Exhibit 8.

A number of comments about this example will be helpful:

- we have set 26 nodes, possibly an unusually large number;
- we have branched at node 5 (which is therefore also node 12), at node 7 (which is also node 10) and, twice, at node 14 (which is thus also node 17 and node 25);
- because node 16 was, at first, the end of a branch, it might seem that we did not have to assign it a new node number when we continued on from it, but since we may need to make either of 2 continuations it is convenient to have both numbers (node 16 and node 21);
- the lower part of the display, of which we give two versions, shows that (first version) we are at node 23, OR (second version) that we are at the second subnode down the trail from node 23 to its successor, node 24.

Exhibit 8

Display of nodes, and when called for, Subnodes
(underscores represent emphasis on currently
occupied nodes or subnodes)

NODES (node numbers within the emphasized segment)

(1) — (2) — (3) — (4) — (5, 12) — (6) — (7, 10) — (8) — (9)
(10) — (11)
(12) — (13) — (14, 17, 25) — (15) — (16) — (21) — (22) — (23) — (24)
(17) — (18) — (19) — (20)
(25) — (26)

SUBNODES (subnode numbers following the emphasized node, two examples)

(2̲3) — ((1)) — ((2)) — ((3)) — ((4)) — ((5)) — ((6)) — ((7)) — ((8)) — (24)

OR

(23) — ((1)) — ((2̲)) — ((3)) — ((4)) — ((5)) — ((6)) — ((7)) — ((8)) — (24)

Note: 23-1, 23-2, 23-3, etc., is probably a better numbering scheme, since subnode numbers restart at each node.

The necessary control in the trail tree can be included in a single menu as shown in Exhibit 9.

R. A. Becker suggests that, when a node is explicitly created, there should be an opportunity to attach to it a brief text string, such as "three groups" or "almost linear."

23. MORE DETAILED LINK DISPLAY

Those who focus much more on where we have come from (initial coordinates) than would correspond to a pristine focus on "geometric objects" may rightfully feel that it may help to know more about what has happened, say, during the links that join one node to another. If we

Exhibit 9

The control menu related to Exhibit 8
Menu Name = NODES+

Segment and node information of Exhibit 8 always displayed, unless PEek enabled an odd number of times when corresponding view is seen (exit only by another press of PEek)

Name & cue		(Skips to)	Function
GOTONO	NN	(INSWIT)	(PIck chooses node)
GO-ON1		(self)	(TA moves on 1 node number)
CRAWLO	TA	self	Steps along branch to next subnode (each use at a node only adds the subnode information, which holds until a new node is called)
WORKER	TA	(over)	Transfers to work tree (menu = WORKER), setting a tentative new node as needed (*)
DENODE	KK	(self)	(Deletes node number KK, linking up adjacent subnodes and relabelling subnodes)
DEBRAN	KK	(self)	(Deletes node number KK and all its immediate successors)
DETRAI		(over)	Deletes entire trail and goes over to WORKER

Note:

* If we are at a node which already has a follower, we set a tentative number for the possible branch at that node. If we are at a *sub*node, we also set a possible node number for that subnode. In addition, of course, we set a tentative node number for the next node down our branch.

can call this information up, we might have a useful basis for trying to understand better what has been done. And we may well want to compare what happens on 2 or 3 such paths.

The sort of display that might be helpful is shown in Exhibit 10.

The control of such expanded information could use a menu like Exhibit 11.

Exhibit 10

Example of expanded information for link paths

23		17	
	ROTATE 1, 2, 7; 2, 4, 9		ROTATE 1, 3, 4; 2, 6, 7
23.1		17.1	
	TWIST 2, 7, 9		G-SHOW 14
23.2		17.2	
	ROTATE 7, 9, 11		TWIST 3, 6, 8
23.3		17.3	
	G-MAKE 29		ROTATE 3, 6, 8
23.4		18	
	G-BOOL 43 from 3, 7, 29		G-SHOW 11
23.5		18.1	
	G-SHOW 43		ROTATE 3, 6, 8
23.6		18.2	
	ROTATE 3, 4, 6		TWIST 2, 5, 4
23.7		19	
	TWIST 3, 4, 6		G-SHOW 9
24		19.1	
	G-SHOW ALL		ROTATE 2, 4, 5
24.1		19.2	
	G-SHOW 21		TWIST 2, 4, 5
24.2		20	
	G-SHOW ALL		ROTATE 2, 4, 5
24.3		21	
	ROTATE 1, 7, 8; 3, 4, 6		
25			

Notes:

1) Number triples after ROTATE are *current* coordinates involved in rotation; more than 1 triple appears when more than one rotation occupation is concatenated into a single link. (Order in triples is irrelevant.)

2) Number triples after TWIST are *current* coordinates involved in twisting; order is A, F, T.

3) Numbers after G-MAKE and G-SHOW are serial numbers of groups made or shown (ALL means all.)

4) Numbers after G-BOOL are near group serial numbers "from" its constituents.

Exhibit 11

Control menu for expanded link information
(Menu Name EXPAND; callable from NODES+)

Name & cue	(jump)	(TA function)
COLUMN KK	(self)	(sets column number, initial default 1, later defaults increase by 1)
INIT-N KK	(TERM-N)	(sets number of initial node)
TERM-N KK	(COLUMN)	(sets number 1 lower than final node; initial value of KK continues from INIT-N)
SCROLL	(ENDLOG)	(use of PIck scrolls all columns up and down)
ENDLOG UP	(UP to NODES+)	(gives up (forgets) the information displayed, and goes to NODES+)

Note:

If only a path from one node to its successor is desired, once KK is set *two* pushes of TA do the needful, and another such push sets the next column ready for KK setting.

24. MORE VISUAL TRAILS

It has been argued that one needs other aids than the expanded link identification just outlined. Two sorts of view displays have been proposed to this end.

The simplest one would allow one to go through all the nodes, and their accompanying displays, and "star" those which have retained interest. It would then be made possible to scan through the cycle of starred views either manually step by step, or automatically. The necessary items, to be added to NODES+, could be:

(Name & cue)	(jump)	
STAR-N NN	(self, NN+1)	(displays view corresponding to node NN, TAke attaches a "star" to this node)
S-VIEW NN	(self)	(shows views for starred nodes *with* their node and group numbers; TAke advances one step; PIck normally active; starts at first star)

SCAN-S NN (self) (scan shows each starred view, and its node and group number, in turn, for a few seconds; opens at first star; TAke starts this scan, which continues cyclically until SKip is pushed)

This should not be hard to implement and is likely to be quite helpful.

A second suggestion, which is less clearly useful, is to keep a small zoomable image of the views corresponding to the last (say) ten nodes on the margin of the data screen at all times. This would seem to be material only appropriately placed on the auxiliary screen. (If this means a high-resolution auxiliary screen and a low-resolution main screen, so be it.)

25. WORKING STASHES

Besides the record, what do we need to keep at hand so that the work tree or the peek tree can function properly?

At the instant after an action, which is also the instant before the next action, we need to have the data table stashed either explicitly or, possibly, if the last actions were linear, implicitly (as an earlier data table and a linear modification matrix).

In the middle of an action, which is thus still tentative, we need additional working stashes.

This is because of our basic structure, which we may have in two or more layers.

* tentativeness and working-stash multiplicity *

If we have j levels of tentativeness ($j = 2$ for present planning, because (a) the whole of any menu is tentative and (b) the current occupation of every item is tentative), we need $j + 1$ working stashes.

This is because we need (a) one for where we are at the instant, (b) one for the beginning of this occupation of this item, and (c) one for the beginning of this occupation of this menu. SKip transfers us from the first to the second, while TAking NULL-R takes us from the second to the third.

One option is to keep 3 complete working stashes. This is possible; but will tend to push us to segmentation (see Part G below). More reasonable alternatives are to keep either one-and-two-fractions or one-and-a-fraction working stashes.

In these latter schemes, we keep one complete stash — corresponding to entry to the menu now occupied — and values for variables that have been modified during that particular occupation. For menus like ROTATE and TWIST, only 3 variables (H, V, B or A, F, T) are involved in modification, so we need only store the new values of such variables in our fractions.

The logic of the situation might run as follows:

1) the basic access to variables is a basic list of, say, 20 variable numbers, with space for two flags;
2) the first flag says that the variable in question appears in the first fraction;
3) the second flag says that the variable in question appears in the second fraction;
4) there are two short lists, one of first flagged variables and another of second flagged variables;
5) the basic stash (at menu entry) contains values for all the variables in the basic list;
6) the fractional stashes (at item entry, at present state) contain values for all the variables in the corresponding short list;
7) if the number of stashes is changed, assembly of variables from two or more stashes may be needed.

To illustrate (7), suppose that variables 1, 2, 4, 8, 9, 10, 11, 12, 13, 19 are available in the basic stash (menu entry), that variables 4, 8 and 12 are H V B (are in an item-entry fraction) and that the item is (..HV..) so that only variables 4 and 8 are in the current fraction. Consider now the effect of various controls.

Pressing SKip destroys the current fraction.

Pressing TAke transfers variables 4 and 8 from the current fraction to the item-entry fraction, and then destroys the current fraction. After this, TAking RETURN transfers all 3 variables (4, 8 *and* 12) from the item-entry fraction to the menu-entry fraction. At the same point, TAking NULL-R only destroys the item-entry fraction.

Thus Take (item), SCan (to RETURN), TAke (RETURN) assembles a whole stash in which variables 4 and 8 came from the current fraction, variable 12 came from the item-entry fraction, and the remainder, variables 1, 2, 9, 10, 11, 13, 19, came from the menu-entry whole.

* side issues *

Doing this means that the matrix combination for successive rotations should occur in the host computer, since we want the fraction to involve only 3 variables in such a case. (Really we are treating each occupation of a rotation as a sublink, and transmitting to the host computer accordingly.)

The acceptance of a non-null nonlinear modification will of course, often set two subnodes, one before and one after, as well as putting status into a link stash. (If there was already a subnode just before the nonlinear action, only one new subnode need be set.)

* and the peek tree *

Speed of communication between memory used for working stashes (say one-and-two fraction style) and the memory behind this one is of great importance as to how we organize "peeking." If this communication is fast enough, when we peek (by pressing PEek) we will transfer the whole of this working-stash memory to a buffer memory in the backup memory, planning to retrieve all of it when PEek is pushed again. If such transfers are too slow, we have little alternative but to hold the one-and-two-fractions WORKER working stash alongside an equally large one-and-two-fractions PEEKER working stash. (This again drives us toward segmentation.)

* one-and-a-fraction *

If transfer is fast enough, we can consider keeping a menu-entry (whole) stash in back and only an item-entry *whole* and a current-position *fraction* in front. Acceptance of an item's occupation with TAke would (a) blend the current-position fraction into the time-entry whole and (b) destroy the current-position fraction. SCanning to RETURN would prepare for, and TAke would carry out a blending of those variables in the item-entry whole into the menu-entry whole which have indeed been modified (which have first flags set).

While it needs to be thought through carefully, the one-and-a-fraction logic is attractive, and may not require too high a transfer speed.

* the peek tree again *

There is a real and important sense in which our discussion of the needs of the peek tree correspond to increasing j from 2 to 5. When we are in the peek tree, there are 5 nested instances of tentativity, namely:

1) the WORKER item-entry;
2) position in the WORKER item;
3) the peek tree itself (stashed as PEEKER menu entry);
4) the PEEKER item-entry;
5) position in the PEEKER item.

We are always ready, in such a situation, to decline (5) or (3) immediately, to decline (4) after (5) and an item rechoice, to decline (2) immediately after declining (3), and to decline (1) after declining (2) and an item rechoice.

This degree of tentativity is important for fluent and convenient control. The relatively heavy demands on working-stash space follow inevitably.

G. SEGMENTS

26. PURPOSE

Close-up memory limitations, exacerbated by the need to keep at least parts of several data tables ready for immediate access, seem often likely to require us to have nearby data tables with fewer coordinates (fewer variables) than our backup data tables.

This leads to a further, large-scale, breakdown of action trails into *segments*, where a segment consists of a tree of consecutive nodes for all of which the same coordinates, specifically the coordinates with the same numbers, are those in the nearby data tables.

Exhibit 12

Possible display of segments,
nodes, and when called for, subnodes
(underscores represent emphasis on currently
occupied segments, nodes, and subnodes)

SEGMENTS (*variable numbers*)

(1,2,3,4,5,6,7,8) — (1,2,3,4,9,10,11,12) — (1,2,3,4,
13,14,15,16) — (1,2,3,4,7,9,15,16) — (1,2,3,4,8,11,12,16)

NODES (*node numbers within the emphasized segment*)

(1) — (2) — (3) — (4) — (5,12) — (6) — (7,10) — (8) — (9)
(10) — (11)
(12) — (13) — (14,17,25) — (15) — (16) — (21) — (22) — (23) — (24)
(17) — (18) — (19) — (20)
(25) — (26)

SUBNODES (*subnode numbers following the emphasized node*)

(23) — ((1)) — ((2)) — ((3)) — ((4)) — ((5)) — ((6)) — ((7)) — ((8)) — (24)

OR

23 — ((1)) — ((2)) — ((3)) — ((4)) — ((5)) — ((6)) — ((7)) — ((8)) — (24)

Note:

The chosen segment (1,2,3,4,7,9,10,16) is shown by underlining in the upper portion.

Exhibit 13

The menu for segment control
Menu Name = SEGMEN

(Segment and node information of Exhibit 12 always displayed)

Name & cue	(SKip)	(Functions of TAke)
CHOOSE ##	Self	On first TAke, establishes new segment, as well as choosing 1st variable. Later TAkes choose variables 2 to 8. (SKip after TAke is to WORKER unless the same segment was previously established when SKip is to DUPLIC)
DUPLIC DN	(NODES+)	(Information prompt only, goes next — on either SKip or TAke — to NODES+ for duplicated segment)
WORKER WO	(side)	Goes to WORKER (in work tree)
ANNULL	CHOOSE	Blanks what has been done in the current engagement (with this segment)
PEEKER	(side)	Goes to PEEKER (in peek tree)
NEWSEG SS	(NODES+)	Changes emphasized segment under PIck control
NODES+	(down)	Transfers to NODES+
DELETE SS	(CHOOSE)	Deletes segment pointed to

Note:

SS = segment number

27. CONTROL

To handle display and control, we have to add an upper stage to our trail display as in Exhibit 12 and an upper control menu (SEGMENT) as indicated in Exhibit 13.

Within a segment we need only stash partial data tables involving that segment's variables. When we change segments, however, we may need to embed each of these partial data tables, copying the rest of the full data table as it was when the segment began.

H. POSTSCRIPT ABOUT RECOVERY

28. A PROBLEM

A very important problem — recognizing where one is at (once at a nice — presumably simple — picture) in terms of where one began (in initial coordinates) — is only addressed very indirectly above. (Enhanced subnode information (Exhibit 10) was a way, probably a very poor way, to begin detailing this problem.)

We try here to make a good beginning on its solution, using first projection pursuit regression as developed by Friedman and Stuetzle [1981].

* projection-pursuit regression *

We sketch the idea of such regression, leaving all details to the paper cited (and its supporting computer program).

Given data points made up of a y and several x's, say $x_1, x_2, \ldots, x_k$ we can try lincoms (linear combinations) of the x's, say

$$x_A = a_1 x_1 + a_2 x_2 + \ldots + a_x x_x$$

by the computer equivalent of

1) plotting y vs. x_A;
2) smoothing y against x_A to find a smooth function $y_A = \psi(x_A)$;
3) looking at the reduction in sum of squares going from y to $y - y_A$;
4) optimizing A to maximize this reduction.

Having done this for A, we can then do

$$y \to y - y_A \qquad x_A \to x_B$$

and find a second linear combination, and a second smoothing, such that y is still more closely approximated by $y_{AB} = \psi_A(x_A) + \psi_B(x_B)$.

The process is then continued to as many terms as may be needed. We will notate the answer as

$$y = \psi(x_1, \ldots, x_k) = \Sigma \, \psi_Q(x_Q) \, .$$

(Possible refinements or shortenings of this process will not be discussed here.)

29. A ONE-DIMENSIONAL EXAMPLE

The simplest recognition problem arises when we start with a curve (a piece of a curved one-dimensional manifold) near which our points are concentrated, and end with a straight-line segment, S, near which the modified points are concentrated. (This could occur in the 3-body final state problem, if we started with poor coordinates.)

We started with x_1, x_2, x_3, x_4 and ended with y_1, y_2, y_3, y_4, where, for the given data points $b \leq y_1 \leq c$ and the other y's are nearly constant (we may as well suppose nearly zero). We begin by propugressing (projection pursuit regressing) each of y_1, y_2, y_3, y_4 on x_1, x_2, x_3, x_4, obtaining, to sufficient accuracy:

$$y_1 = \psi_1(x_1, x_2, x_3, x_4)$$

$$y_2 = \psi_2(x_1, x_2, x_3, x_4)$$

$$y_3 = \psi_3(x_1, x_2, x_3, x_4)$$

$$y_4 = \psi_4(x_1, x_2, x_3, x_4) .$$

If $x_{10}, x_{20}, x_{30}, x_{40}$ and $y_{10}, y_{20}, y_{30}, y_{40}$ correspond to a data point, we can solve, in as small steps for θ as need be, for $0 \leq \theta \leq 1$,

$$y_{10} = \psi_1(x_1(\theta), x_2(\theta), x_3(\theta), x_4(\theta))$$

$$(1-\theta)y_{20} = \psi_1(x_1(\theta), x_2(\theta), x_3(\theta), x_4(\theta))$$

$$(1-\theta)y_{30} = \psi_1(x_1(\theta), x_2(\theta), x_3(\theta), x_4(\theta))$$

$$(1-\theta)y_{40} = \psi_1(x_1(\theta), x_2(\theta), x_3(\theta), x_4(\theta))$$

thus associating, for $\theta = 1$, a point $y_{11}, 0, 0, 0)$ on the line segment, S, close to $y_{10}, y_{20}, y_{30}, y_{40})$ with an original coordinate point $x_{11}, x_{21}, x_{31}, x_{41}) = (x_1(1), x_2(1), x_3(1), x_4(1)$ on — or close to — the original curve segment.

We now do propugression (which here reduces close to ordinary regression) of each x_{j1} on y_{10}, with results

$$x_{11} = \psi_1(y_{10})$$
$$x_{21} = \psi_2(y_{10})$$
$$x_{31} = \psi_3(y_{10})$$
$$x_{41} = \psi_4(y_{10})$$

giving a parametric representation of the curve segment with which we started. (The relationship between this operation, and those proposed in CAKEMAKE, see Friedman, Tukey, and Tukey, 1979, deserves thought.)

It is natural to argue that we have described an unduly complex procedure for finding a parametric form. Perhaps this is true in the simplest instance, but the occurrence of weak "rods" (weak resonances) in the 3-body final state case that are parallel to strong ones suggests that we might find curves this way that were much less detectable in the original picture.

30. A TWO-DIMENSIONAL EXAMPLE

Let us now turn to 4-body final state data, where we have $(x_1, x_2, \ldots, x_7)$ and expect to see a collection of pieces of curved and twisted two-dimensional manifolds, of which we may suppose PRIM-81's grouping facilities will let us cut out one. And suppose that we have been able to make this one flat, which means

a) $b \leq y_{10} \leq c$ and $d \leq y_{20} \leq e$
b) all other y's — $y_{30}, \ldots, y_{70}$ — nearly zero.

If we do propugression of each y_{j0} on all the x_{k0}'s, we will get

$$y_{j0} = \psi_j(x_{10}, \ldots, x_{70})$$

and can solve, for θ ranging from 0 to 1,

$$y_{10} = \psi_1(x_{10}(\theta), \ldots, x_{70}(\theta))$$
$$y_{20} = \psi_2(x_{10}(\theta), \ldots, x_{70}(\theta))$$
$$\theta y_{j0} = \psi_j(x_{j0}(\theta), \ldots, x_{70}(\theta)) \quad 3 \leq j \leq 7$$

associating each data point $(x_{10}, x_{20}, \ldots, x_{70})$ with a nearby point $(x_{10}(1), x_{20}(1), \ldots, x_{70}(1))$ on (or very near) the two-dimensional manifold.

We can then proceed to propugression of each $x_j = x_{j0}(1)$ on y_{10} and y_{20} to find

$$x_{j0}(1) = \Psi_j(y_{10}, y_{20}) \quad 1 \leq j \leq 7$$

which gives a parametric account of the original piece of a 2-manifold in terms of y_{10} and y_{20}.

31. CONTINUATIONS AND IMPROVEMENTS

* nonsingulations *

The coordinates y_{10} and y_{20} are, really, undetermined up to a nonsingulation (a nonsingular linear transformation). Thus we could take the last form, put

$$[y_{10}, y_{20}] = \begin{bmatrix} f g \\ h j \end{bmatrix} (z_{10}, z_{20})$$

which has an inverse

$$(z_{10}, z_{20}) = \begin{bmatrix} f g \\ h j \end{bmatrix} (y_{10}, y_{20})$$

so that we can do the propugression of the x_{j0}'s on z_{10}, z_{20} instead of on y_{10}, y_{20}. It is then in order to look for those z's, of the given class, that make these propugressions (collectively for all 7 x's) as simple as possible. This should give us a pretty good parametric form.

* coordinate comparisons *

If we are actually dealing with 4-body final-state data, and not just an analog of this, there is another thing we can do. There are on the order of 40 natural coordinates in the seven-dimensional space of the reduced momenta, corresponding to intrinsic masses, etc., etc. We can take each of these

$$t_m(x_1, x_2, \ldots, x_7)$$

in turn, and propugress

$$t_m(x_{10}, x_{20}, \ldots, x_{70}) \text{ on } (y_{10}, y_{20}).$$

Then we can pick out of the roughly $\binom{40}{2}$ pairs those whose (two) propugressions are simplest. We can then, if we desire, complete them to a coordinate system with 5 others that relate reasonably naturally, and, finally, look at our data in this new coordinate system, where our piece of a two-manifold may well be reasonably flat.

32. EXTENSIONS

Extensions to 5- and more-body final states are clear in principle, and should be feasible if we work up slowly.

Extensions to other simple figures, like circles, etc., seem also relatively clear.

33. DEMANDS ON PRIM-81

Our discussion in Section 16 indicated a plan for some re-expression. We ought probably to plan for more flexibility of this sort.

34. PROPUMORPHISM

We have just described an approach to relating $(x_1, x_2, \ldots, x_n)$ to $(y_1, y_2, \ldots, y_n)$ when these alternative sets of coordinates are used to describe approximately lower dimensional clouds of points, less simple for the x's and more simple for the y's. (Our explicit discussion was for the case where "simple = flat" but extensions to "simple = spherical" and the like were noted to be easy.) We now turn to the case where the point cloud is of full dimension, so that we naturally want to relate the whole x-space with the whole y-space, preferably in as reversible a way as possible.

* n = 2 *

For convenience, let us begin with $n = 2$. Let y_A be a lincom (linear combination with constant coefficients) of y_1 and y_2. M-propugression of y_A on x_1, x_2 selects an x_A such that

y_A is as well-fitted by $\psi_A(x_A)$ as possible .

Here M-propugression refers to a modification of FS (Friedman and Stuetzle)-propugression that makes $\psi_A(\ )$ monotone, and hence easily invertible. (We return to such modifications below.)

If we take $y_A = \tau_A(y_1 \cos \theta_A + y_2 \sin \theta_A)$, choosing τ_A to scale y_A to a chosen size according to some at least moderately robust size measure, we can plot the degree of fit of $\psi(x_A)$ as a function of θ_A and seek out local maxima. We will want to choose two maxima which are (a) high and (b) not too close together. (Clearly one can be the highest of all.) Suppose θ_B and θ_C are the two selected. Then,

$$y_B/\tau_B = y_1 \cos \theta_B + y_2 \sin \theta_B$$

$$y_C/\tau_C = y_1 \cos \theta_C + y_2 \sin \theta_C$$

so that

$$y_1 = \frac{\sin \theta_C}{\sin(\theta_B - \theta_C)} \frac{y_B}{\tau_B} - \frac{\sin \theta_B}{\sin(\theta_B - \theta_C)} \frac{y_C}{\tau_C} = a_{1B} y_B + a_{1B} y_C$$

$$y_2 = \frac{\cos \theta_C}{\sin(\theta_B - \theta_C)} \frac{y_B}{\tau_B} - \frac{\cos \theta_B}{\sin(\theta_B - \theta_C)} \frac{y_C}{\tau_C} = a_{2B} y_B + a_{2C} y_C$$

and

$$y_B \text{ is closish to } \hat{y}_B = \psi_B(x_B) = \psi_B(d_{B1}x_1+d_{B2}x_2)$$

$$y_C \text{ is closish to } \hat{y}_C = \psi_C(x_C) = \psi_C(d_{C1}x_1+d_{C2}x_2)$$

so that

$$y_1 \text{ is closish to } \hat{y}_1 = a_{1B}\hat{y}_B + a_{1C}\hat{y}_C$$
$$= a_{1B}\psi_B(d_{B1}x_1+d_{B2}x_2) + a_{1C}\psi_C(d_{C1}x_1+d_{C2}x_2)$$

$$y_2 \text{ is closish to } \hat{y}_2 = a_{2B}\hat{y}_B + a_{2C}\hat{y}_C$$
$$= a_{2B}\psi_B(d_{B1}x_1+d_{B2}x_2) + a_{2C}\psi_C(d_{C1}x_1+d_{C2}x_2).$$

We can now replace (x_1, x_2) by $(\hat{y}_1, \hat{y}_2)$ for every data point, and repeat the process, eventually approximating y by the concatenation of a chain of reversible steps, each of the symbolic form

$$L'ML''$$

where the L's are nonsingular matrices and M is a diagonal invertible (monotone) function. The resulting morphism, since

$$L_5' M_5 L_5'' \, L_6' M_6 L_6'' = L_1 M_5 L_2 M_6 L_3$$

is of the form

$$L_k M_{k-1} L_{k-2} M_{k-3} \ldots M_4 L_3 M_2 L_1$$

whose inverse is

$$L_1^{-1} M_2^{-1} L_3^{-1} M_4^{-1} \ldots M_{k-3}^{-1} L_{k-2}^{-1} M_{k-1}^{-1} L_k^{-1}$$

with the usual significance of L_i^{-1} and with M_j^{-1} being the diagonal nonlinear operation whose diagonal elements, $\psi_{jk}^{-1}(\ )$, are the inverse functions to those, $\Psi_{jk}(\ )$, of M_j.

* important comments *

Such mappings seem to be a natural n-dimensional generalization of monotone functions.

The extent to which such composites of L's and M's can be used to approximate general homeomorphisms (among squares, cubes, or parallelpipedons) remains unclear. Abstractly, this is quite an interesting problem.

If we consider a different sort of reversible transformation, which we shall call Q (perhaps for "queer") which we shall only define in the two-dimensional case, namely those with

$$(v_1, v_2) = Q(u_1, u_2),$$

where,

$$v_1 = f_\lambda(u_1),$$

$f_\lambda(\cdot)$ monotone,

$$\lambda = g(u_2)$$

$g(\cdot)$ continuous

$$v_2 \equiv u_2$$

it can be shown that sequences alternating L's and Q's can be used to approximate any homeomorphism of a square into another.

* implementation *

To implement such a calculation for $n = 2$, we require (i) an M-propugression (see below) and (ii) an algorithm to choose θ_B and θ_C. Both will require trial, and, probably, improvement, but reasonable starts for both can already be suggested. For (ii), we could start with, after crudely orthogonalizing y_1 and y_2, θ_B minimizing, for all θ_B,

$$\text{residual sum of squares for } y_B - \psi_B(x_B)$$

and θ_C minimizing, for all θ_C,

$$\frac{Q}{[\sin(\theta_B - \theta_C)]^R} + \frac{\text{residual SS for } y_C - \psi_C(x_C)}{\text{residual SS for } y_B - \psi_B(x_B)}$$

where we might start with $R = 1$ and $Q = 1/4$.

* larger n *

Except for choice of algorithms, the extension of this calculation to larger n is clear in principle. We explore all lincoms of the y's for M-propugression on x, finding n local maxima not too close to one another. (Exploration should probably be by hill-climbing from $P > n$ starts to locate $N \geq n$ local maxima.) Then we find the nonsingular matrices relating the initial y's to these lincoms and the propugression x's to the initial x's. We now have a step in the form

$$\hat{y} = L'ML''x$$

and we can go ahead, chaining as many of these together as necessary.

* M-propugression *

Two approaches to M-propugression, one more flexible (MF-propugression), the other with more simply describable result (MS-propugression) deserve to be outlined.

The technologies described in Tukey 1982 can be used for any choice of y_A and x_A to provide:

a) first a possibly bumpy monotone smooth (by combining median smoothing and tie condensation);
b) then a smoother description (by selecting points for retention á la smelting);
c) then a fitted functional form (by identifying near straightness of an appropriate plot).

Going only as far as (a) or (b) will produce an effective MF-propugression procedure.

To produce an effective MS-propugression procedure, we will need to go either to (c) or to one or another of its following extensions:

d) extension of identification to more simple, invertible functions (see TM-82-112-3 for a possible list, which should probably be shortened to keep monotonicity);
e) going to an explicit concatenation process, perhaps with "rocking" as discussed below, in which the desired simple function is obtained as the concatenation of elementary simple ones.

* "rocking" a concatenation *

Whenever we approximate a given relationship by the concatenation of elementary simple relationships, it is natural to include "rocking" in the calculation. Schematically, with E's elementary simple operators, with inverses, E^{-1}, and the relation of y to x as the target, it is easy to outline a rocking scheme.

Let us start by choosing E, to make

$$y^* = E_1 x \quad \text{close to} \quad y$$

and then choose E_2, to make

$$y^{**} = E_2 y^* \quad \text{close to} \quad y$$

and then find E_{11}, to make

$$y^* = E_{11}x \quad \text{close to} \quad E_2^{-1}y$$

so that

$$y^{**} = E_2 E_{11}x \quad \text{is close to} \quad y.$$

If we made any appreciable improvement by modifying E_1 to E_{11}, we probably should go on to modify E_2 to E_{21} so that

$$y^{**} = E_{21}y^* = E_{21}E_{11}x \quad \text{is closer to} \quad y.$$

We may continue cycling between E_1 and E_2 as long as we get substantial improvement. (We continue to write E_{11} and E_{21} for the result.)

Then we can press on, choosing E_3 to make

$$y^{***} = E_3 y^{**} = E_3 E_{21} E_{11} x \quad \text{close to} \quad y$$

and revise, first, E_2 to E_{22}, making

$$E_{22}y^* = E_{22}E_{11}x \quad \text{closer to} \quad E_3^{-1}y$$

and hence, we hope,

$$E_3 E_{22} E_{11} x \quad \text{closer to} \quad y.$$

Then we can revise E_{11} to E_{12} and E_3 to E_2 (and, if worthwhile, continue cycling at this level).

The extension to an E_4 and beyond should now seem clear.

Algorithms to shorten such a calculation will be always welcome — and sometimes necessary. What we are doing here is only to show how such things *can* be done, not how they should be done.

I. CLOSE

35. AN UNLIKELY EXTENSION

In the unlikely event that we want to know what we have done to the whole n-dimensional region, and not just to the data points, we might want to add some "invisible points" to the data before starting our modifications, and then restoring them at the end. (It seems unlikely that this would help very often, if at all.)

The facilities proposed for PRIM-81, however, provide the necessary means for doing this, whether we remember or forget.

If we remember, we can add the necessary field to the initial data, going to $n + 1$ dimensions with a 0-or-1 variable to separate the two kinds of points. Grouping can then be used to suppress the field points from view during all of our modifications. At the end we can give up the grouping and work, as above, with all points.

If we forget that we are likely to want field points — or if including field points would increase the computing or memory load too far, we can do something else. We need only separate all status information for nodes and subnodes as a process, and pass the additional field points through this process, eventually combining data points *and* field points for both initial and final coordinates. We have not been explicit above about these last two facilities, but if PRIM-81 is to be able to work effectively with larger data sets, as it should, they will be essential.

Adequate coverage of field points may well be feasible, even if grids of field points are likely to be quite impractical. The basic idea would be to move toward orthogonal arrays rather than grids. (A two-value grid in 40 dimensions, requiring $2^{40} \approx 10^{12}$ points is quite impractical, but an orthogonal array of strength 5 will require at most a few thousand points.)

36. CLOSING COMMENTS

It is to be hoped that the examples set out above

1) sufficiently illustrate the proposed general philosophy of control, prompting, and trail-maintenance; and
2) offer specific guidance to assist the initial implementation of PRIM-81.

37. ACKNOWLEDGMENTS

The significant contributions of Jerome Friedman, Mary Ann Fisherkeller and Werner Stuetzle were mentioned in the first paragraph. The tone and content of this account owes much to comments by R. A. Becker and R. Gnanadesikan, and especially to those by P. A. Tukey. The reader owes them more thanks than he or she is likely to realize.

REFERENCES

Fisherkeller, M. A., Friedman, J. H., and Tukey, J. W. (1974). PRIM-9: An interactive multidimensional data display and analysis system. A. E. C. Scientific Computer Information Exchange Meeting, May 2-3, 1974 and Film Review. (Film). Film Review in: (1976) *Proceedings of the 4th International Congress for Stereology.* Gaithersburg, MD: National Bureau of Standards Special Publication 431.

Friedman, J. H. and Stuetzle, W. (1981). "Projection pursuit regression," *Journal of the American Statistical Association,* **76**: 817-823.

Friedman, J. H., Tukey, J. W., and Tukey, P. A. (1980). "Approaches to analysis of data that concentrate near intermediate-dimensional manifolds," *Data Analysis and Informatics,* E. Diday, L. Lebart, J. T. Pagès, and R. Tomassone, eds., 3-13. Amsterdam: North-Holland.

Tukey, J. W. (1977). *Exploratory Data Analysis.* Reading, MA: Addison-Wesley.

Tukey, J. W. (1982). "The use of smelting in guiding re-expression," *Modern Data Analysis,* A. F. Siegel and R. Launer, eds., 83-102. New York: Academic Press.

7

ORION I: INTERACTIVE GRAPHICS FOR DATA ANALYSIS

John Alan McDonald
Stanford University

ABSTRACT

This paper reports on work in a new branch of research in statistical methods: applications of interactive computer graphics. It describes the Orion I workstation, an experimental computer graphics system built at the Stanford Linear Accelerator Center in 1980-81. Orion I has been used to experiment with new methods for data analysis—made possible by advances in computer technology—that use color, real-time motion, and interaction to discover structure in many-dimensional data.

1. INTRODUCTION

This paper reports on work in a new branch of research in statistical methods: applications of interactive computer graphics. The concrete results of this research are the Orion I workstation—an experimental computer graphics system built at the Stanford Linear Accelerator Center in 1980-81—a several programs written for it. These programs implement

---

This research was supported in part by the Department of Energy under contracts DE-AC03-76SF00515 and DE-AT03-81-ER10843, the Office of Naval Research under contract ONR N00014-81-K-0340, and the U.S. Army Research Office under contract DAAG29-82-K-0056.

From *Dynamic Graphics for Statistics*, William S. Cleveland and Marylyn E. McGill, eds., copyright © 1988 by Wadsworth, Inc. All rights reserved.

new methods for data analysis made possible by advances in microprocessor and computer graphics technology. These methods can be used on relatively inexpensive and soon to be widely available machinery—the next generation of personal computers.

Systems like Orion I are primarily intended for exploration and description of data, that is, for discovery, summary, and understanding of *structure* or *patterns* in a data set. They are particularly useful for the discovery and interpretation on *unanticipated kinds of structure* in data.

Computer graphics can quickly expose many views of a data set to human perception, so that a data analyst can detect a variety of patterns. The analyst can interpret apparent structure using knowledge of the data set's context and can select directions for further exploration using results obtained so far. Problems that are extremely difficult to solve by automatic, objective methods often become much simpler with interactive methods—because of the addition of human intelligence.

A written description cannot really do justice to an interactive graphics system, so we have made two films that demonstrate some of the techniques described below: *Exploring Data with the Orion I workstation*, [McDonald, 1982b], and *Projection Pursuit regression with the Orion I workstation* [McDonald, 1982c].

2. THE PRIM SYSTEMS

Orion I is the youngest descendant of a graphics system called PRIM-9, which was developed at the Stanford Linear Accelerator Center (SLAC) in 1972 [Fisherkeller, Friedman, and Tukey, 1975].

PRIM-9 was used to explore data in up to 9 dimensions. It used motion graphics to draw three-dimensional scatter plots. Through a combination of *projection* and *rotation*, PRIM-9 could display an arbitrary three-dimensional subspace of the 9-dimensional data. *Isolation* and *masking* were used to divide a data set into subsets.

The computing for PRIM-9 was done in a large mainframe computer (an IBM 360/91) and used a significant part of the mainframe's capacity. A Varian minicomputer was kept busy transferring data to the IDIIOM vector drawing display.

Successors to PRIM-9 were built at the Swiss Federal Institute of Technology in 1978 (PRIM-S) and at Harvard in 1979-80 (PRIM-H) [Donoho, Huber, and Thoma, 1981; Donoho *et al.*, 1982]. PRIM-S used a DEC-10, a PDP-11/34, and Evans and Sutherland Picture System 2 as analogs of the IBM 360/91, the Varian, and the IDIIOM. PRIM-H is based on a VAX 11/780 computer and an Evans and Sutherland Picture System 2. Computation for rotations is done by hardware in the Evans and

Sutherland. PRIM-S and PRIM-H had and have slightly greater graphical capabilities than PRIM-9. A significant advantage of PRIM-H is the inclusion of a flexible, interactive statistical package called ISP [Donoho et al., 1982].

3. EXPLORING MANY-DIMENSIONAL DATA

Although they have many other uses, a principal application of the PRIM and Orion systems and the emphasis of this paper is the graphical analysis of many-dimensional data.

Histograms and scatter plots—for one- and two-dimensional data—have been part of a data analyst's tool kit for centuries. A *three-dimensional scatter plot* is a natural extension. There are several ways of drawing three-dimensional scatter plots [Nicholson and Littlefield, 1982; Littlefield and Nicholson, 1982]; real-time rotations give a convincing and accurate perception of three-dimensional shape from the apparent parallax in the motion of the points in the scatter plot. Rotating three-dimensional scatter plots were used in all the PRIM systems. They are demonstrated in the PRIM-9 film [PRIM-9, 1973] and in the Orion I films [McDonald, 1982a, 1982b, 1982c].

With Orion I, we have gone beyond three-dimensional scatter plots to experiment with ways of looking at more than three variables at a time. There is no completely satisfactory method. Three basic approaches to viewing many-dimensional structure, described briefly below, are discussed in more detail in the following sections:

3.1 Higher-Dimensional Views

The idea is to represent as many variables as possible in a single picture.

In a simple scatter plot, observations are represented by featureless points. We add dimension to the picture by replacing points with objects that have features, such as color, size, and shape. These features represent variables in addition to those represented by a point's position in the scatter plot. These *featurefull* objects are sometimes called *glyphs* [Gnanadesikan, 1977; Tukey and Tukey, 1981].

3.2 Projection Pursuit

The basic problem with adding dimension to a single picture is that the picture quickly becomes too complicated to understand. The alternative is to restrict our picture to a few (≤ 3) dimensions and then select low-dimensional pictures that capture interesting aspects of the multivariate structure in our data. This is the basic idea of projection pursuit, which originated with PRIM-9 [Kruskal, 1969; Fisherkeller, Friedman, and Tukey, 1975; Friedman and Tukey, 1974; Friedman and Stuetzle, 1981, 1982a; Friedman, Stuetzle, and Schroeder, 1981; Huber, 1981, 1985; Chen and Li, 1981; McDonald; 1982a, 1983; Diaconis and Freedman, 1984; Diaconis and Shahshahani, 1984].

3.3 Multiple Views

Usually no single low-dimensional view will capture everything that is interesting about a many-dimensional data set. Therefore we will need to look at several views of the data. It may be possible to understand more about the many-dimensional structure in the data if we look at several views *simultaneously* and *connect* the contents of the views in some way [McDonald, 1982a, 1982b; Kolata, 1982; Tukey and Tukey, 1981; Diaconis and Friedman, 1984].

4. THREE- (AND HIGHER-) DIMENSIONAL SCATTER PLOTS

This section describes programs which reproduce most of the functions of the PRIM-9 system [Fisherkeller, Friedman, and Tukey, 1975], with the addition of some elementary uses of color. We describe parts of a *user model* which refers to a set of abstract concepts that make it possible to understand what a program is doing, ignoring unnecessary internal details [Foley and van Dam, 1982; Newman and Sproull, 1979]. The user model defines the kinds of *objects* that a program acts on and the *actions* or *commands* that affect objects.

4.1 Using Orion I

A user of Orion I (as opposed to a programmer) controls the system using an input device called a *trackerball* while looking at pictures displayed on a video screen.

The *trackerball* [Foley and van Dam, 1982; Newman and Sproull, 1979], is a hard plastic ball about 3 inches in diameter set into a metal box so that the top of the ball sticks out. The ball can be easily rotated by hand. The position of the ball determines the values of two coordinates. Also mounted on the metal box are six switches, which are used for discrete input to programs.

In order to get Orion I to do something, a user *executes commands*, either by using a switch on the trackerball housing that is identified with a particular command, or by using the trackerball and switches to position a cursor and pick an item from a menu of choices listed on the screen.

The pictures produced by Orion I are displayed on a 19 inch high resolution video monitor. Orion I uses a *raster graphics device* which means that the picture is constructed from a *raster* or rectangular array of—in this case—1280 (horizontal) by 1024 (vertical) little dots, called *pixels*. Each pixel can have any color in a list of 256 colors called the *color map*. The 256 colors in the color map are subset chosen by each program from a pallette of 2^{24} different colors the graphics device can display.

The screen is divided into three regions (see Plate 1a). A strip on the bottom of the screen, one character high, shows six labels that indicate the current functions of the six switches on the trackerball housing. A region approximately 16 characters or 200 pixels wide on the right side of the screen is reserved for the display of assorted information about the current state of the program. The remaining area, approximately 1000×1000 pixels, is used for the display of pictures of data. Occasionally, the pictures of data will be temporarily overwritten with a menu of commands or of options for a particular command.

4.2 Data

Simple data sets are represented as a two-dimensional arrays of *observations* versus *variables*.

Euclidean variables take on continuous, ordered values which can be reasonably represented by real numbers. *Categorical variables* take on discrete values, called *categories*. Usually the values of a categorical variable are assumed to be unordered. It is convenient to be able to treat ordered, discrete variables as either euclidean or categorical.

Suppose a data set contains p_e euclidean variables and p_c categorical variables. The p_e euclidean variables may be thought of as the canonical orthogonal basis of a p_e-dimensional real inner product space, called the *euclidean data space*. The p_c categorical variables form the *categorical data space*, a less familiar object. The cartesian product of the p_e-dimensional euclidean data space and p_c-dimensional categorical data space is the *data space*.

4.3 Three-Dimensional Scatter Plots

The picture we see in a three-dimensional scatter plot is drawn in the *view space*. The view space is spanned by three (basis) vectors: screen-x, screen-y, and screen-z. *Screen-x* is horizontal in the plane of the display screen; *screen-y* is vertical in the plane of the display screen; *screen-z* is perpendicular to the plane of the screen, pointing out.

There are two steps in mapping the data space to the view space. First, the data space is mapped onto a three-dimensional space called the *world space*, by a *projection*. Second, the world space is mapped onto the view space by the *viewing transformation*.

Although it is not obvious why it should be, small, rapidly repeated, changes in the viewing transformation are seen as continuous motion of a rigid object—the point cloud. We automatically see the three-dimensional shape of the point cloud, using the unconscious human ability to perceive *shape from motion* [Marr, 1982].

The Projection

The projection is restricted to be an orthogonal projection from the euclidean data space to the world space. It is chosen with either of two commands:

- *Get a new projection.*
  To do this, three euclidean variables are chosen with the switches on the trackerball housing. The new projection maps the data onto their linear span. Mechanically, this is accomplished by using x, y, and z switches to increment the index of the three x, y, and z basis variables.
- *Update the projection.*
  One euclidean variable is chosen by picking an item from a menu.

The screen plane (screen-x vs. screen-y) corresponds at any moment to some two-dimensional linear subspace of the euclidean data space. The updated projection maps the euclidean data space onto the linear span of this plane and the basis vector corresponding to the chosen variable. This process can be thought of as throwing away the one-dimensional subspace of the world space corresponding to screen-z and replacing it by one of the euclidean variables.

The update command is designed this way, which may seem unnatural at first, to let us mimic the search strategy of a Rosenbrock method for numerical optimization [Rosenbrock, 1960]. The update command is used for interactive projection pursuit, which is discussed below.

Updating the projection allows us to see three-dimensional subspaces of the euclidean data space which are not simply the linear span of three variables. This method of updating does not allow us to select any three-dimensional subspace, but it does allow us to display any two-dimensional subspace on the screen. For the projection pursuit applications considered so far an arbitrary two-dimensional subspace is good enough.

The Viewing Transformation

In general, a viewing transformation is composed from translation, scaling, rotation, the perspective transformation, clipping, and hidden object elimination [Foley and van Dam, 1982; Newman and Sproull, 1979].

In the programs discussed here, only translation, scaling, and rotation are performed. Only the rotation can be controlled by the user. The translation and scaling are determined so that the point cloud fits on the screen, no matter what the angle of rotation.

The trackerball provides the angles of the rotation. When the trackerball is rotated, the viewing transformation is changed and the points on the screen appear to rotate in the same way. The user also has the option of automatic rotation. Then the effect of the last motion of the trackerball is repeated continuously, giving smoother motion than can be achieved by hand.

Coordinate Axes

We need to be able to tell which three-dimensional subspace of the euclidean data space we are looking at. We also need to see how the point cloud is oriented in that space. To satisfy these needs we draw, in a corner of the screen, an object called the *coordinate axes*. This object was called the *dreibein* (German for tripod) in previous PRIM systems [Fisherkeller, Friedman, and Tukey, 1975] and is sometimes referred to as the *gnomon* in the computer graphics literature [Foley and van Dam, 1982].

If we are looking at a world space (see Plate 1a), spanned by three variables, then we see the three coordinate axes as projected on the plane of the screen in the current rotation. As we rotate the data, the coordinate axes rotate also, but about a point in the corner rather than about the point in the center of the screen used by the point cloud. By comparing the motion of the point cloud to the motion of the axes we can determine the orientation of the point cloud in three dimensions.

When the world space is a more general three-dimensional subspace of the euclidean data space (see Plate 1b)—as happens in projection pursuit—we will see a collection of vectors radiating from a common origin, each labeled with the index of the variable it represents. Each vector is the projection of a unit vector in the direction of the corresponding variable onto the three-dimensional world space. In this case, the coordinate axes are not as easy to interpret.

4.4 Identify

The *Identify* command (see Plate 1b) shows us which observation is represented by a given point on the screen. A cursor is positioned on the screen with the trackerball. The point on the screen nearest the cursor is highlighted by changing its color and drawing a circle around it and the index of the corresponding observation is written nearby.

The highlighted point can also be temporarily deleted from the data set by moving the delete switch. Deleted points take no part in any actions and are unaffected by them, until restored. This feature is used, for example, in the regression program discussed below. Unusual observations can be deleted from the data set temporarily, so that they do not interfere with the construction of a regression model.

4.5 Adding Dimension with Color

We use color to represent the values of an additional variable in a two- or three-dimensional scatter plot. By default, color is used to represent the values of categorical variables. Euclidean variables can be represented in color using the *Discretize* command (see Plate 1a), which converts a euclidean variable into an ordered discrete variable.

All points in the scatter plot should appear equally bright, so we should choose a spectrum of colors varying in hue and/or saturation [Newman and Sproull, 1979; Foley and van Dam, 1982]. Two default coloring schemes are provided—which can be modified interactively. One default scheme is for unordered variables; it consists of white and six saturated hues—red, yellow, green, cyan, blue, magenta. For ordered variables, we provide a perceptually ordered range of seven saturated hues, from blue through magenta to red. (We arbitrarily assume no categorical variables have more than seven categories.)

4.6 Grouping

The *Group* command is used to partition a data set by creating or modifying a categorical variable (see Plate 1c).

The Group command modifies a categorical variable with a series of moves. Each move is made by defining a rectangular region on the screen and moving all the points inside the rectangular region into a chosen category. The rectangular region is defined using the trackerball to position a cursor on the screen and to mark the rectangle's two corners.

The ability to modify existing categorical variables is useful because it lets us define a partition based on several views of the data. We first find a view that shows structure that suggests a natural partition of the data set. To summarize the structure, we create a new categorical variable that records the partition. We can then project and rotate to a new view that suggests a refinement or other modification of the partition and modify the categorical variable accordingly. This interactive clustering is illustrated in our film: *Exploring data with the Orion I workstation* [McDonald, 1982b].

We imitate the Isolation and Masking functions of PRIM-I using Group and a command called *Select Category*. The user can set categories of the color variable *active* or *inactive*. Observations in inactive categories are invisible.

5. PROJECTION PURSUIT

The basic idea of projection pursuit is to find low-dimensional views of a data set that capture aspects of its many-dimensional structure [Kruskal, 1969; Fisherkeller, Friedman, and Tukey, 1975; Friedman and Tukey, 1974; Friedman and Stuetzle, 1981, 1982a; Friedman, Stuetzle, and Schroeder, 1981; Huber, 1981, 1985; Chen and Li, 1981; McDonald, 1982a, 1983; Diaconis and Freedman, 1984; Diaconis and Shahshahani, 1984; Henry, 1983]. By temporarily reducing the dimension of the data, we can sometimes replace a problem that is difficult or impossible to solve with one or several *smaller* problems that are more manageable. This is especially true for graphics. It is hard to understand pictures that show many variables at once; it is easy to understand histograms and scatter plots.

To reduce dimension we *project* the data from the many-dimensional data space to a low-dimensional view. In general, the projection could be any mapping from the many-dimensional data space to a low-dimensional view. The projection pursuit methods that have been developed so far restrict the mappings considered to orthogonal projections of the data onto one-, two-, or three-dimensional subspaces of the euclidean data space.

Except for PRIM-9, previous work on projection pursuit has concentrated on automatic algorithms. In automatic projection pursuit, an optimization algorithm chooses the projection to maximize a numerical criterion of interesting structure.

The first projection pursuit algorithm [Friedman and Tukey, 1974] used a numerical optimizer to search for a one- or two-dimensional projection that maximized a *clottedness* index (related to entropy [Huber, 1981, 1985]).

Automatic projection pursuit methods have been developed for: regression [Friedman and Stuetzle, 1981; Friedman and Stuetzle, 1982a], classification [Friedman and Stuetzle, 1982a], and density estimation [Friedman and Stuetzle, 1982a; Friedman, Stuetzle, and Schroeder, 1981;

Huber, 1981; Henry, 1983]. These techniques construct a model that summarizes the apparent dependence of a *response* on some *predictors*. In regression the response is one of the euclidean variables; in classification the response is a categorical variable; in density estimation, the response is the *mass* of the observations.

Projection pursuit models are built up from several low-dimensional views of the data, following the *projection pursuit paradigm* [Friedman and Stuetzle, 1982a]:

Choose an initial model.

Repeat

    Projection Pursuit:
        Find the projection that shows
        the greatest deviation from
        the current model, indicating
        previously undetected structure.

    Model Update:
        Change the model to incorporate
        that structure.

Until the model is accurate in all projections.

Projection pursuit methods have a natural connection to interactive graphics. As long as we project on a subspace of dimension no greater than three, we can look at a picture of the result. Automatic versions of projection pursuit choose the projection using an optimization algorithm to maximize an objective numerical criterion. On a system like Orion I, however, an analyst can interactively search for a projection using perception, judgement, and a knowledge of context of the data, instead of relying on a single, to some degree arbitrary, numerical measure of what constitutes an interesting view.

5.1 Non-parametric Regression

In regression, we distinguish one of the p_e euclidean variables as the *response variable*, y_i. The remaining $p_e - 1$ euclidean variables form a vector of *predictor variables*, $\vec{x}_i$.

One way of looking at the regression problem is to say that we are looking for a function or *model*, $f(\cdot)$, to summarize the apparent dependence of y_i on $\vec{x}_i$. We want two things from the regression model:

- The model, $f(\cdot)$, should be *accurate*.

In other words $y_i \approx f(\vec{x}_i)$ or the residuals, $r_i = y_i - f(\vec{x}_i)$, should be small. For objective model fitting, we must choose a numerical criterion, such as $\sum r_i^2$. In an interactive graphics environment, the accuracy of the model can be assessed using several numerical criteria simultaneously and, more importantly, using human visual perception.

- The model, $f(\cdot)$, should be *simple*.

In parametric regression, a model is simple if it uses few parameters.

In non-parametric regression, we require the model to satisfy a more vague constraint of *smoothness*. Smooth is taken to mean that observations that are close in the predictor space have similar values of $f(\cdot)$. The function, $f(\cdot)$, should vary slowly relative to the spacing of the observations in the predictor space.

5.2 The PPR Model

Projection Pursuit Regression (PPR) models a response variable as the sum of general smooth functions of linear combinations of the predictor variables:

$$f(\vec{x}_i) = \sum_{k=1}^{K} g_k\left(<\vec{u}_k, \vec{x}_i>\right).$$

$\vec{u}_k$ is a unit vector—a direction in predictor space. $g_k(\cdot)$ is a general smooth function. The inner product, $<\vec{u}_k, \vec{x}_i>$, is the projection of $\vec{x}_i$ on the direction $\vec{u}_k$.

The model is constructed, iteratively, following the projection pursuit paradigm [Friedman and Stuetzle, 1982a]. We subtract the current model from the responses to get a set of residuals, r_{ik}. We then search for a new direction, $\vec{u}_k$, and a new smooth function, $g_k(\cdot)$, to model the residuals.

The functions, $g_k(\cdot)$, are found by smoothing r_{ik} as function of $z_{ik} = <\vec{u}_k, \vec{x}_i>$. We use a smoothing algorithm based on running linear fits (Cleveland, 1979; Friedman and Stuetzle, 1982b).

In the original, automatic version of PPR, the directions, $\vec{u}_k$, are found using a numerical optimizer, usually a modified Rosenbrock search [Rosenbrock, 1960]. The basic idea is to minimize a function of several variables by optimizing over one variable, while holding all others fixed. The method we actually use is slightly simpler in some respects than Rosenbrock's original proposal and is modified to restrict the search to the unit sphere.

For a complete description see Friedman and Stuetzle [1981, 1982a].

5.3 Interactive Projection Pursuit Regression

The interactive projection pursuit regression program (IPPR) permits an analyst to modify or completely take over functions handled by automatic procedures in the original projection pursuit regression. For simplicity of notation, we will describe how Orion I is used to find a one term PPR model:

$$f(\vec{x}_i) = g\left[<\vec{u}, \vec{x}_i>\right].$$

More general models are fit by iterating the process on the residuals.

As mentioned above, it is very difficult to present an adequate verbal description of an interactive graphics system. For a better demonstration of IPPR, see our film *Projection Pursuit Regression with the Orion I Workstation* [McDonald, 1982c].

Tactics of Manual Search

The *Interactive Projection Pursuit Regression (IPPR)* command on Orion I provides facilities that let a data analyst carry out a Rosenbrock search manually, using visual perception and judgement in context for guidance, in addition to one or several numerical criteria.

In IPPR, the projection and the viewing transformation are restricted so that screen-y (vertical) always corresponds to the response variable. Therefore we can only rotate in the screen-x—screen-z plane, that is, we can only rotate about the vertical, screen-y axis.

The screen-x—screen-z plane corresponds to a two-dimensional subspace of the predictor space, P. The screen-x direction corresponds to the current choice of $\vec{u}$, which must therefore lie in P. By rotating, we can look at any $\vec{u}$ that lies in the current P. We can get general two-dimensional subspaces, P, using combinations of the commands, *Get a new projection* and *Update the projection*, described above.

The screen shows, at each instant, a scatter plot of the response variable, y_i, versus the data projected on the current direction in the space of predictor variables, $z_i = <\vec{u}, \vec{x}_i>$ (see Plate 2a). We may choose to display the curve of the smooth of y_i as a function of z_i. Also at our option, a vertical bar, whose height indicates the value of an objective criterion may be drawn on the bottom of the screen. The default objective criterion is the percentage of variance in the response explained by the smooth.

As we rotate, the direction, $\vec{u}$, corresponding to screen-x changes. The smooth is recomputed in real-time as the direction changes. On Orion I, a data set with 500 hundred points can be rotated and re-smoothed and the picture erased and redrawn about 3 times a second.

The speed of iteration, in this system, is limited by the graphics device; the computation alone takes less than 1/10 of a second.

The vertical bar indicating the value of the objective criterion is drawn at a horizontal position depending on the angle of rotation. As the data is rotated, the vertical bar traces out a graph of the value of the objective criterion as a function of the angle of rotation. This makes it easy to manually rotate to the best direction in the current predictor plane. We can now update the projection with one of the predictor variables to choose a new plane, P. The iteration of rotating to the best direction in the current plane, updating the plane, rotating again, etc., is essentially a manual version of the Rosenbrock search used by the original PPR.

Strategy of Manual Search

Two aspects of searching can benefit from human judgment. First, we can sometimes shortcut a numerical optimizer. Prior knowledge of the data may suggest good starting values of the direction. Prior knowledge may also suggest which directions of search may be most useful. Second, the automatic search is limited by the need for a numerical objective criterion to optimize. We can search for a direction that is *good* in a subjective—as well as an objective—sense.

Subjective aspects of the relationship between the current direction and the response variable can influence our choice of a direction for the next term in the model. For example, a direction involving a small number of variables may be preferred because it is easier to interpret, even though it has a slightly higher sum of squared residuals. In another case, a certain sub-optimal direction may be preferred because the shape of the smooth is simple. A combination of a *simple* direction with a *nicely* shaped smooth may be especially interesting because it suggests a parametric model for the dependence of the response variable on the predictors. The advantages of this interactive model construction are illustrated in *Projection Pursuit Regression with the Orion I workstation* [McDonald, 1982c].

Grouping

Another way in which IPPR benefits from interaction is by combining IPPR with the Group and Select Category commands discussed above. The Group command allows us to partition the data set into subsets. We may set some subsets inactive with the Select Category command and proceed to fit the regression model to the remaining, active observations. This provides a convenient way to compare models fit on all the data to models fit on various subsets.

Assessing the Variability of the Smoother

When we search manually for a direction, we often sacrifice some amount of the objective, numerical criterion of *goodness of fit* to improve some subjective impression. For example, we may prefer a direction that has non-zero coefficients of as few predictor variables as possible. To achieve a parsimonious model, we may be willing to sacrifice a few percent of variance explained.

To judge how important a few percent of variance is, we need some indication of the natural variability of our regression model. In a non-interactive setting, it is appropriate to try to assess the variability of the full projection pursuit algorithm by bootstrapping [Efron 1979, 1981].

In an interactive program, we need a way of assessing variability that can be done rapidly, in real time, and that has a natural graphical representation. A reasonable solution is to bootstrap the smoother only. That is, we hold the direction, $\vec{u}$, fixed so that have a two-dimensional scatter plot of y_i vs. $z_i = <\vec{u}, \vec{x}_i>$. We then bootstrap the smooth of y_i as a function of z_i.

When the bootstrap command is chosen, the program chooses a bootstrap sample, smooths it, and adds the new smooth curve to the picture. This process is executed about 5 times a second for a data set with 500 observations. Thus 100 bootstrap replications will be seen in about 20 seconds. The curves of all replications of the smooth accumulate on the screen. The original smooth is redrawn each time in a distinct color, so that it can be compared with the bootstrap replications. The smooths of the bootstrap samples soon fill in a *confidence band* about the original curve, and give an indication of the variability of the smoothed value for any value of z (see Plate 2b).

6. MULTIPLE VIEWS

In projection pursuit, we develop a model of many-dimensional structure from a sequence of low-dimensional views. To interpret a projection pursuit model, we need to understand the relationships between the contents of two or more views in the sequence. The methods described in this section allow us to do this.

This approach is related to *M-and-N-plots* of Diaconis and Friedman [1983]. Diaconis and Friedman draw a four-dimensional picture by drawing two two-dimensional scatter plots side-by-side. The two scatter plots show four variables from a given data set. To display four-dimensional structure, they connect pairs of corresponding point with line segments. Any structure is obscured if all pairs of corresponding points are connected, so Diaconis and Friedman offer an algorithm for thinning the line segments.

A simple version of our procedure works as follows (see Plate 2c and Kolata [1982]):

On the screen there are two scatter plots, side by side, showing four variables. There is a cursor on the screen in one of the two scatter plots. The scatter plot that the cursor is in is the *active scatter plot*. We position the cursor in the active scatter plot by moving the trackerball. Points near the cursor in the active scatter plot are red. Points at an intermediate distance from the cursor are purple. Points far from the cursor are blue. A point in the non-active scatter plot are given the same color as the corresponding point in the active scatter plot. The colors are continuously updated as the cursor is moved. We can also move the cursor from one scatter plot to the other, changing which scatter plot is active.

Our technique is not as precise at showing pairs of corresponding points as M-and-N-plots. On the other hand, it is somewhat easier to see regional relationships between the two scatter plots. Also, we can look at connections between more than two scatter plots at once, as is demonstrated in *Exploring data with the Orion I workstation* [McDonald, 1982b].

7. ORION I HARDWARE

This section is a technical description of the Orion I workstation. For exposition of the terminology, see McDonald and Pedersen [1983b], a survey paper on raster graphics aimed at statisticians by Beatty [1983], the texts on computer graphics by Foley and van Dam [1982] and Newman and Sproull [1979], and standard texts on microcomputer systems, such as Wakerly [1981].

The Orion I workstation was designed and built in 1980-81 by Jerome H. Friedman, David Parker, and Werner Stuetzle in the Computation Research Group at the Stanford Linear Accelerator Center [Friedman and Stuetzle, 1981]. The architecture of Orion I is fundamentally different from the PRIM systems.

The PRIM systems consisted of mainframe computers with associated graphics terminals. This arrangement is limited in the amount of computing power available to the user of the graphics system, who must share the mainframe with many other users, and is also usually limited in the speed and flexibility of communication between the mainframe and the graphics terminal.

Orion I, in contrast, was based on a graphics workstation, that is, a powerful computer, with an integrated high-resolution display, used by a single person [McDonald and Pedersen, 1983, 1985]. It was, in consequence, both less expensive and more powerful than the PRIMS.

The total cost for hardware in Orion I in January, 1982 was less than $60,000. Equivalent color graphics hardware would today (August, 1983) cost less than $20,000; hardware for a black and white system with the capabilities of the earlier PRIM systems would cost less than $10,000 (McDonald and Pedersen, 1983, 1985). For comparison, the hardware in PRIM-9 (essentially, SLAC's computing center) cost millions of dollars. The PRIM-H system cost hundreds of thousands of dollars.

For speed of computation, Orion I is equivalent to a large mainframe computer (say one half of an IBM 370/168 or three times the VAX 11/780 used in PRIM-H) and is used by a single person. For comparison, this is equivalent to the computing power of a typical university computer center five to ten years ago.

The important parts of the hardware are:

7.1 The Master Processor

The master processor controls the action of the other parts of the system and handles the interaction with the user. Orion I uses a prototype of the central processor board for the SUN microcomputer, which was developed by the Stanford Computer Science Department for Stanford University Network (Bechtolsheim, 1982a, 1982b, 1982c; Bechtolsheim and Baskett, 1980; Bechtolsheim, Baskett, and Pratt, 1982; Pratt, 1982). The SUN microcomputer is based on the Motorola MC68000 microprocessor (Motorola, 1982) with a MULTIBUS architecture (Boberg, 1980). Our version of the SUN board runs the processor at 8 mhz with no wait states, which makes it roughly equivalent, for non-floating point operations, to a VAX 11/750. The board has 128k of RAM (random access memory), two (RS-232) serial ports and a memory mapping and protection mechanism.

The only system software is a rudimentary monitor, stored in ROM (read only memory), that is used to download object code from a host computer. The monitor is written in the programming language C. The graphics programs discussed below were written mostly in Pascal; a few time-critical routines were written in MC68000 assembly language. All of these programs were edited and *cross-compiled* or *cross-assembled* on a mainframe host computer, and then downloaded for execution on the SUN board.

7.2 The Graphics Device

The graphics device is a Lexidata 3400 bitmap display, which stores and displays the current picture. The picture is determined by a *raster* of

1280 (horizontal) by 1024 (vertical) colored dots, called *pixels*. There are 8 bits of memory for each pixel which, through a look up table, determine the setting (from 0 to 255) of each of the three color guns (red, green, blue). Thus, at any time, there may be 256 different colors on the screen, from a potential palette of 2^{24}.

The Lexidata 3400 contains a display processor that is used for drawing vectors, circles, and characters. Neglecting the overhead in communication with the master processor, the Lexidata can draw 500,000 pixels per second in single pixel writes and about 300,000 pixels per second in vector drawing. It communicates with the SUN board through a 16-bit parallel port that is interfaced to the MULTIBUS. The communication between the master processor and the frame buffer is much slower than that found in more recent graphics workstations; the speed of communication has been the major limitation in the use of Orion I for real-time graphics.

7.3 The Arithmetic Processor

The arithmetic processor is a special purpose *bipolar, bit-slice* processor called a 168/E [Kunz, 1976, 1978]. The 168/E emulates the cpu of an IBM 370/168 without I/O and interrupt capabilities. It was developed by SLAC engineers for the processing of particle physics data and has about half the speed of the true 370/168. Because it has no input/output facilities, its communication with the outside world must be handled by a master processor, in this case, the SUN computer.

In Orion I, the 168/E serves as a programmable high-speed floating point processor. It increases the speed of the system in floating point operations by a factor of 50–100.

The 168/E is programmed in FORTRAN. As with the SUN board, the programs are edited and cross-compiled on the host and then downloaded for execution.

7.4 The Host

Our host computer is SLAC's IBM 3081 mainframe. The host is used only for software development and long term data storage; the programs described below are executed by the Orion I workstation alone and do not use the host. The programs were written on the 3081 and then cross-compiled or cross-assembled for execution on the SUN board and the 168/E. Data sets are also prepared on the 3081. Programs and data are downloaded to the SUN board and the 168/E through a high speed serial interface that connects the MULTIBUS and the 3081.

8. CONCLUSION

Our original intent in constructing the Orion I workstation was to show that the functions of the PRIM systems could implemented with inexpensive hardware, using a microprocessor-based computer and a color bitmap display. We exceeded this goal; the final version of Orion I was much more powerful than any of the PRIM systems. Because of this, we were able to experiment with new methods (IPPR, Multiple Views) that were not previously possible.

In addition to the methods described above, Orion I has been used for:

- Projection pursuit classification (T. Hastie).
- Interactive multidimensional scaling (Buja, 1983).
- Periodic smoothing of time series (McDonald, 1983).
- The Grand Tour (of multivariate data) (Asimov, 1983).
- Interactive non-linear transformation of multivariate data (Twisting) (J. W. Tukey and M. Fisherkeller; see Friedman, Tukey, and Tukey, 1980).
- Exploration of seismic data (G. Lin and J. A. McDonald).
- Interactive analysis of heart X-rays (K. Oyama and J. A. McDonald).
- Static display of three-dimensional scatter plots using perspective, shading, and hidden surface removal (J. Chu and J. A. McDonald).

The major drawback of Orion I was that it was a special purpose machine; critical parts were designed and built at SLAC. Therefore the same system cannot be easily reproduced anywhere else. Furthermore, Orion I was very difficult to program; the total system contained six processors programmed in about a dozen languages. Orion I was used for developing and experimenting with selected techniques for interactive graphical data analysis, but it was not feasible to write an integrated statistical package for it.

Fortunately, in the two years since Orion I was built, several companies have begun to sell graphics workstations that equal or surpass Orion I (McDonald and Pedersen, 1983, 1985). These workstations range in price from about $15,000 to $150,000. A reasonable configuration for statistical applications will cost about $50,000.

Commercial graphics workstations are complete computers, including a large amount of supporting software that was missing from Orion I. It is therefore now possible to begin to develop integrated statistical packages for these machines that include the type of methods described here. One example is ISP which is being developed in the PRIM-H project at Harvard [Donoho, Huber, and Thoma, 1981; Donoho et al., 1982].

REFERENCES

Asimov, D. (1983). "The grand tour," Technical Report, SLAC-PUB-3211. Sanford, CA: Stanford Linear Accelerator Center.

Beatty, J. C. (1983). "Raster graphics and color," *American Statistician*, **37**: 60-75.

Bechtolsheim, A. (1982a). "SUN graphics board," available from SUN Microsystems, 2550 Garcia Ave., Mountain View, CA 94043.

Bechtolsheim, A. (1982b). "SUN 68000 board," available from SUN Microsystems, 2550 Garcia Ave., Mountain View, CA 94043.

Bechtolsheim, A. (1982c). "SUN 3-mbit ethernet board," available from SUN Microsystems, 2550 Garcia Ave., Mountain View, CA 94043.

Bechtolsheim, A. and Baskett, F. (1980). "High performance raster graphics for microcomputer systems," *Computer Graphics, SIGGRAPH '80 Conference Proceedings*, **14**(3): 43-47. New York: The Association for Computing Machinery.

Bechtolsheim, A., Baskett, F., and Pratt, V. O. (1982). "The SUN workstation architecture," Report No. CSL-TR-229. Stanford, CA: Stanford University ERL Computer Systems Lab.

Boberg, R. W. (1980). "Proposed microcomputer system 796 bus standard," *IEEE Computer*, **13**: 89-105.

Buja, A. (1982). "Interactive multi-dimensional scaling". Produced by the author at Stanford Linear Accelerator Center, Stanford, CA. (16mm film—available from the author).

Chen, Z. and Li, G. (1981). "Robust principal components and dispersion matrices via projection pursuit," Research Report PJH-8. Cambridge, MA, Harvard University, Department of Statistics.

Cleveland, W. S. (1979). "Robust locally weighted regression and smoothing scatter plots," *Journal of the American Statistical Association*, **74**: 829-836.

Diaconis, P. and Freedman, D. A. (1984). "Asymptotics of graphical projection pursuit," *Annals of Statistics*, **12**: 793-815.

Diaconis, P. and Friedman, J. H. (1983). "M and N plots," *Recent Advances in Statistics: Papers in Honor of Herman Chernoff on His Sixtieth Birthday*, M. H. Riszi, J. Rustagi, and D. O. Siegmund, eds., 425-447. New York: Academic Press.

Diaconis, P. and Shahshahani, M. (1984). "On non-linear functions of linear combinations," *SIAM Journal on Scientific and Statistical Computing*, **5**: 175-191.

Donoho, D. L., Huber, P. J., Ramos, E., and Thoma, H. M. (1982). "Kinematic display of multivariate data," *Proceedings of the Third Annual Conference and Exposition of the National Computer Graphics Association*, **1**: 393-398. Fairfax, VA: National Computer Graphics Association.

Donoho, D., Huber, P. J., and Thoma, H. (1981). "The use of kinematic displays to represent high dimensional data," *Computer Science and Statistics: Proceedings of the 13th Symposium on the Interface*, 274-278. New York: Springer-Verlag.

Efron, B. (1979). "Bootstrap methods: another look at the jackknife," *Annals of Statistics*, **7**: 1-26.

Efron, B. (1981). "Nonparametric estimates of standard error: the jackknife, the bootstrap, and other methods," *Biometrika*, **68**: 589-599.

Fisherkeller, M. A., Friedman, J. H., and Tukey, J. W. (1975). "PRIM9: an interactive multidimensional data display and analysis system," *Data: Its Use, Organization, and Management*, 140-145. New York: The Association for Computing Machinery.

Foley, J. D. and van Dam, A. (1982). *Fundamentals of Interactive Computer Graphics*. Reading, MA: Addison-Wesley.

Friedman, J. H. and Stuetzle, W. (1981). "Projection pursuit regression," *Journal of the American Statistical Association*, **76**: 817-823.

Friedman, J. H. and Stuetzle, W. (1982a). "Projection pursuit methods for data analysis," *Modern Data Analysis*, R. L Launer and A. F. Siegel, eds., 123-147. New York: Academic Press.

Friedman, J. H. and Stuetzle, W. (1982b). "Smoothing of scatter plots," Technical Report Orion 3. Stanford, CA: Stanford University, Department of Statistics.

Friedman, J. H. and Stuetzle, W. (1983). "Hardware for kinematic statistical graphics", *Computer Science and Statistics: Proceedings of the 14th Symposium on the Interface*, 163-169. Amsterdam: North-Holland.

Friedman, J. H., Stuetzle, W., and Schroeder, A. (1981). "Projection pursuit density estimation," Technical Report Orion 2. Stanford, CA: Stanford University, Dept. of Statistics.

Friedman, J. H. and Tukey, J. W. (1974). "A Projection Pursuit Algorithm for Exploratory Data Analysis," *IEEE Transactions on Computers*, **C-23**: 881-890.

Friedman, J. H., Tukey, J. W., and Tukey, P. A. (1980). "Approaches to analysis of data that concentrate near intermediate-dimensional manifolds," *Data Analysis and Informatics*, E. Diday, L. Lebart, J. T. Pagès, and R. Tomassone, eds., 3-13. Amsterdam: North-Holland.

Gnanadesikan, R. (1977). *Methods for Statistical Data Analysis of Multivariate Observations*, New York: Wiley.

Henry, D. H. (1983). "Multiplicative models in projection pursuit," Technical Report Orion 25. Stanford, CA: Stanford University, Department of Statistics.

Huber, P. J. (1981). "Density estimation and projection pursuit methods," Research Report PJH-7. Cambridge, MA: Harvard University, Department of Statistics.

Huber, P. J. (1985). "Projection pursuit," *Annals of Statistics*, **13**: 435-475.

Kolata, G. (1982). "Computer graphics comes to statistics," *Science*, **217**: 919-920.

Kruskal, J. B. (1969). "Toward a practical method which helps uncover the structure of a set of multivariate observations by finding a linear transformation which optimizes a new 'index of condensation'," in *Statistical Computation*, R. C. Milton and J. A. Nelder, eds., 427-440. New York: Academic Press.

Kunz, P. F. (1976). "The LASS hardware processor," *Nuclear Instruments and Methods*, **135**: 435-440.

Kunz, P. F. et al. (1978). "The LASS hardware processor," *Proceedings of the 11th Annual Microprogramming Workshop, SIGMICRO Newsletter*, **9**.

Littlefield, R. J. and Nicholson, W. L. (1982). "Use of color and motion for the display of higher dimensional data," *Proceedings of the 1982 DOE Statistical Symposium*.

Marr, D. (1982). *Vision*. New York: W. H. Freeman.

McDonald, J. A. (1982a). "Interactive graphics for data analysis," Technical Report Orion 11. Stanford, CA: Stanford University, Department of Statistics.

McDonald, J. A. (1982b). "Exploring data with the Orion I workstation," a 25 minute, 16mm sound film, which demonstrates programs described in McDonald (1982a). It is available for loan from: Jerome H. Friedman, Computation Research Group, Bin #88, SLAC, P.O. Box 4349, Stanford, CA. 94305.

McDonald, J. A. (1982c). "Projection pursuit regression with the Orion I workstation," a 20 minute, 16mm color sound film, which demonstrates programs described in McDonald (1982a). It is available for loan from: Jerome H. Friedman, Computation Research Group, Bin #88, SLAC, P.O. Box 4349, Stanford, CA 94305

McDonald, J. A. (1983). "Periodic smoothing of time series," Technical Report Orion 17. Stanford, CA: Stanford University, Department of Statistics.

McDonald, J. A. (1986). "Periodic smoothing of time series," *SIAM Journal on Scientific and Statistical Computing*, **7**: 665-688.

McDonald, J. A. and Pedersen, J. (1983). "Graphics workstations for data analysis," *American Statistical Association 1983 Proceedings of the Statistical Computing Section*, 74-82. Washington, DC: American Statistical Association.

McDonald, J. A. and Pedersen, J. (1985). "Computing environments for data analysis II: hardware," *SIAM Journal on Scientific and Statistical Computing*, **6**: 1013-1021.

Motorola, Inc. (1982). "MC68000, 16-bit microprocessor," User's Manual, Third Edition.

"PRIM-9" (1973). Produced by Stanford Linear Accelerator Center, Stanford, CA. (S. Steppel, ed.) Bin 88 Productions. (Film)

Newman, W. M. and Sproull, R. F. (1979). *Principles of Interactive Computer Graphics*. New York: McGraw-Hill.

Nicholson, W. L. and Littlefield, R. J. (1982). "The use of color and motion to display higher dimensional data," *Proceedings of the Third Annual Conference and Exposition of the National Computer Graphics Association*, **I**: 476-485. Fairfax, VA: National Computer Graphics Association.

Pratt, V. O. (1982). *The SUN Workstation*, a preprint, delivered at The Office Automation Conferences, San Francisco, April 5-7, 1982.

Rosenbrock, H. H. (1960). "An automatic method for finding the greatest or least value of a function," *Computer Journal*, **3**: 175-184.

Tukey, J. W. and Tukey, P. A. (1981). "Graphical display of data sets in 3 or more dimensions," *Interpreting Multivariate Data*, V. Barnett, ed., 189-275. Chichester: Wiley.

Wakerly, J. F. (1981). *Microcomputer Architecture and Programming*. New York: Wiley.

8

BRUSHING SCATTERPLOTS

Richard A. Becker
William S. Cleveland
AT&T Bell Laboratories

ABSTRACT

A dynamic graphical method is one in which a data analyst interacts in real time with a data display on a computer graphics terminal. Using a screen input device such as a mouse, the analyst can specify, in a visual way, points or regions on the display and cause aspects of the display to change nearly instantaneously. Brushing is a collection of dynamic methods for viewing multidimensional data. It is very effective when used on a scatterplot matrix, a rectangular array of all pairwise scatterplots of the variables. Four brushing operations—highlight, shadow highlight, delete, and label—are carried out by moving a mouse-controlled rectangle, called the brush, over one of the scatterplots. The effect of an operation appears simultaneously on all scatterplots. Three paint modes—transient, lasting, and undo—and the ability to change the shape of the brush allow the analyst to specify collections of points on which the operations are carried out. Brushing can be used in various ways or on certain types of data; these usages are called brush techniques and include the following: single point and cluster linking, conditioning on a single variable, conditioning on two variables, subsetting with categorical variables, and stationarity probing of a time series.

Reprinted from (1987) *Technometrics*, **29**: 127-142, by permission of the American Statistical Association.

From *Dynamic Graphics for Statistics*, William S. Cleveland and Marylyn E. McGill, eds., copyright © 1988 by Wadsworth, Inc. All rights reserved.

1. INTRODUCTION

1.1 High-Interaction Graphics

In the late 1960s and early 1970s, most statistical graphics was through systems that produced printer plots in batch output that took from 15 minutes to days to come back. The time and effort for a data analyst to look at a graph, react to it, and change the graph to see information in a new improved way, were expensive. This was indeed *low-interaction* graphics, or in many cases, *no-interaction* graphics. Many current statistical graphics systems have *medium-interaction* capabilities. A graph can be studied and a new one made within seconds or minutes. For example, in many current systems a data analyst can make a scatterplot of y_i against x_i, for $i = 1$ to n, decide that it is hard to see the form of the dependence of y on x, and issue a command to add to the graph a nonparametric regression curve of y on x to better see the dependence. Typically, the execution of the commands takes less than a minute.

The newest generation of computer graphics hardware and software has given us, at moderate cost, the capability to develop *dynamic* graphical methods for data analysis. By using a screen input device such as a mouse, the data analyst can point to elements of a graphical display and manipulate them, and can cause aspects of the display to change continuously in real time. It would be hard to overemphasize the importance of this new medium for data display, one which differs significantly from the traditional medium of static graphical displays that have been employed for statistical graphics in the past. The goal now is to invent graphical methods that exploit this dynamic, dynamic medium and that are useful for data analysis.

1.2 The Evolution of High-Interaction Methods

Interactive computer graphics applications began to be developed in the early 1960s [Newman and Sproull, 1979]. The earliest dynamic graphical method in statistics of which we are aware is a probability plotting procedure developed by Fowlkes [1969, 1971]. By turning a knob the user could continuously change a shape parameter for the reference distribution on a probability plot displayed on a screen; the knob could be turned to make the pattern of points as nearly straight as possible. The data could be transformed by a power transformation, $(y+c)^p$, with c and p each changed by a knob. Points could be deleted by positioning a cursor on them, and graphs could be saved on files and brought up later on the screen for further dynamic development. The system demonstrated that dynamic graphical methods had the potential to be important tools for data analysis.

Another early system was PRIM 9 [Fisherkeller, Friedman, and Tukey, 1975], a set of dynamic tools for projecting, rotating, isolating, and masking multidimensional data in up to nine dimensions. The central operation was user-controlled rotation, which allows the data analyst to study three-dimensional data by showing the points rotating on the computer screen; this provides one type of three-dimensional scatterplot. PRIM 9 also had isolation and masking, features that allowed point deletion in a lasting or in a transient way.

Dynamic methods were not widely used in the early years of their development, and invention moved ahead slowly, because special-purpose, expensive hardware was needed. Today, however, statistical computing environments have evolved to the point where the hardware and software to carry out dynamic graphics are available at low cost. This will be discussed further Section 4.

1.3 Brushing

Brushing is a collection of new dynamic graphical methods for analyzing data in two, three, and higher dimensions. The data are displayed by a scatterplot matrix showing all pairwise scatterplots of the variables [Chambers et al., 1983; Cleveland, 1985]. Figure 1 has an example; the data, which will be described in Section 2.4, are 111 measurements of four variables. The numbers in the corners of the boxes along the main diagonal in Figure 1 are the limits of the scales of the variables. For example, the scale for temperature goes from 57 to 97.

The central object in brushing is the *brush*, a rectangle that is superimposed on the screen, as illustrated in Figure 2. (The data in the figure, which will be described in Section 2.3, consist of 30 measurements of three variables.) The data analyst moves the brush to different positions on the scatterplot matrix by moving a mouse. There are four basic *brushing operations*—highlight, shadow highlight, delete, and label. Each operation is carried out in one of three *paint modes*—transient, lasting, and undone. At any time, the data analyst can stop the brushing and change one or more of the features—the shape of the brush, the operation, or the paint modes—and then resume the brushing. The brushing methodology provides a medium within which a data analyst can invent data analytic methods, which we call *brush techniques*. This framework for invention is analogous to having, say, a Cartesian graph as the medium for inventing

204 BRUSHING SCATTERPLOTS

static displays such as normal probability plots, plots of residuals against fitted values [Anscombe and Tukey, 1963], sharpening [Tukey and Tukey, 1981], and many other graphical methods [Chambers et al., 1983].

Thus the salient features of brushing are visual manipulation of the graph by the analyst, instantaneous change, the movement of the brush to different positions on the scatterplot matrix, the ability to change the shape of the brush, four operations, three paint modes, and brush techniques built up from the previous capabilities. In the next two sections, these features are described and illustrated by several applications.

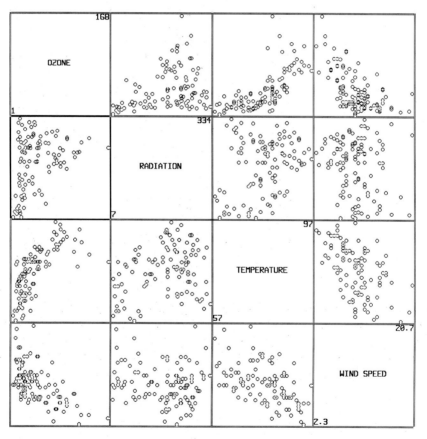

Figure 1. Ozone and Meteorological Data. A scatterplot matrix, all pairwise scatterplots in a rectangular array, shows four-dimensional data.

2. THE HIGHLIGHT AND SHADOW HIGHLIGHT OPERATIONS, THE THREE PAINT MODES, AND SEVERAL BRUSH TECHNIQUES

2.1 The Highlight Operation

Figure 3 is a scatterplot matrix of 275 measurements of three variables, which will be described shortly. On the (1,2) panel we can see that for low values of $\log_2 n$ there are several outliers—values of $\log_2 w$ that are large. Consider the three largest outliers, which occur at the third and fourth values of $\log_2 n$; by scanning across to the (1,3) panel we can

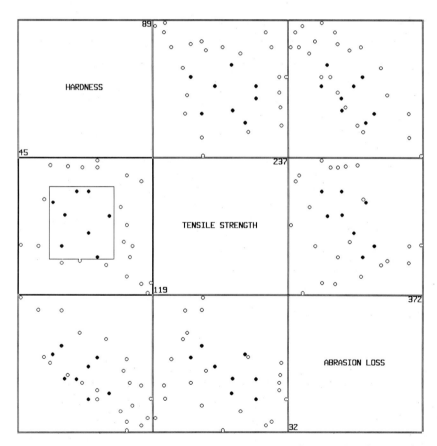

Figure 2. Abrasion Loss Experiment. The central object in brushing is the brush, the rectangle in the (2,1) panel of the scatterplot matrix.

see that these three outliers have large values of f. Consider, however, the next five largest outliers, which occur at the two lowest values of $\log_2 n$. When we scan across to the (1,3) panel, we cannot tell whether these five outliers correspond to points with high values of f or to points lying in the dense spire in the middle of the (1,3) panel. We cannot *link* the points across the two panels.

The operation *highlight* allows us to solve the linking problem and to carry out a number of other brush techniques. Highlight is illustrated in Figure 2. One panel of the scatterplot matrix is selected as the *current panel*. All points which are both inside the brush and inside the current panel are highlighted, as are the points on other panels that correspond to these points.

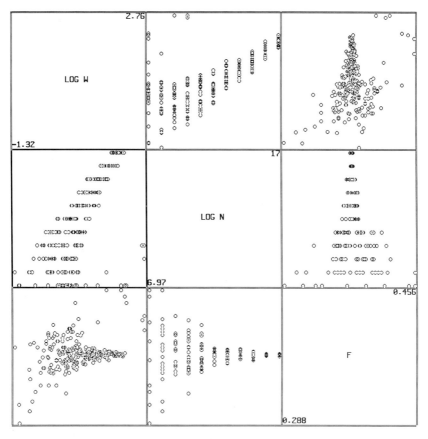

Figure 3. Bin Packing Data. The scatterplot matrix shows three-dimensional data. It is not possible to visually link all of the outliers in the upper left corner of the (1,2) panel to the corresponding points in the (1,3) panel.

The data of Figure 3 are measurements of a computer bin packing algorithm [Bentley et al., 1983]. First, n weights were generated as random numbers, uniform on the interval 0 to .8. The weights were ordered from largest to smallest and then packed, in order, into bins of size 1. Bin 1 is tried; if there is no room, bin 2 is tried, and so forth until a bin with enough room for the weight, perhaps a completely empty bin, is found. As many bins are available as are needed to pack the weights. The data in the figure are for 275 independent runs of the bin packing; n was set at 11 different values ranging from 125 to 128,000 and there were 25 runs for each value of n.

The variable w (for white space) is the amount of empty space in bins with at least one weight. The (1,2) panel in Figure 3 shows that $\log_2 w$, not surprisingly, increases as $\log_2 n$ increases. But in addition the variability decreases, and here are several outliers for low values of $\log_2 n$. What causes these outliers? The mathematical theory of bin packing suggested that the weights above .5 might be the cause; for this reason f, the fraction of weights above .5, was computed for each run and added as an explanatory variable. In Figure 4 the outliers on the (1,2) panel for the first two levels of $\log_2 n$ have been highlighted by brushing. Now we can see clearly from the (1,3) panel that these outliers, as well as the three largest, do indeed correspond to high values of f and are not in the dense spire. The (2,3) panel in Figure 4 provides further insight: The law of large numbers insists that f not vary by much from its expected value of .375 when $\log_2 n$ is large; thus f cannot, with nonnegligible probability, be large and cause outliers in $\log_2 w$.

2.2 The Transient Paint Mode

When the brush is in the *transient* paint mode, the following happens: As the brush is moved, the new points that come inside the brush are highlighted, but points no longer inside the brush are no longer highlighted. This is illustrated in Figures 5 to 7; the brush starts out on the left of the active panel, moves to the center, and then moves to the right.

2.3 Conditioning on a Single Variable

Suppose the goal is to see how one variable depends on the others. This is the case for the data in Figures 5 to 7, which are measurements of tensile strength, hardness, and abrasion loss (the amount of material rubbed off by an abrasive material) for 30 rubber specimens [Davies, 1958, p. 210]. The purpose of the experiment that generated the data was to determine how abrasion loss depends on tensile strength and hardness.

208 BRUSHING SCATTERPLOTS

In the original analysis, abrasion loss was modeled as a linear function of hardness and tensile strength.

Conditioning on a single variable is a brush technique that provides a simple, direct way to fix the values of one variable to a certain range and see the behavior of the other variables. This technique is achieved by selecting the highlight operation and the transient paint mode and by making the brush long and narrow. This is illustrated in Figure 5 in which the brush has been shaped and positioned in the (2,1) panel so that all points with low values of hardness are highlighted. Brushing only affects the active panel, so the large brush can be used without fear that points on surrounding panels will be affected. The highlighted values on

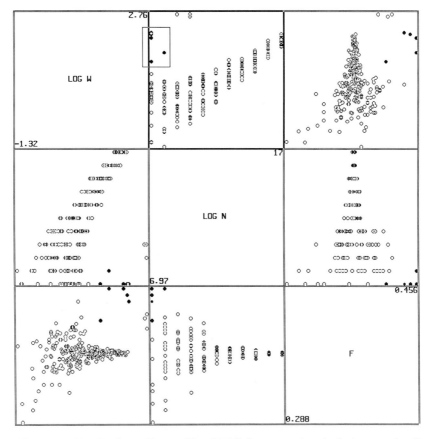

Figure 4. Bin Packing Data. The highlight operation is being used; all points inside the brush are highlighted as well as points on the other panels that correspond to these points. Now we can see the points of the (1,3) panel that correspond to the outliers of the (1,2) panel.

the (3,2) panel are a plot of abrasion loss against tensile strength for hardness held fixed to low values; the configuration of points looks something like a hockey stick. In Figure 6 the brush has been moved to the right on the (2,1) panel, so we are now conditioning on middle values of hardness, and the (3,2) panel again shows a hockey stick effect. In Figure 7 we have conditioned on the high values of hardness; the hockey stick effect on the (3,2) panel is still apparent, although there appear to be one or two unusual points in the lower left corner. Thus this brushing analysis suggests that there is a nonlinear dependence of abrasion loss on tensile strength and that the linearity of the original model is inadequate. These suggestions have been confirmed through fitting a regression model that is nonlinear in tensile strength.

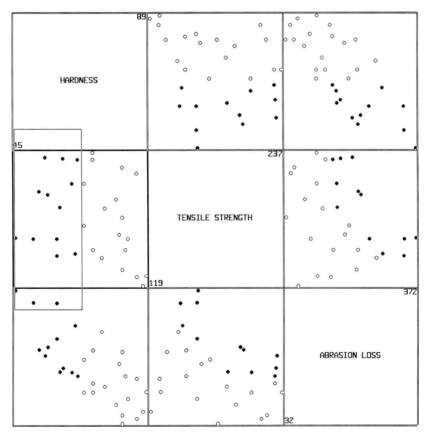

Figure 5. Abrasion Loss Experiment. Figures 5 to 7 illustrate conditioning on a single variable. In this figure, low values of hardness are highlighted by the brush in the (2,1) panel. The highlighted points of the (3,2) panel show the dependence of abrasion loss on tensile strength for low values of hardness.

2.4 Conditioning on Two Variables and the Shadow Highlight Operation

We can condition on two variables by making the brush a square and moving it around an active panel that graphs the two variables against one another. We will illustrate this technique using the variables of Figure 1. The data are 111 daily measurements of ozone, temperature, wind speed, wind direction, and solar radiation at various sites in the New York City metropolitan region [Bruntz et al., 1974]. (A table of the data is given by Chambers et al., 1983, p. 347-9.) The goal of the analysis is to see how ozone depends on the other variables. Solar radiation is certainly an explanatory variable, since it is a requisite for the photochemical reactions that produce ozone. The (1,3) panel in Figure 1 shows that ozone and temperature measurements are correlated, but photochemical

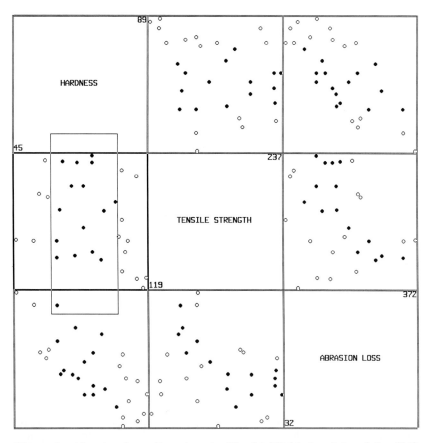

Figure 6. Abrasion Loss Experiment. The highlighted points of the (3,2) panel show the dependence of abrasion loss on tensile strength for middle values of hardness.

theory suggests that temperature itself is not responsible for greater ozone generation, since the rates of the reactions change negligibly over the range of temperatures in the data set. Temperature, however, is correlated with the meteorological stagnation that can lead to high ozone; when stagnation is high, temperature tends to be high. For example, in the (3,4) panel of Figure 1 we can see that temperature tends to be high when wind speed is low.

Is temperature a necessary surrogate for stagnation in this data set, or is there little explanatory power beyond solar radiation and wind speed? Figure 8 helps us to see. We have conditioned on both wind speed and solar radiation in the (2,4) panel. The operation is now shadow highlight. This behaves like highlight except that only the points that are highlighted on the active panel appear on the other panels. The

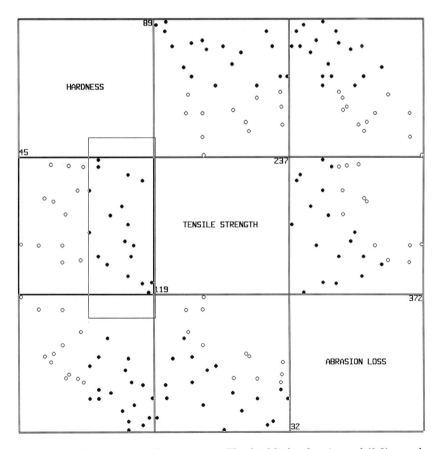

Figure 7. Abrasion Loss Experiment. The highlighted points of (3,2) panel show the dependence of abrasion loss on tensile strength for high values of hardness.

principal use of shadow highlight is in situations in which, because there are a large number of points, highlighted points cannot be easily visually detected because of overlap. This is the case for the data of Figure 8.

Figure 8 suggests temperature is needed. Panel (1,3) shows a relationship between temperature and ozone. The (1,2) and (1,4) panels show no relationship; this assures us that the temperature variable provides explanatory power beyond what is provided by wind speed and solar radiation. When we move the brush around the (2,4) panel a similar temperature-ozone relationship always appears, but when the brush is in a region of low solar radiation the temperature-ozone slope tends to decrease, which suggests a temperature-radiation interaction.

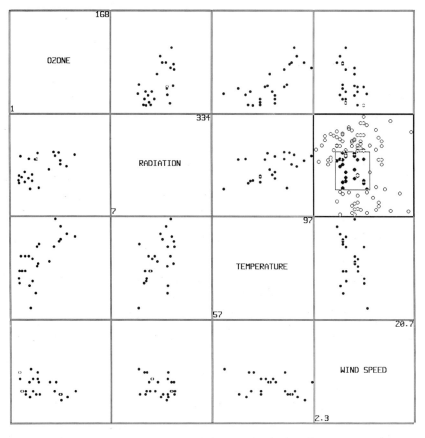

Figure 8. Ozone and Meteorological Data. The figure illustrates conditioning on two variables and the shadow highlight operation. The points of the (1,3) panel show the dependence of abrasion loss on temperature for radiation and wind speed held fixed to the ranges determined by the brush in the (2,4) panel.

2.5 Subsetting Categorical Variables

Often, multidimensional data can be broken up into groups, which can be described by a categorical variable. We can incorporate such a categorical variable by defining a numerical variable that takes on values from 1 to k, where k is the number of groups, and add it to the list of variables shown by the scatterplot matrix; this enables us to study the structure of each group.

An added categorical variable is illustrated in Figure 9. The data are from a survey of graphs in scientific publications [Cleveland, 1984]. One quantitative variable is the base 10 logarithm of the ratio of legend area to graph area for 50 scientific journals; another quantitative variable

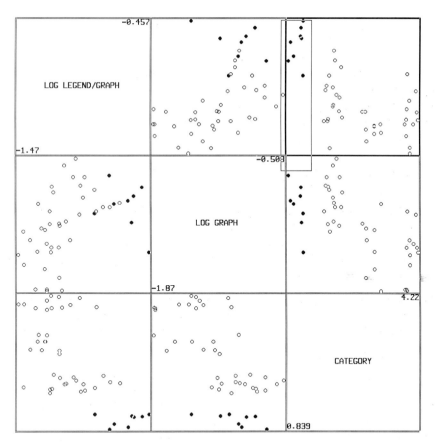

Figure 9. Journal Data. Brushing can be used to show a breakup of the data into subsets by a categorical variable. In this example the brush is highlighting the observations in the first of four categories.

is the base 10 logarithm of the fraction of space devoted to graphs (not including legends) in 50 scientific journals. The second variable is a measure of the amount of use of graphs in a journal, and the first variable is a rough measure of the average amount of explanation a graph receives in the legends. The four journal categories are as follows:

1. Biological—biology, medicine
2. Physical—physics, chemistry, engineering, geography
3. Mathematical—mathematics, statistics, computer science
4. Social—psychology, economics, sociology, education.

A categorical variable was defined by assigning one of the numbers 1 to 4 to each journal, according to its category, and then adding a random number from the interval $-.25$ to $.25$. This *jittering* [Chambers et al., 1983] is done to alleviate overlap, but in such a way that each journal's category can be correctly identified. The jittered categorical variable is included in Figure 9. The brush on the (1,3) panel is highlighting the first group, which is biological journals. We can see that these journals have a greater amount of explanation than the other categories and that the amount of explanation per unit area of graphs for this category does not depend on the fraction of space devoted to graphs. Using the brush, we can quickly shift to the other journal groups and also study their behavior.

2.6 The Lasting and Undo Paint Modes and Irregular Cluster Linking

When the paint mode is transient, we can quickly move from highlighting one rectangular region to highlighting another; this is important for the three brushing techniques discussed in the previous three sections. But how can we perform cluster linking when the cluster region on a panel is irregular and cannot be enclosed in a rectangle without including other points? In other words, how can we highlight any subset of points on the panel?

We solve this problem of irregular cluster linking by the *lasting* paint mode; points inside the brush remain highlighted even after the brush no longer covers them. (In our implementation the analyst invokes the lasting paint mode by pressing button 1 of a 3-button mouse.) With this capability we can highlight an irregular region much as we would paint it; this is illustrated in Figure 10 for jittered values of the famous four-dimensional iris data [Anderson, 1935]. A subset of points has been highlighted by painting the cluster in the upper left of the (1,2) panel, which is a plot of sepal width and length. Now we can readily see from panel (3,4) that this corresponds to the cluster of flowers with large petal lengths and widths.

The *undo* paint mode can be used to remove highlighting. When the brush is in this mode, any lasting highlighting inside the brush is undone, so if we moved the brush over the highlighted cluster in Figure 10 in this mode, we would remove the highlighting. (In our implementation the analyst invokes the undo paint mode by pressing button 2 of the mouse.) This undo operation allows us to separately study a variety of clusters.

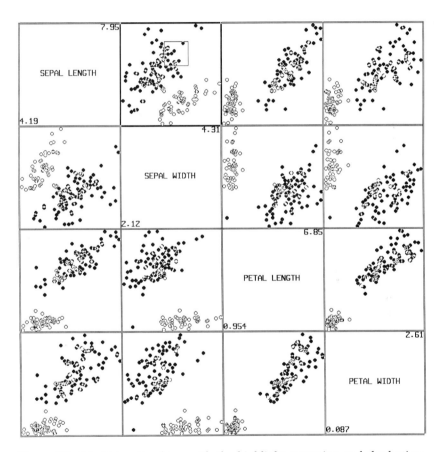

Figure 10. Iris Data. Brushing with the highlight operation and the lasting paint mode (points once highlighted remain highlighted) has been used to highlight the points of an irregular cluster in the (1,2) panel. The lasting highlighting can be removed by the undo operation.

2.7 Other Variables in the Scatterplot Matrix

It is important to emphasize that variables other than the original measurements can be displayed by the scatterplot matrix and brushed. For example, if we are involved in a regression, putting residuals in the scatterplot matrix can be helpful. One possibility is to replace the dependent variable by the residuals from regressing the dependent variable on all variables; in this case the variables of the matrix would be the residuals, together with the explanatory variables. If some explanatory variables have been decided upon as important and necessary in the regression and r additional variables are under consideration, then the variables of the scatterplot matrix can be the set of residuals from regressing the dependent variable on the entered explanatory variables and r sets of residuals from regressing each of the r not-entered variables on the entered ones.

3. THE DELETE AND LABEL OPERATIONS AND MORE BRUSH TECHNIQUES

3.1 The Delete Operation

Another brushing operation is *delete*. It works in the same way as highlight except that points are deleted instead of highlighted. The transient, lasting and undo paint modes also operate in the same way; button 1 makes the deletion lasting rather than transient, and the deletion can be undone by holding down button 2 and painting the region of deleted points.

3.2 Stationarity Probing of a Time Series

In the usual approach to analyzing a time series, $x(t)$, the autocovariance is estimated as if stationary; that is, the autocovariance

$$c(k) = \text{cov}(x(t), x(t+k))$$

is assumed independent of t. But in real life $c(k)$ can vary appreciably through time. It can slowly drift or, if $x(t)$ has a strong seasonal component, $c(k)$ can be cyclic.

Suppose we want to investigate the autocovariance at lag ℓ. Brushing can be used to probe nonstationarity by defining three variables: $x(t)$,

$x(t-l)$, and $d(t)$, where $d(t)$ might be t if we want to check for drift, or $d(t)$ might be t mod p if we want to check for a cycle of period p in $c(k)$.

Figure 11 is a scatterplot matrix where the three variables are $x(t)$, $x(t-2)$, and t. The (1,2) panel is a graph of $x(t)$ against $x(t-2)$ and the (1,3) panel is a graph of $x(t)$ against time. The time series is the residuals of monthly atmospheric CO_2 concentrations [Keeling et al., 1982] after a seasonal component and trend component, computed using the SABL procedures [Cleveland, Devlin, and Terpenning, 1982], were subtracted. The (1,2) panel suggests that there is correlation at lag 2; indeed, the autocorrelation coefficient at lag 2 is .44.

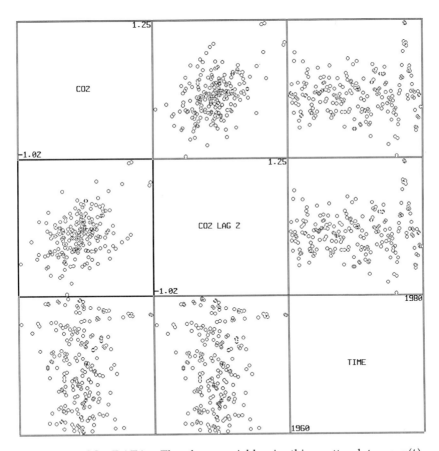

Figure 11. CO_2 DATA. The three variables in this scatterplot are $x(t)$, $x(t-2)$, and t for a time series, x. The (1,2) panel suggests that $x(t)$ and $x(t-2)$ are positively correlated.

Brushing, however, shows that the autocorrelation at lag 2 is not a stationary phenomenon. This is illustrated in Figure 12. Using the delete operation we have removed all data from a narrow time slice; with these points removed, the autocorrelation disappears. Figure 13 shows a shadow highlight; it is clear that the points in the narrow time slice have a high lag 2 autocorrelation and that they are the sole contributor to the autocorrelation of all of the data.

3.3 The Label Operation

Very frequently in the analysis of measurements of two or more variables, each observation of the variables has a name, or label, associated with it. An example is shown in Figure 14. The data are values of

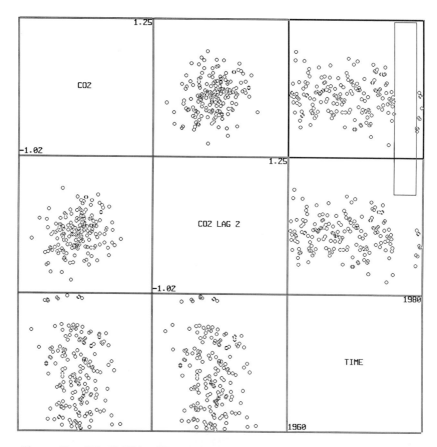

Figure 12. CO_2 DATA. The points in a narrow time interval have been removed by the brush in the (1,3) panel and the delete operation. Now the points in the (1,2) panel appear to have no correlation.

four variables for the 48 contiguous U.S. states [*The World Almanac*, 1983, pp. 116, 126]—per capita income in 1975 and 1979, per capita state government debt in 1980, and per capita state taxes in 1980. In this case each observation (the four values of the variables) has a label, the name of the state.

In most situations where the observations are labeled, we find ourselves wanting to know some or all of the labels of the points on graphs portraying the data. Usually, however, we cannot display all labels simultaneously since the overlap of the labels produces an uninterpretable mess; this would be the case for the graph in Figure 14.

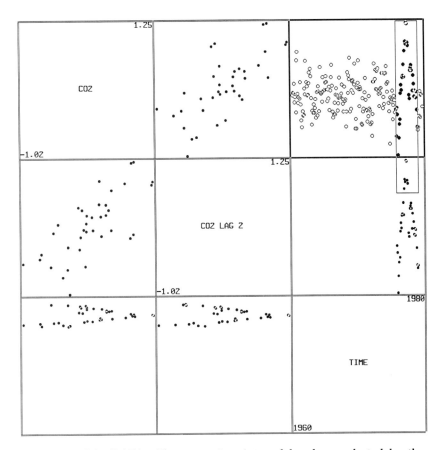

Figure 13. CO_2 DATA. The same time interval has been selected by the brush as in Figure 12, but now the operation is shadow highlight. The points in the (1,2) panel have a strong correlation. Thus the deletion and shadow highlight of Figures 12 and 13 show the correlation in $x(t)$ and $x(t-2)$ results only from the data in a narrow time interval.

220 BRUSHING SCATTERPLOTS

Brushing, however, provides a convenient mechanism for studying the labels, because we can use the brush to turn labels on and off. If the *label operation* is used, the effect of brushing is to cause the names to appear, as in Figure 14. As with the highlight, shadow highlight, and delete operations, label can be used in the transient, lasting, or undo paint modes; brushing in the lasting mode can be used to turn on a collection of labels and brushing in the undo mode can be used to turn them off.

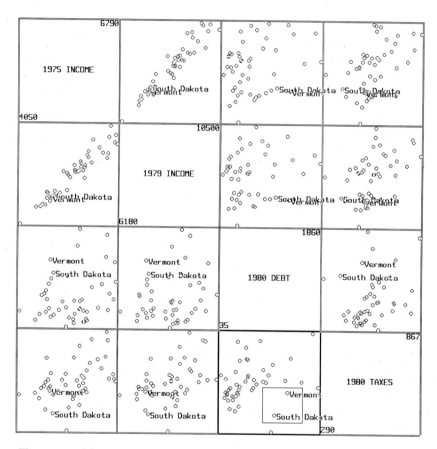

Figure 14. CO_2 DATA. The label operation is being used. All points inside the brush and on the active panel have their labels displayed as well as corresponding points on other panels.

4. DISCUSSION

4.1 Linking

There is a long history of attempts to design static graphical displays that allow the data analyst to carry out linking of corresponding points on different scatterplots [Chambers et al., 1983]. In fact, the scatterplot matrix is a static method that arose, in part, because it provides partial linking.

The first dynamic method for linking of which we are aware was developed by MacDonald [1982]. In his method a cursor is moved around one plot. Points close to the cursor on the plot are red, points intermediate in distance are purple, and far points are blue. All points corresponding to the same observation on any other scatterplots have the same color. MacDonald did not apply this method to the scatterplot matrix, although one could, of course, do so. This method of linking gives quite striking results in certain cases, however, the method does not allow the highlighting of many regions that are of interest in data analysis—for example, irregular clusters, such as the one in Figure 10, and the long and narrow rectangular regions of conditioning on a single variable. The aspects of brushing that provide the needed generality are the highlight operation, the three paint modes, and a rectangular brush whose shape can change.

4.2 Other Methods and Extensions

Brushing is just one of many dynamic methods that are available to data analysts. Others are reviewed by Becker, Cleveland, and Wilks [1987]. Moreover, brushing can be extended to more general configurations of graphical displays than the scatterplot matrix. For example, one can brush a scatterplot and have the result appear also on a plot of three variables shown by rotation [Becker, Cleveland, and Weil, 1988; Stuetzle, 1987] or on histograms of the variables [Stuetzle, 1987].

Brushing by itself is not, of course, a sufficiently rich environment for data analysis. For the methodology to be helpful in analyzing data, it must be embedded in a software environment that includes data management and standard statistical methodology, both graphical and numerical. For example, our implementation is within the S system for data analysis and graphics [Becker and Chambers, 1984] and the Wang and Gugel [1988] implementation is callable from SAS [SAS Institute, 1985]. Several other dynamic capabilities can enhance brushing as a data analytic tool.

For example, after points have been deleted it is often helpful to rescale the scatterplots of the matrix so that the remaining points fill up the plotting regions of the panels. It is also helpful to have a locate capability, which is the opposite of label; a name is chosen from a pop-up menu (controlled by the mouse) and the points of the scatterplot matrix that have this name are highlighted [Becker and Cleveland, 1985].

4.3 Brushing and Point-Cloud Rotation

An important consideration is how brushing a scatterplot matrix compares with rotation, since both are oriented toward analyzing multidimensional data. Our experience in applications is that rotation is better suited for certain tasks and brushing is better suited for others. Thus both are worthwhile tools to have available for data analysis.

Suppose there are three variables; that is, the data are in three dimensions. Point-cloud rotation can be useful for locating groups of points that cluster. Many clusterings are relatively easy to discover with rotation but not brushing. One example is the *randu* random numbers; detecting their planar clusterings [Marsaglia, 1968] appears impossible with brushing. If, however, the goal is to determine how one variable depends on the other two, brushing is usually far more informative than rotation. One reason is perceptual. Perceiving dependence on a two-dimensional scatterplot can be difficult; it is even more difficult on a three-dimensional point cloud because our perception of depth is not nearly as effective as our perception of position along the horizontal or along the vertical. Furthermore, it is quite easy to become befuddled in viewing a rotating point cloud and lose a sense of the axes. Another reason for the better performance of brushing is that the brushing technique of conditioning on a single variable is such a powerful method for understanding dependence in three-dimensional data.

4.4 Computer Implementation

Brushing can be implemented in many different computing environments. Our original implementation was on an AT&T Teletype 5620 graphics display terminal together with a host machine running under the control of the UNIX operating system. Details of this implementation are given in [Becker and Cleveland, 1984]. Since then we have implemented brushing on an IRIS graphics workstation [Becker, Cleveland, and Weil, 1988], Tierney [1986] has implemented it on a Macintosh, Stuetzle [1987] has implemented it on a Symbolics workstation, and Wang and Gugel [1988] have done a color implementation using the DI-3000 graphics system.

In developing high-interaction software it is important to maintain uniformity in the operation of the mouse. Without general concepts that cut across many operations, the data analyst can become confused by the mouse buttons. For example, in our implementation the operation of the four brushing operations is the same; with no button depressed brushing is transient, with button 1 depressed it is lasting, and with button 2 depressed it is undone.

Space is a problem for the scatterplot matrix on the computer screen; many plotting symbols need to be squeezed into a small area. To give the matrix as much room as possible on the screen, we put the variable names inside the main-diagonal panels, instead of around the sides; we omitted tick marks and put extreme scale values in the corners of the main-diagonal panels; and we squeezed the panels as close as possible, allowing just enough space to have good visual separation of the panels.

Acknowledgements

Jon Bentley, John Chambers, Brian Kernighan, and Allan Wilks gave us many helpful comments on an earlier version of this manuscript.

REFERENCES

Anderson, E. (1935). "The irises of the Gaspé peninsula," *Bulletin of the American Iris Society*, **59**: 2-5.

Anscombe, F. J. and Tukey, J. W. (1963). "The examination and analysis of residuals," *Technometrics*, **5**: 141-160.

Becker, R. A. and Chambers, J. M. (1984). *S: An Interactive Environment for Data Analysis and Graphics*, Belmont, CA: Wadsworth.

Becker, R. A. and Cleveland, W. S. (1984). "Brushing a scatterplot matrix: high interaction graphical methods for analyzing multidimensional data," Technical Memorandum. Murray Hill, NJ: AT&T Bell Laboratories.

Becker, R. A., Cleveland, W. S., and Weil, G. (1988). "The use of brushing and rotation for data analysis," *Dynamic Graphics for Statistics*, W. S. Cleveland and M. E. McGill, eds., 247-275. Pacific Grove, CA: Wadsworth & Brooks/Cole.

Becker, R. A., Cleveland, W. S., and Wilks, A. R. (1987). "Dynamic graphics for data analysis," *Statistical Science*, **2**: 355-395.

Bentley, J. L. and Johnson, D. S., Leighton, T., and Cole McGeogh, C. (1983). "An experimental study of bin packing," *Proceedings, Twenty-First Annual Allentown Conference on Communication, Control, and Computing*, 51-60. Urbana-Champaign, IL: University of Illinois.

Bruntz, S. M., Cleveland, W. S., Kleiner, B., and Warner, J. L. (1974). "The dependence of ambient ozone on solar radiation, wind, temperature, and mixing height," *Symposium on Atmospheric Diffusion and Air Pollution*, 125-128. Boston: American Meteorological Society.

Chambers, J. M., Cleveland, W. S., Kleiner, B., and Tukey, P. A. (1983). *Graphical Methods for Data Analysis.* Monterey, CA: Wadsworth.

Cleveland, W. S. (1984). "Graphs in scientific publications," *The American Statistician,* **38**: 261-269.

Cleveland, W. S. (1985). *The Elements of Graphing Data.* Monterey, CA: Wadsworth.

Cleveland, W. S., Devlin, S. J., and Terpenning, I. J. (1982). "The SABL seasonal and calendar adjustment procedures," *Time Series Analysis: Theory and Practice 1,* O. D. Anderson, ed., 539-564. Amsterdam: North-Holland.

Davies, O. L., ed. (1958). *Statistical Methods in Research and Production,* London: Oliver and Boyd.

Fisherkeller, M. A., Friedman, J. H., and Tukey, J. W. (1975). "PRIM9: an interactive multidimensional data display and analysis system," *Data: Its Use, Organization, and Management,* 140-145. New York: The Association for Computing Machinery.

Fowlkes, E. B. (1969). "User's manual for a system for interactive probability plotting on graphic-2," Technical Memorandum. Murray Hill, NJ: AT&T Bell Telephone Laboratories.

Fowlkes, E. B. (1971). "User's manual for an on-line interactive system for probability plotting on the DDP-224 computer," Technical Memorandum. Murray Hill, NJ: AT&T Bell Laboratories.

Keeling, C. D., Bacastow, R. B., and Whort, T. P. (1982). "Measurement of the concentration of carbon dioxide at Mauna Loa Observatory, Hawaii," *Carbon Dioxide Review: 1982,* W. C. Clark, ed., 377-385. New York: Oxford University Press.

Marsaglia, G. (1968). "Random numbers fall mainly in the planes," *Proceedings of the National Academy of Sciences,* **61**: 25-28.

McDonald, J. A. (1982a). "Interactive graphics for data analysis," Technical Report Orion 11. Stanford, CA: Stanford University, Department of Statistics.

Newman, W. M. and Sproull, R. F. (1979). *Principles of Interactive Computer Graphics.* New York: McGraw-Hill.

SAS Institute Inc. (1985). *SAS User's Guide: Basics,* Version 5 Edition. Cary, NC: SAS Institute Inc.

Stuetzle, W. (1987). "Plot windows, *Journal of the American Statistical Association,* **82**: 466-475.

Tierney, L. (1986). "Unpublished software for the Apple Macintosh," Minneapolis, MN: University of Minnesota, School of Statistics.

Tukey, J. W. and Tukey, P. A. (1981). "Graphical display of data sets in 3 or more dimensions," *Interpreting Multivariate Data,* V. Barnett, ed., 189-275. Chichester: Wiley.

Wang, C. M. and Gugel, H. W. (1988). "A High-performance color graphics facility for exploring multivariate data," *Dynamic Graphics for Statistics,* W. S. Cleveland and M. E. McGill, eds., 379-389. Pacific Grove, CA: Wadsworth & Brooks/Cole.

The World Almanac and Book of Facts 1983. New York: Newspaper Enterprise Association.

9

PLOT WINDOWS

Werner Stuetzle
University of Washington

ABSTRACT

Personal workstations with bitmap displays and window systems are becoming a widely used computing tool. This article describes some of the capabilities of an experimental package for drawing scatter plots and histograms that has been implemented on such a workstation. An example illustrates how those capabilities can be used in the graphical analysis of a data set.
KEY WORDS: Workstation; Scatter plot; Desk top; Object-oriented programming; Window system.

1. INTRODUCTION

Most data analysis in the future will be done on personal workstations. A typical workstation executes a million instructions per second, has a megabyte of memory, and a bitmap display with a resolution of

---

Werner Stuetzle is Associate Professor, Department of Statistics, University of Washington, Seattle, WA 98195. The author wishes to thank Andreas Buja, Catherine Hurley, and John McDonald for their careful reading and criticism of an earlier draft of this article. Special thanks go to John McDonald for many valuable suggestions and generous help in programming and debugging. This research was supported by National Science Foundation Grant DMS-8504359 and U. S. Department of Energy Grant DE-FG06-85-ER25006.

Reprinted from (1987) *Journal of the American Statistical Association,* **82**: 466-475, by permission of the American Statistical Association.

From *Dynamic Graphics for Statistics,* William S. Cleveland and Marylyn E. McGill, eds., copyright © 1988 by Wadsworth, Inc. All rights reserved.

1,000 × 1,000 pixels. It is used by one person at a time. The screen usually can be divided into several windows assigned to different tasks. A graphical input device such as a mouse controls the location of a cursor on the screen and thus allows graphical input. Machines that fit the description are offered by a number of computer companies [see McDonald and Pedersen, 1985a,b].

The goal of this article is to illustrate how the capabilities of a personal workstation can be used for better statistical graphics. It draws extensively on two areas of research. The first one is research on human interfaces and screen organization, done mostly at the Xerox Palo Alto Research Center (PARC). The ideas developed at PARC form the basis of the Macintosh user interface and have achieved considerable popularity. The second line of ancestry can be traced to the work of Fowlkes [1971] on interactive choice of data transformations, the pioneering PRIM-9 system [Fisherkeller, Friedman, and Tukey, 1974], and later systems for graphical data analysis, such as PRIM-H [Donoho et al. 1982], ORION [Friedman, McDonald, and Stuetzle, 1982], and MacSpin [Donoho, Donoho, and Gasko, 1986]. These systems have demonstrated the use of dynamic, interactive graphics for the visual inspection and interpretation of multivariate data.

Plot Windows is a package of functions for drawing scatter plots and histograms that has been implemented on a Symbolics 3600 Lisp workstation. This article is intended to describe some of its functionality. Implementation techniques and design issues were discussed in Stuetzle [1986]. Reasons for the choice of a Lisp environment as a basis for the implementation can be found in McDonald and Pedersen [1988]. All figures are snapshots of the Symbolics screen.

Systems with capabilities similar to those of *Plot Windows* will undoubtedly be written for other workstations. The *Data Desk* package for the Macintosh [Velleman, 1986] represents a step in this direction.

2. THE DESKTOP CONCEPT

Most statistical languages now in use, such as Minitab [Ryan, Joiner, and Ryan, 1976], ISP [PC-ISP, 1985] and S [Becker and Chambers, 1984], were originally designed for keyboard interaction from terminals connected to a computer through relatively slow (10-1,000 characters per second) serial lines. The user types commands; the system scrolls the screen and displays its response. This means that text and plots, once displayed, disappear as more commands are entered and answered. Unlike data, functions, or macros, plots are not persistent objects. The only way to save a plot is by making a hardcopy. Typically, plots also support little or no interaction — they are merely images that can be looked at.

On a personal workstation we can do better than that. The screen is divided into a number of rectangular *windows*, as illustrated in Figure 1. The large *Editor* window in white on the lower left is the window at which we type commands to have them executed. Commands, in our case, are simply statements in the Lisp language, and the *Editor* window comes as part of the Symbolics Lisp environment. Whatever we type is stored in the editor buffer, which can be saved if we wish to keep a record of the commands issued during a session. We can scroll the buffer, cut and paste, go back to commands typed some time ago and select them for evaluation, and so forth. The ability to execute commands out of an editor is a major convenience. It also makes it very easy to package frequently used command sequences as new functions that can be invoked by a single command.

Commands produce two different kinds of results. First, every command is a Lisp expression and upon evaluation returns a value, which can be a number, an array, or a more complex object, such as a scatter plot. This value or, more precisely, a printed representation thereof is automatically displayed in the *typeout pane* at the bottom of the *Editor* window. The typeout pane scrolls like a conventional terminal screen. Second, commands can have side effects. A command to draw a scatter plot not only returns the scatter plot as its value; it also creates a new window on the screen, in which the plot appears. Similarly, a command to perform a regression analysis might result in the creation of several new windows, displaying the regression coefficients, residual plots, and so forth. In any case, results do not appear interspersed with the commands themselves.

Apart from the command window, Figure 1 shows three more exposed windows, a scatter plot (*Plot1*) at the top right, the *Data-Inspector* in the middle right, and a folder (*All-Windows*) at the bottom right. Those types of windows will be discussed further. Other windows are partially visible in the gray area of the screen. Let us ignore those.

We can think of this whole arrangement as a desk top. Each window and, in particular, each plot corresponds to a sheet of paper on the desk top. Generating a new plot places a new sheet of paper on the desk.

3. SHUFFLING PAPER ON THE DESK

We can perform the same operations on windows — virtual sheets of paper — on a virtual desk that we are used to when working with real sheets of paper on a real desk, plus some more.

Moving Windows. Windows can be moved by pointing at the double arrow in the *title pane* and clicking the left mouse button. (The Symbolics mouse has three buttons.) After the click, an outline of the window will follow the movements of the mouse. Another left click deposits the window at the newly chosen location.

Reshaping Windows. Windows can be reshaped by clicking left on one of the arrows in the top right corner or top left corner of a window. After the click, the chosen corner will follow the movements of the mouse. Clicking left once more deposits the corner and makes the window change to the new outline. Figure 2 shows the screen after we have moved and enlarged Plot1. Note that the plot has automatically rescaled itself to fill up the larger window.

Burying Windows. A window can be buried by clicking left on the circular icon in the title pane. Burying a window means moving it to the bottom of the pile of paper on the desk, without changing its location.

Uncovering Windows. If a window is partially visible, we can expose it by pointing at its visible part with the mouse and clicking left. A completely hidden window can be retrieved by sequentially burying all the windows covering it up. Using folders, however (as described in Sec. 4), is usually a more efficient method for recovering a hidden window than simply digging through the pile.

Other Operations. There are several less frequently used actions that we want to perform on a window, like throwing it away or putting it into a folder. Other operations are specific to what the window shows. If it shows a scatter plot, for example, we want to add or remove coordinate axes. A list of such actions is presented in a menu that we obtain by clicking right on the title of the window. We can then point at an action with the mouse and have it executed by clicking left.

4. FOLDERS

On the simplest level, a folder is just what we are used to — a place to collect (virtual) sheets of paper (i.e., windows). The title of every window in the folder is displayed in the folder's window. The text in a folder window can be scrolled up and down under mouse control if the window is too small to have room for all of the titles of the windows that it contains.

Folders have two important properties, which have no exact parallel in the world of real paper and cardboard. First, a folder can contain other Lisp objects besides windows, such as data sets or regression models. This is useful for organizing a data analysis and keeping track of the results.

Second, an object can be in more than one folder. Strictly speaking, a folder in our sense does not contain objects — it contains pointers to objects. This was done partly to avoid the notion of "copying."

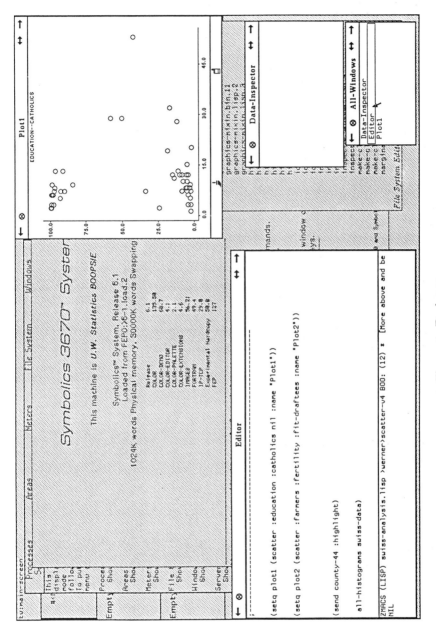

Figure 1. Desk top.

230 PLOT WINDOWS

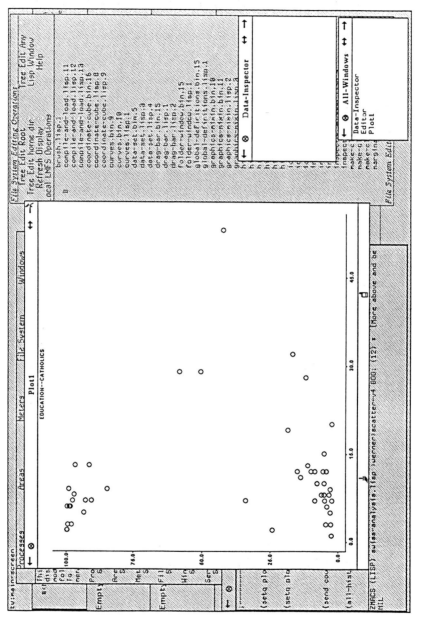

Figure 2. Desk top after moving and enlarging the scatter plot.

The system maintains a list of folders, to which newly opened windows are automatically added. The user can create folders and add them to the list or remove folders from the list. This makes it easy to organize work and keep logically connected plots together. One folder, *AllWindows*, at the bottom right in Figures 1 and 2, is defined by default. It contains all of the windows that are currently in existence. Windows that are thrown away are automatically removed from all folders.

Clicking left on the title of a window in a folder alternately exposes and buries the window. On the Symbolics the response to the clicks is instantaneous. This makes it convenient to search for windows that got lost in the shuffle, providing an alternate to constantly inventing and remembering new, meaningful names. Clicking right on the name of a window in the folder brings up the same menu that we would obtain by clicking right on the name in the window's title pane.

5. A DATA SET

For the remainder of the article I will discuss scatter plots and histograms. To illustrate their capabilities, we will look at a set of demographic data for 47 Swiss counties in 1888. The data set was collected by Francine van de Walle of Princeton University to study the demographic transition occurring when Switzerland changed from a medieval agricultural society to a modern industrial society. It was reported by Mosteller and Tukey [1977]. The variables are the following:

- *education* — proportion of population with education beyond primary school
- *catholics* — proportion of Catholics
- *infant mortality*
- *fertility* — a measure of fertility
- *farmers* — proportion of inhabitants engaged in agriculture
- *fit draftees* — proportion of draftees receiving the highest mark on the army examination.

The *Plot1* window in Figures 1 and 2 shows a scatter plot of *catholics* on the vertical axis versus *education* on the horizontal axis. Internally, the data set is represented as a list of records, with one record for each observation. The fields of each record in our example are :*education*, :*catholics*, and so forth. In addition, there are fields specifying the label of the observation and the way in which the observation is to be drawn on the screen.

6. SCATTERPLOT WINDOWS

A scatter plot in Plot Windows differs from scatter plots produced by most standard statistical packages in three important aspects:

1. It can display an arbitrary number of 2-dimensional or 3-dimensional objects, such as point clouds, polygons, and coordinate axes.
2. It supports interactive operations.
3. It always reflects the current state of the objects that it displays. If these objects change, the plot automatically updates itself.

I will now discuss these three properties in more detail.

6.1 Displaying Objects

To give a feel for the type and level of commands, here is the Lisp statement that we have to type at the command window in order to create a new scatter plot window:

> (setq plot2 (scatter :farmers :fertility :fit-draftees
> :name "Plot2"))

This statement assigns the value returned by the *scatter* function to the symbol *plot2*; *setq* is the Lisp assignment operator, analogous to := in Pascal. The value returned by *scatter* is a Lisp data structure representing the scatter plot. (The *:name* parameter can be omitted, in which case the system makes up a default name.)

Figure 3 shows the screen after execution of the command. The new plot appears in a default size in the upper right corner. The proportion of inhabitants engaged in agriculture is drawn on the horizontal axis; *fertility* is drawn on the vertical axis. The third variable is not visible, because the third coordinate initially is orthogonal to the screen plane. To perceive the 3-dimensional structure we have to rotate the point cloud, as described in Section 6.2.1.

In our example the *scatter* function is called with four arguments, the names of the three variables, and the name of the plot. We also could specify the name of the data set, in this case *swiss-data*. If no data set is specified, the default is used, which had been set to *swiss-data* previously. By giving additional arguments to *scatter*, the user can specify the locations of the top left and bottom right corners of the window or indicate that they are to be positioned with the mouse.

We can at any time add additional objects to a scatter plot, such as another point cloud, a curve, or coordinate axes, or we can remove objects from the plot. By default, scaling is automatic: the viewing transformation is chosen so that all objects in the display list are completely visible inside the window.

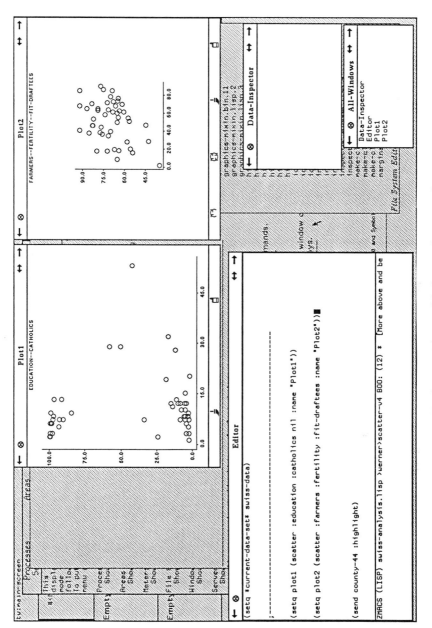

Figure 3. Desk top with an added second scatter plot.

6.2 Interactive Operations

Scatter plot windows support a number of interactive operations. They are similar to the ones that were available in earlier systems for dynamic viewing of multivariate data, such as PRIM-9 and ORION.

6.2.1 Rotation

When a 3-dimensional object is shown in a scatter plot window, we initially see the projection of the object onto the first two coordinates. We can perceive the 3-dimensional structure by making the object spin and watching the projection on the screen change. Rotation is controlled either through the menu pane at the bottom of a scatter plot window or by typing commands at the *Editor* window.

Pointing at the leftmost icon in the menu pane and holding the left mouse button down makes the objects in the scatter plot spin around the screen's x axis. Rotation starts slowly and then accelerates to some maximum speed. Releasing the button stops the rotation; pressing it down again resumes the rotation, but in the opposite direction. In an analogous fashion, the middle and right buttons control rotation around the screen y and z axes, respectively. (The screen's z axis sticks out of the screen; rotation of a point cloud around the z axis moves the points in circles on the screen.)

Pointing at the next icon to the right and clicking left makes the objects spin around the x axis; in contrast to what was described before, however, the spinning continues after the button is released. Clicking left again stops the motion, and clicking left once more starts rotation in the opposite direction. Any command sent to a scatter plot window, either through menu selection or by typing at the command window, will also stop the rotation.

We can also make a scatter plot spin under program control. Evaluating the expression

(send plot2 :rotate-around :y-axis)

makes the objects spin around the y axis. Direction of rotation can be specified by providing an additional optional argument. The rotation can be stopped by sending the plot the *:stop-rotation* message. As a general rule, whatever can be done through menu selection can also be done under program control.

Of course we can have more than one scatter plot window on the screen at any time, and any number of windows can spin simultaneously. Spinning will also continue while new commands are executed or the user interacts with another plot. It should become apparent that this is a useful feature.

6.2.2 Identification and Highlighting

One of the most useful features of *Plot Windows* is the ability to establish a correspondence between observations and their representations on the screen. Our example does not illustrate this fact very well, because the observations have no meaningful labels — we do not know the names or locations of the 47 counties. Thus we labeled the observations simply by their sequence numbers.

To locate County 44 in the two plots, we issue the command

(send county-44 :highlight).

Figure 4 shows the screen after execution of the command. In each plot the icon representing County 44 has been highlighted. The command does nothing more than changing the *:intensity* field of the record representing County 44. This is enough to highlight the observation in all plots, because, by construction, plots always reflect the current state of the objects they display.

To get the label of any observation on the screen, we can click left on the pointing hand icon in the scatter plot menu and move the cursor inside the window. Pointing at any observation and clicking the left button highlights this observation, prints its label next to it, and displays the values of all of its variables in the *Data-Inspector*. Figure 5 shows the screen after labeling one of the observations in *Plot1*.

6.2.3 Painting

We select the painting operation by clicking left on the paintbrush icon in the menu pane of a scatter plot window and moving the cursor inside the window. The arrow will be replaced by a little square, the brush, which we can move with the mouse. Painting with the left button held down highlights the observations that we brush over, whereas painting with the middle button held down removes highlights. The painting operation is terminated by moving the cursor out of the window.

The system offers a choice between different brush shapes, selectable from a menu. Apart from the square brush, there is a thin and high brush, stretching over the entire vertical range of the window, and a flat and wide brush, stretching over the entire horizontal range. We can also define our own brush by selecting the corresponding menu item and then positioning the upper left and lower right corners of the brush with the mouse.

Painting allows us to interactively define subsets of a data set. We simply paint the observations that we want in the subset and then evaluate a Lisp function that loops over the data set and collects into a list all those observations for which the *intensity* field has the value *:highlighted*.

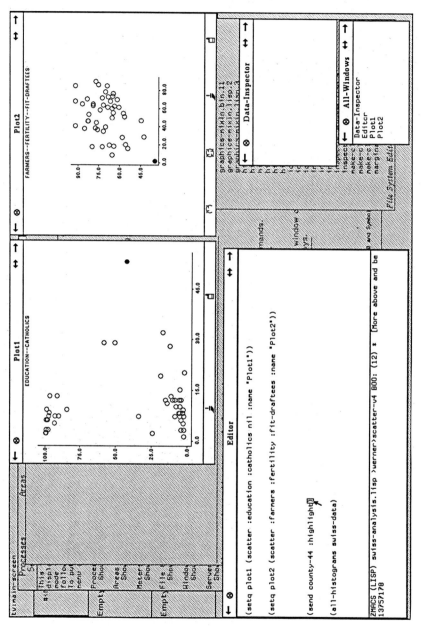

Figure 4. Desk top after highlighting observation 44.

SCATTERPLOT WINDOWS   237

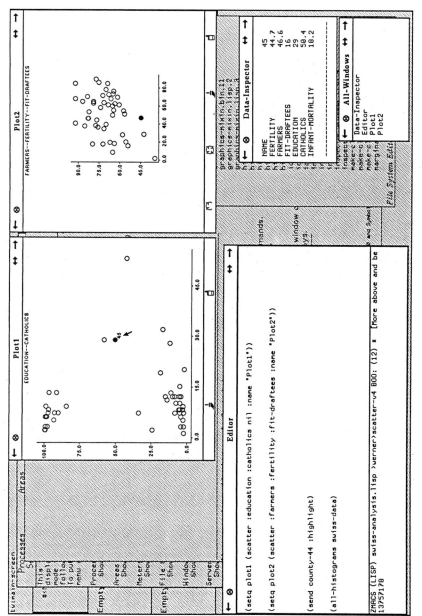

Figure 5. Desk top after identifying observation 45.

6.3 Automatic Updating

A scatter plot always automatically reflects the latest state of the objects it displays. If the objects change, the scatter plot changes. This is a far-reaching concept. I will discuss here one particular application of this concept, connected scatter plots.

6.3.1 *Connected Scatter Plots*

To illustrate the idea of connected scatter plots, let us return to Figure 3. It shows a plot of *catholics* versus *education* and a plot of *fertility* versus *farmers*. These two plots, however, do not give a complete insight into the distribution of the data in the 4-dimensional space spanned by the 4 variables involved; we cannot reconstruct the 4-dimensional coordinates of the points from the 2-dimensional projections. For the reconstruction we also need to know which point in one scatter plot corresponds to which point in the other — we need to be able to identify pairs of points corresponding to the same observation. This connection can be established through painting and automatic updating. We select the painting operating in one of the scatter plots (the *active plot*) and highlight or downlight observations. All other scatter plots automatically keep themselves up to date. Thus the intensity with which an observation is drawn is always the same in all plots showing the observation.

Automatic updating of scatter plots to show connections between points was first suggested and implemented by McDonald [1982]. He used transient highlighting based on distance — at any time points within a certain distance from the cursor and the corresponding points in the other plots were highlighted. The idea of painting, that is, the persistent highlighting described in Section 6.2.3, and of using brushes of various shapes, is due to Becker and Cleveland [1984].

6.3.2 *Illustration*

To illustrate the usefulness of connected scatter plots, let us have a closer look at our data set. We first highlight the three isolated points in *Plot1* by painting them with the left mouse button held down. Figure 6 shows that those three districts are also extreme in *Plot2* — they have the lowest values of *fertility*.

We would also like to know where the two clusters with a high proportion of *catholics* and a low proportion of *catholics* fall in *Plot2*. We first erase the highlights by painting over the highlighted observations with the middle button held down. Figure 7 shows the screen after highlighting the cluster with a high proportion of *catholics*.

We see in *Plot2* that districts with a high proportion of *catholics* also have a high proportion of *farmers* and high *fertility*. Spinning the cloud in *Plot2* shows that they also tend to have a low proportion of *fit draftees*. Note that we can spin one scatter plot while painting another; this ability is useful, allowing us (at least in theory) to completely reconstruct the geometric configuration of 5-dimensional data. Districts with a low proportion of *catholics* (the lower cluster in *Plot1*) tend to have lower *fertility* than the districts with a high proportion of *catholics*. The three outliers do not follow the rule — they are about 50% Catholic, but they have the lowest values of *fertility*.

6.3.3 Engineering Aspects of Painting

Painting taxes the computer hardware. A substantial number of plots besides the active plot can be exposed on the screen, and they all have to be updated constantly. (We will see an example with eight plots.) Processor time has to be divided between moving the brush and changing the intensities in the active plot, and bringing the other plots up to date. Experience shows that it is crucial for the brush to follow smoothly and continuously the mouse — jerky brush movement is highly disturbing. This was resolved by halting any other activity in the system during painting: While the mouse is moved with either the left or the middle button held down, other plots are not updated, all rotations stop, and so forth. As soon as painting is interrupted (all mouse buttons released), the other plots are updated, rotations start again, and system activity reverts to normal. The fact that other plots are momentarily out of date and rotations are stopped does not seem to matter, because during painting one is forced to concentrate on the active plot; it is very hard to look at two plots simultaneously.

There are two slightly different ways of bringing the other plots up to date, once painting is interrupted. Either we can erase them totally and then redraw them from scratch, or we can incrementally update the icons representing the observations whose intensities have been changed. Both ways were tried, and it was found that the second one is clearly preferable. Suppose we have two scatter plots on the screen, and *Plot1* is the active plot. We will want to see which icons in *Plot2* change intensity, once we interrupt painting. This is difficult if *Plot2* is completely erased and then redrawn. Our memory of the picture apparently does not survive the minute period during which the window is blank.

The preceding remarks illustrate the fact that small details in the implementation of interactive methods can make a big difference in ease of use. Therefore, it is important to build prototype systems in an environment that allows one to make changes quickly and easily.

240 PLOT WINDOWS

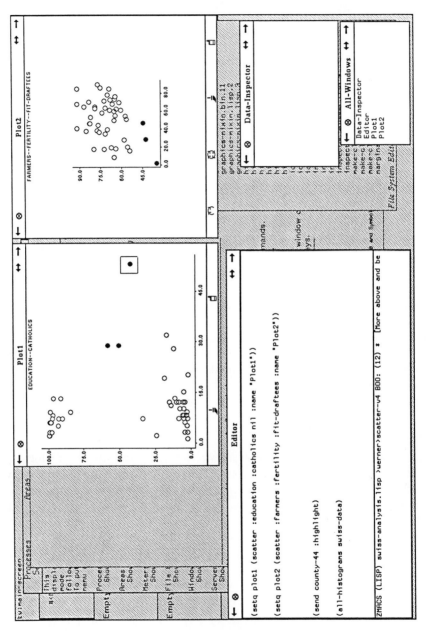

Figure 6. Desk top after painting of the outliers in Plot1.

SCATTERPLOT WINDOWS 241

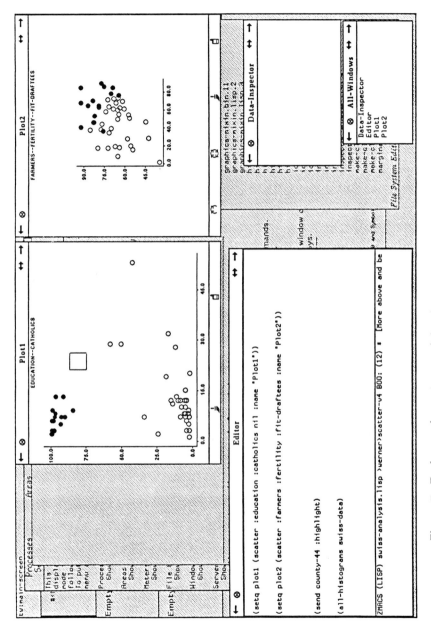

Figure 7. Desk top after painting of the cluster with a high proportion of Catholics.

7. HISTOGRAMS

Discovering structure in a data set helps in our understanding only if we can describe what we see in terms of the underlying variables. To give a concrete example, suppose we spin a 3-dimensional scatter plot and we see a cluster. We then need to come up with a statement like "These districts have medium proportion *catholics*, low *fertility*, and high *education*." Extracting such a statement, based on a 3-dimensional coordinate system drawn into the plot, requires considerable mental effort. Depending on the orientation, the axes can crisscross the point scatter, cluttering it up, and it is always difficult to see whether they point into the screen or out of the screen.

As a first step in describing structure, it is useful to be able to characterize an observation or group of observations in terms of the original variables, such as "The three outliers correspond to districts with the lowest *fertility*, medium proportion *catholics*, highest *education*, the highest proportion of *fit draftees*; they are not special in terms of *infant mortality*." As long as each of the variables is represented in at least one scatter plot, we have the necessary information on the screen. We just have to paint the observations in question and then see where they fall in the other plots. Note, however, that the preceding statement concerns only 1-dimensional marginals, and it seems a waste of mental effort to extract it from plots of 2-dimensional marginals.

These two empirical observations suggest implementing histograms. A histogram in *Plot Windows* is special in three ways:

1. Each bar of the histogram is composed of the icons representing the observations that fall into the corresponding class. In this aspect, the histogram resembles a stem-and-leaf plot.
2. The intensities of the observations are automatically kept up to date. The highlighted observations in each class are always drawn at the bottom of the corresponding bar, and the others are drawn at the top. This arrangement makes it easy to see what fraction of the observations in a particular bar is highlighted.
3. Like a scatter plot, a histogram supports the *identify* and *paint* operations. We can paint outliers and clusters apparent in a 1-dimensional marginal and see where those observations fall in the other variables. We can also easily answer questions like "What goes with a high proportion of *catholics*?" by simply painting the right tail of the *catholics* histogram.

Figure 8 shows the screen with the two scatter plots and histograms of the six variables. Note that the location of the Catholic districts in each of the variables is now readily apparent.

HISTOGRAMS 243

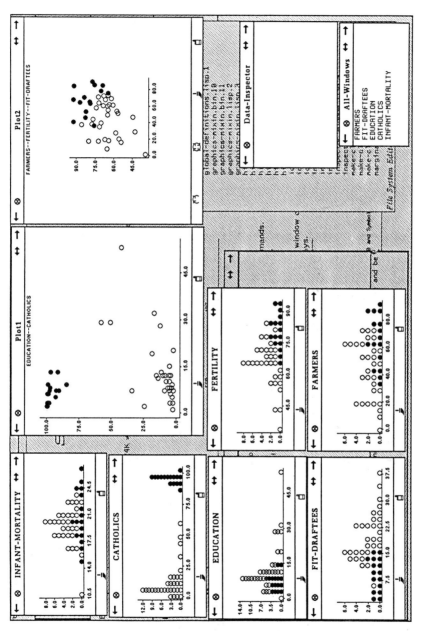

Figure 8. Desk top with marginal histograms.

8. DISCUSSION

The *Plot Windows* system — which allows the data analyst to view, manipulate, and store a large number of plots — offers a major increase in the ability to dissect and interpret multivariate data. Painting of connected scatter plots and histograms appears to be more useful for detecting and understanding structure than rotation, because it is not limited to three dimensions. There have been many attempts at displaying more than three dimensions in a single plot. A rather natural idea for showing additional variables is to encode them in the shape, color, or intensity of the icons representing the observations. But extracting a statement like "high values of variables 1 and 3 go with low values of variables 2 and 4" from a spinning cloud of icons is not nearly as immediate as grasping and describing a spatial point pattern.

A different approach to presenting higher-dimensional data is the *grand tour*, a dynamic display of the projections of a high-dimensional point set on a continuous sequence of 2-dimensional planes [Asimov, 1985; Buja and Asimov, 1986]. Watching a spinning 3-dimensional point cloud means watching a special grand tour of 3-dimensional space: one of the vectors spanning the projection plane remains fixed; the other one rotates. In more than three dimensions, however, we are not able to synthesize the sequence of views into a mental image of the configuration; our ability to visualize high-dimensional space is too limited. The *grand tour* is a useful tool for detecting clusters and outliers in higher dimensions; it appears to be less suitable for deriving statements about the associations between variables.

Luckily, in an environment like the one described here we do not have to decide whether we want to look at marginal histograms, at scatter plots, or at a grand tour — we can have it all simultaneously, and we can connect what we see in the various plots through painting. Higher dimensional views show us features of the data that might not be visible in lower dimensions; lower-dimensional views, on the other hand, are easier to interpret. We can have the advantages of both at the same time.

REFERENCES

Asimov, D. (1985). "The grand tour: a tool for viewing multidimensional data," *SIAM Journal on Scientific and Statistical Computing*, **6**: 128-143.

Becker, R. A. and Chambers, J. M. (1984). *S: An Interactive Environment for Data Analysis and Graphics*, Belmont, CA: Wadsworth.

Becker, R. A. and Cleveland, W. S. (1984). "Brushing a scatter plot matrix: high interaction graphical methods for analyzing multidimensional data," Technical Memorandum. Murray Hill, NJ: AT&T Bell Laboratories.

Buja, A. and Asimov, D. (1986). "Grand tour methods: an outline," *Computer Science and Statistics: Proceedings of the 17th Symposium on the Interface*, 63-67. Amsterdam: Elsevier.

REFERENCES

Donoho, A. W., Donoho, D. L., and Gasko, M. (1986). *MACSPIN: A tool for dynamic display of multivariate data*. Monterey, CA: Wadsworth & Brooks/Cole.

Donoho, D. L., Huber, P. J., Ramos, E., and Thoma, H. M. (1982). "Kinematic display of multivariate data," *Proceedings of the Third Annual Conference and Exposition of the National Computer Graphics Association*, **1**: 393-398. Fairfax, VA: National Computer Graphics Association.

Fisherkeller, M. A., Friedman, J. H., and Tukey, J. W. (1974). "PRIM-9: An interactive multidimensional data display and analysis system," SLAC-PUB-1408. Stanford, CA: Stanford Linear Accelerator Center.

Fowlkes, E. B. (1971). "User's manual for an on-line interactive system for probability plotting on the DDP-224 computer," Technical Memorandum. Murray Hill, NJ: AT&T Bell Laboratories.

Friedman, J. H., McDonald, J. A., and Stuetzle, W. (1982). "An introduction to real time graphics for analyzing multivariate data," *Proceedings of the Third Annual Conference and Exposition of the National Computer Graphics Association*, **1**: 421-427. Fairfax, VA: National Computer Graphics Association.

McDonald, J. A. (1982). "Interactive graphics for data analysis," PhD thesis. Stanford, CA: Stanford University.

McDonald, J. A. and Pedersen, J. (1985a). "Computing environments for data analysis, Part 1: Introduction," *SIAM Journal on Scientific and Statistical Computing*, **6**: 1004-1012.

McDonald, J. A. and Pedersen, J. (1985b). "Computing environments for data analysis, Part 2: Hardware," *SIAM Journal on Scientific and Statistical Computing*, **6**: 1013-1021.

McDonald, J. A. and Pedersen, J. (1988). "Computing environments for data analysis, Part 3: Programming environments," *SIAM Journal on Scientific and Statistical Computing*, **9**: 380-400.

Mosteller, F. and Tukey, J. W. (1977). *Data Analysis and Regression*, Reading, MA: Addison-Wesley.

PC-ISP (1985). *Interactive Scientific Processor User's Guide and Command Descriptions*. London: Chapman & Hall.

Ryan, T. A., Joiner, B. L., and Ryan, B. F. (1976). *Minitab Student Handbook*, North Scituate, MA: Duxbury Press.

Stuetzle, W. (1986). "Design and implementation of plot windows," *American Statistical Association 1986 Proceedings of the Statistical Computing Section*, 32-40. Washington, DC: American Statistical Association.

Velleman, P. F. (1986). "Graphics for specification: a graphic syntax for statistics," *Computer Science and Statistics: Proceedings of the 17th Symposium on the Interface*, 59-62. Amsterdam: Elsevier.

10

THE USE OF BRUSHING AND ROTATION FOR DATA ANALYSIS

Richard A. Becker
William S. Cleveland
Gerald Weil
AT&T Bell Laboratories

ABSTRACT

Brushing and rotation are two dynamic graphical methods for studying multidimensional data. In this paper we describe: (1) the IRIS implementation; (2) features of the methods that we have found useful during data analysis; (3) data analytic tasks for which each method has particular strengths. We implemented the methods on a high-performance graphics workstation—a Silicon Graphics IRIS-2400T—in order to study their properties by using them in the analysis of a large number of data sets.

NOTE FOR READERS

The figures contain five stereograms. One way to view them is to look between the images and then defocus the eyes, bringing the two images together. When the images are on top of one another, the 3-D effect will appear. Some people have trouble doing this, while others can do it but get a splitting headache after a short time. Another way is to use the stereo viewer included in this book. Such a viewer makes fusing the stereogram considerably easier.

From *Dynamic Graphics for Statistics*, William S. Cleveland and Marylyn E. McGill, eds., copyright © 1988 by Wadsworth, Inc. All rights reserved.

1. INTRODUCTION

Brushing [Becker and Cleveland, 1984, 1987] and rotation [Fisherkeller, Friedman, and Tukey, 1975] are two dynamic graphical methods that are being implemented in many graphics software systems and that are likely to be widely-used tools for data analysis in the future [Becker, Cleveland, and Wilks, 1987]. Because both methods are used to analyze multidimensional data—observations of several variables whose relationship is to be studied—the two methods are sometimes viewed as competitors. In fact, as we shall see, they complement one another. In this paper, we describe: (1) features of the methods that we have found useful during data analysis; (2) the data analytic tasks for which each method is useful. Our comments are based on implementing brushing and rotation on a high-performance graphics workstation and then using them in the analysis of a variety of data sets.

1.1 Implementation on an IRIS Workstation

To provide a fair test of the data-analytic differences between brushing and rotation, we had to provide a no-compromise implementation of each technique. For the implementation, we chose a Silicon Graphics IRIS-2400T workstation [Becker, Cleveland, and Weil, 1987]. The IRIS has a powerful "Geometry Engine"—special purpose hardware to assist in dynamic displays. In particular, the IRIS performs rotation so efficiently that thousands of points can be rotated under mouse control with no perceptible lag from the time the mouse is moved to the time the points move.

The IRIS has a high-resolution (1024×768) color display which can accommodate relatively large scatterplot matrices or multiple displays on the screen at once. Up to 32 bit-planes of graphic memory allow flexible choices of color maps, double buffering, and synchronization with external stereo glasses.

The IRIS provides a comfortable programming environment for developing software for dynamic displays. It utilizes the UNIX operating system, so the basic tools needed for program development are provided. Our graphics programs were written in the C language and make use of a library of graphics routines that easily accomplish rotations, segment control, color manipulation, and other important tasks.

In our system one of two types of data display is shown on the screen—a scatterplot matrix or a rotating point cloud.

A scatterplot matrix consists of all pairwise scatterplots of the p variables arranged in the form of a matrix. An example is shown in Plate 3 where the data on the states of the U.S. are plotted. The matrix arrangement allows the user to see one variable graphed against each of the others in a particularly simple and highly integrated way—by visually scanning along a row or a column. For example, in Plate 3, the bottom row has scatterplots of income against all other variables; the same is true of the leftmost column.

The rotating point cloud is, in a sense, a 3-D scatterplot. The data for three variables are plotted on orthogonal axes and rotated while the screen shows a 2-D projection of the 3-D scatterplot. Since motion is one of the visual cues that the human visual system uses to see in depth, the user can see the point cloud in 3-D. This is suggested by Plate 4. Of course, on a static page we cannot give the feeling of depth achieved by point cloud rotation.

The user interaction in this system is handled almost completely through the use of the mouse. The system is menu driven, with four different types of menus; these are shown in Plates 3 and 4. The high-level interaction is handled through the use of a group of pull-down menus at the upper-left of the screen. A pop-up menu (at the current position of the cursor) is used for the operations that are performed frequently in the brushing mode. Two static menus are used as well. The first is a color palette at the upper-right of the screen, and the second is a scrolling menu containing a list of names on the right of the screen. These two menus will also be described in more detail later.

The mouse attached to the IRIS has three buttons, all of which are used in this system. The middle button is consistently used to perform menu selections, the left to perform an action or decrease a quantity, and the right to do the opposite of the left. The uses of these buttons are kept consistent to help make the system easy to learn and easy to use.

Implementation on the IRIS gave us two benefits: good graphics performance and the ability to test a full complement of features of the methods. This was important for an effective assessment of the methods. For example, had we implemented brushing on a workstation that did not have the capability of providing nearly instantaneous updating when the brush was moved, the assessment would have been affected by the implementation insufficiency. Or, had we implemented rotation on a workstation that could not provide stereo viewing, we would have underestimated the potential usefulness of rotation. Our goal was to assess the intrinsic properties of brushing and rotation.

Is it realistic to assess rotation and brushing on a state-of-the-art graphics workstation such as the IRIS, when most data analysts have computing environments with much less graphics power? The answer is "yes" simply because we expect that in the future, data analysts will have IRIS-like power available at low cost. Our goal is to provide assessments of brushing and rotation that will have lasting value rather than assessments that have a transient present value.

2. ROTATION

Suppose x_{ij} for $i=1$ to n and $j=1$ to p are n measurements of p variables. Rotation allows us to explore the joint structure of any three of the variables. The three-dimensional point cloud is shown rotating on a computer screen. The rotation shows us different views of the points and it produces a 3-D effect, allowing us to see depth. We see depth because the human visual system uses motion parallax—the smaller movement of more distant objects as the observer and the objects move relative to one another—to see depth in real scenes. Thus, the rotation of the cloud allows us to perceive its three-dimensional structure.

2.1 Rotation Control

The rotating cloud can be manipulated by one of two coordinate systems. The first system is determined by the screen: two coordinate axes are in the plane of the screen, one vertical and one horizontal, and the third is perpendicular to the screen. The second coordinate system is determined by the data. The screen axes remain fixed as the point cloud rotates, and the positions of the points change with respect to the screen axes. During rotation, data axes move with respect to the screen axes, but the data-axis coordinates of the points stay fixed.

The data analyst can control the rotation by making the cloud rotate about any one of these six axes. This is illustrated in the left two panels of Figure 1. In the left panel, the rotation is about the x coordinate axis; in the middle panel, the rotation is about the horizontal screen axis. Rotation about the screen axes is useful for getting the cloud into a particular orientation and carrying out general explorations. Rotation about a data axis is particularly convenient when we want to study how one variable depends on the other two; selecting the axis of the dependent variable as the axis of rotation helps our visual system focus on the dependent variable.

Our implementation uses the three-button mouse to provide user control of operations, including rotation. The rotation axis is selected through a pull-down menu and the direction of rotation is selected by pushing the left or right button of the mouse. The velocity of the rotation increases until the button is released; after the release, the velocity remains constant at the final value it attained while the button was pushed.

Another method for controlling the rotation of the point cloud, called *pushing*, gives very natural control of the position of the cloud.

Imagine the data embedded in a clear plastic ball. The mouse gives the analyst a way of pushing the ball, horizontally, vertically, or at some angle. As the ball rotates, the points rotate. We call the direction the mouse moves the *movement direction*. Consider a point on the ball hemisphere that faces the data analyst; it moves in the movement direction so the axis of rotation is perpendicular to the movement direction. The amount of rotation is determined by the amount of movement of the mouse. The feel that the data analysts have from this push control is a very natural one; it is as they reached out and pushed the ball to make it rotate. Pushing gives us rotations about all axes in the screen plane. This is illustrated in the right panel of Figure 1; the mouse has been pushed from the lower right of the pad to the upper left. Clockwise and counterclockwise rotations about an axis perpendicular to the screen can be gotten by left and right button pushes using the method described above. Pushing can also be used with point cloud inertias—when the pushing stops, the cloud continues to rotate.

The hypothetical ball we described above is a mouse-oriented analogue to a tracker ball. McDonald [1982] describes the use of a tracker ball to define a movement direction.

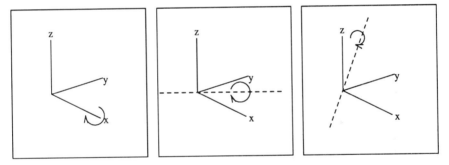

Figure 1. Rotation Control. Measurements of three variables can be graphed by showing the data rotating on a computer screen. There are various methods for controlling the rotation. The axis of rotation can be specified to be one of the coordinate axes of the data (left panel) or one of the screen axes (middle panel), and the direction and speed of rotation can be controlled by mouse buttons. Another method of control is pushing (right panel). When the mouse is moved, the point cloud rotates about an axis perpendicular to the mouse movement direction, and the amount of movement depends on how far the mouse is moved. It is as if the analyst reached out and pushed the cloud in the direction of the mouse movement. With this method the axis of rotation can be any line in the plane of the screen.

2.2 Two Problems with Rotation

There are two disadvantages to rotation, one quite major in its implication for data analysis and the other minor, but worthy of consideration.

The major problem is that if the rotation stops, the 3-D effect disappears. This is unfortunate because it is helpful to stop rotation to get one's bearings with respect to the axes; the continuous movement can make it quite difficult to get these bearings. The scenario often goes as follows: some geometric phenomenon is observed in the data that we want to interpret in terms of the three variables, but the motion makes it difficult to fix one's attention on the axes. Consider, for example, Figure 2, a scatterplot of two variables. The geometric pattern we observe is two clusters of points. One interpretation in terms of the variables is that in the northwest cluster the average x is lower than in the southwest cluster, and the average y is higher. The process of interpreting a pattern after we have perceived it is so simple for most scatterplots that we are hardly conscious that perception and interpretation are distinct. We become painfully aware of the distinction for a rotating point cloud [Huber, 1987]. The reader might get an intimation of the problem by continuously rotating the page about the axis from the eyes to the page, while attempting to study the plot.

The second problem with rotation is that the 3-D effect is less than what it is when we view objects in the real world because the rotating cloud does not have the other visual cues—stereopsis (the two different views of the eyes), receding parallel lines, decreasing sizes and brightnesses of elements with distance, shading, masking of objects in the background and others that contribute to seeing depth in real scenes [Kaufman, 1974].

2.3 Stereo Views with Dynamic Scaling and Dynamic Symbol Size

Both of the problems of the previous section can be reduced in severity by adding stereo viewing. In our implementation this was achieved by showing red and green images of the cloud from two different eye points; we view the cloud using glasses with one red lens and one green lens to get a stereo view. A better version of stereo viewing can be achieved by using glasses with electro-optic shutters in the lenses. Two views, one for each eye, are alternately displayed on the screen. In synchronization with the views, the lenses alternately are clear or opaque. This allows full color stereo viewing. The addition of stereo makes the rotation capability less important for achieving depth; rotation becomes a method for getting different views of the cloud. For this reason pushing becomes a very attractive method of control when stereo is used.

The binocular disparity introduced by stereo viewing provides an element of perspective. Our implementation enhances perspective with monocular depth cues by having the apparent sizes of the plotting symbols decrease with depth and by drawing a frame (edges of a rectangular parallelepiped) in perspective around the point cloud; the frame also serves as an indicator of the data axes. (The rotation origin—that is, the fixed point—is the center point of the frame.)

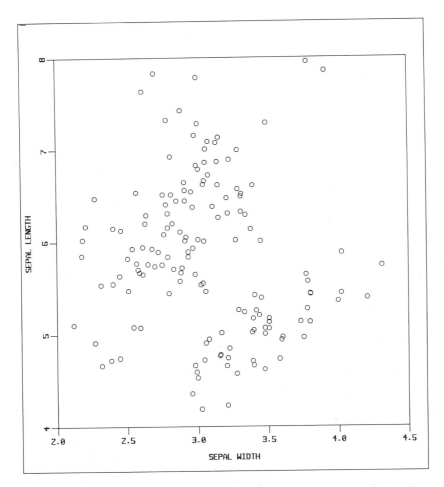

Figure 2. Interpretation of Effects in Rotation. On this scatterplot we can readily identify two clusters of points. The average y value is greater in the northwest cluster than in the southeast one, and the average x value is smaller. Interpretation of effects in terms of the scales in this way is difficult on a rotating point cloud. For this reason stereo is an important adjunct to rotation; with it, the data analyst can stop the rotation when an interesting view is found, still study the cloud in depth, and interpret the pattern. Using pushing with stereo allows the data analyst to delicately probe a point cloud in a highly-controlled way.

An alternative to stereo that still provides depth cues and a stable view for interpretation is *rocking* [Tukey, 1988]. The idea is to have the cloud move slowly with small deviations from the desired view. The motion provides depth cues but does not distract from the contemplation of a particular view.

The length of each side of the frame can be dynamically controlled, that is, changed continuously in real time, which changes the scaling of the variables. The idea for doing this was stimulated by Buja *et al.* [1988]. Also, the size of the plotting symbol, a three-dimensional cross, can be dynamically changed by the data analyst. Finally, the distance of the viewer to the center of the frame can be dynamically changed, which changes the size of the cloud on the screen. (This distance can be

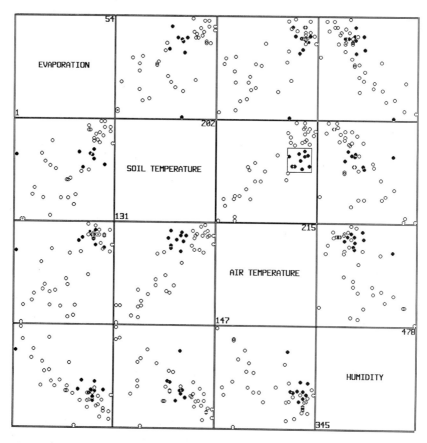

Figure 3. Brushing With the Highlight Operation. The rectangle in the (2,3) panel of the scatterplot matrix is the brush, which the data analyst moves around the screen by moving a mouse. In this figure the operation is highlight: points inside the brush on the active panel are highlighted as well as corresponding points on other panels.

reduced to zero or even be negative. In so doing the data analyst enters the point cloud; by also invoking the rotation control, the analyst can fly around the cloud like a spaceship among the stars. However, we have not yet found an application where this point-cloud penetration is useful.)

2.4 Summary: Features and Performance

In summary, we implemented the following features of rotation: (1) Stereo; (2) Monocular perspective; (3) Dynamic control of scaling and symbol size; (4) Rotation about any data axis or any screen axis; (5) Pushing; (6) Rocking.

3. BRUSHING

Brushing is a dynamic method in which a rectangle, called the brush, is moved around the screen by moving a mouse. Typically, the screen contains scatterplots, often arranged in the form of a scatterplot matrix. Figure 3 shows a scatterplot matrix with a brush in the (2,3) panel. There are three brush operations: *delete*, *label*, and *highlight*. Here, we will describe only the last since its use provides the exploration of data that we will be comparing with rotation; the description is in Section 3.1. There are three paint modes in brushing—*transient*, *lasting*, and *undo*—each of which can be used with any brush operation; they are described in Section 3.2.

3.1 The Highlight Operation

In brushing, one scatterplot panel is chosen as the active panel (or, the active panel can be the panel that contains the center of the brush). In Figure 3 this is the (2,3) panel. When the brush operation is *highlight*, points inside the brush on the active panel are highlighted. The data we are considering consist of n observations, $(x_{i1}, ..., x_{ip})$, of p variables. The highlighted points on the active panel select, in effect, a subset of observations for special treatment. This special treatment is extended to all points on the other panels that correspond to these selected observations; that is, the corresponding points on other panels are also highlighted. This is illustrated in Figure 3.

The highlighting can be done in various ways. If the screen is monochromatic, unfilled circles can be used for points not highlighted and filled circles can be used for highlighted points; this method gives good visual contrast, as illustrated in Figure 3. In the first implementation of brushing, which was on a Teletype 5620 terminal, this method was used [Becker and Cleveland, 1985]. However, when there is substantial overlap of the plotting symbols, it can be difficult to see the highlighted points. In such a case it can be helpful to omit, on all panels that are not active, points that are not highlighted; this *shadow highlight* is illustrated in Figure 4.

If the screen is in color, points can be highlighted by using different colors for highlighted and unhighlighted points. We used this method in our IRIS implementation. Since color provides such excellent visual discrimination, shadow highlight is not needed in an implementation of this type.

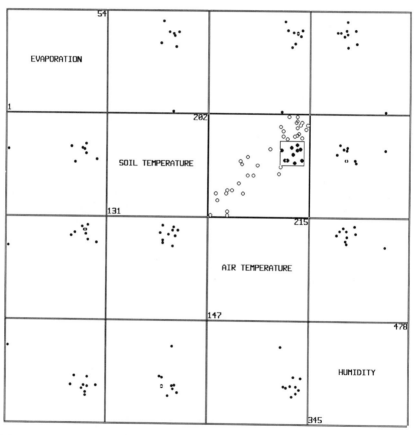

Figure 4. Shadow Highlight. When there is a lot of overlap of the plotting symbols it can be difficult, on screens without color, to see which points are highlighted. In such a case it is helpful to switch to shadow highlight: on all panels except the active one, points that are not highlighted do not appear.

3.2 The Paint Modes and Brush Shape

The position of the brush is controlled by a mouse; when the mouse is moved, the brush moves. The paint mode of brushing can be *transient*, *lasting*, or *undone*. When the mode is *transient*, points that were previously selected but are no longer covered by the brush are no longer highlighted. This is illustrated in Figures 5 and 6. Transient highlighting allows one to move easily from highlighting one subset of the observations to another. To facilitate such change it is important to have a nearly instantaneous updating of the new highlighting and the highlight removal. Smooth continuous change is needed so that the data analyst can produce rapid sequences of brush views, moving quickly from one to another, storing one view in short-term visual memory to compare with

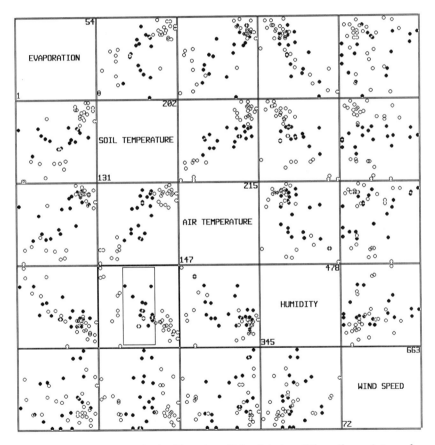

Figure 5. Brushing With the Transient Paint Mode. When the paint mode is transient, points that were previously inside the brush are no longer highlighted. See Figure 6.

258 THE USE OF BRUSHING AND ROTATION FOR DATA ANALYSIS

another. It is particularly disconcerting if performance drops to the point where there is a noticeable time lag between the movement of the mouse and the movement of the brush on the screen.

When the brush mode is *lasting,* points that are selected by the brush remain highlighted when the brush is moved. This allows us to paint arbitrary subsets of points with highlighting, just as one can paint with an actual brush. This is illustrated in Figure 7 where an irregular cluster has been highlighted. When the brush is in *undo* mode, lasting highlight of points selected by the brush is removed.

In our IRIS implementation, transient highlight is carried out by moving the mouse with no buttons depressed, lasting highlight is carried out by moving the mouse with the left button depressed, and highlight

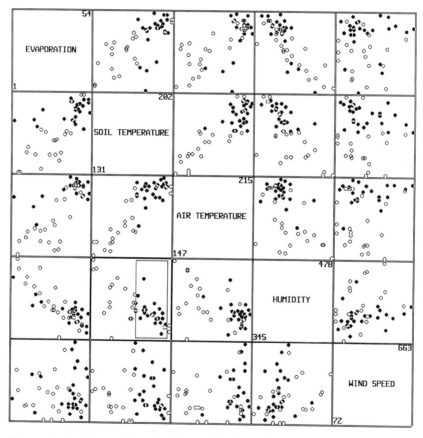

Figure 6. Brushing With the Transient Paint Mode. The brush has been moved to the right; points that were previously inside the brush, but are no longer inside, are not highlighted.

removal is carried out by moving the mouse with the right button depressed. Furthermore, the shape of the brush can be changed by selecting "shape change" from a menu, depressing a button to mark one corner of the rectangle, moving the mouse to sweep out a rectangle, and releasing the button when the desired size and shape is achieved. Finally, separate brushings with different colors can be carried out; the colors are chosen from the color palette in the upper right of the screen.

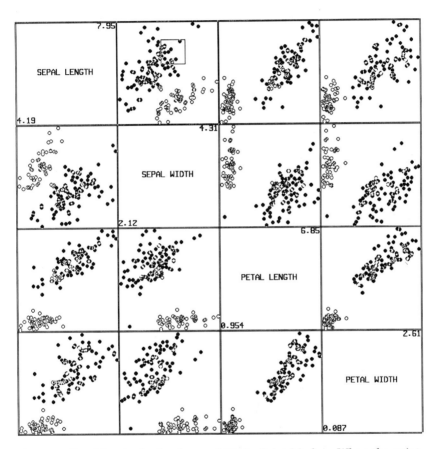

Figure 7. Brushing in the Lasting and Undo Paint Modes. When the paint mode is lasting, points that are highlighted remain highlighted when the brush is moved. This allows the data analyst to highlight any subset of points such as the cluster to the northwest on the (1,2) panel in this figure. When the paint mode is undo, lasting highlight of points inside the brush is removed.

3.3 Summary: Features and Performance

In summary, we implemented the following features of brushing with the highlight operations: (1) A rectangular brush whose size and shape can be easily changed. (2) Instantaneous updating of highlighting on all panels when the rectangle is moved by moving a mouse; (3) Three paint modes: transient, lasting, and undo. (Two other operations, label and delete, were implemented, but are not discussed here.)

4. USES OF BRUSHING AND ROTATION IN DATA ANALYSIS

The previous sections have described implementations of brushing and rotation in a high-performance graphics environment with a full complement of features for each method. In this section we describe the strengths and weaknesses of the methods in analyzing data.

4.1 Three Variables: Exploring the Trivariate Structure

Suppose the data are three-dimensional and the goal is to study the trivariate structure in a symmetric way with respect to the variables; that is, the goal is *not* to see how one specified variable depends on the other two. With such data we tend to look for clusters and outliers and to determine if the pattern of the points can be described by one-dimensional or two-dimensional manifolds.

Rotation (with stereo and perspective) is at its best in this scenario because each single view of the points is a view of the trivariate structure. We can attempt various forms of brushing, moving quickly from one display to the next and trying to integrate the many views in short-term memory, but the result is typically not as effective as rotation.

One example is provided by measurements of glucose intolerance, insulin response to oral glucose, and insulin resistance on 145 subjects. The data were analyzed by Reaven and Miller [1979] and are reported by Andrews and Herzberg [1985]. Rotation shows clearly a central collection of points, which are normal subjects, and two appendages that contain diabetic subjects. Figure 8 shows one 3-D view of this structure by a stereogram. It is hard with brushing to fully appreciate the three-dimensional structure of the central collection and the appendages.

Of course, not every trivariate structure is immediately and easily perceived by rotation with stereo views. We have experimented with simple structures, such as nested hemispheres of data and hemispheres with different scales and orientations. Such features can be difficult to recognize. Donoho *et al.*, [1986] described another structure that causes certain problems—three-variable normal data with a hole cut out of the center. The problem with viewing such data is that the outer points obscure the lack of data in the center. They found that the hole in the middle is not visible by means of rotation or stereo pairs alone; recognition of the hole required rotation combined with stereo.

4.2 Three Variables: Dependence of One Variable on Planar Position of Two Others

Suppose the data are three-dimensional and the goal is to see how one variable depends on the position in the plane of points determined by the other two variables, but it is not of any special interest to see how the dependent variable depends on each of the independent variables for the other held fixed. The major domain of application for this scenario is spatial data: measurements of a variable at geographical locations.

In this scenario, rotation typically gives us more insight into the data than brushing, again, because each single view of the data provides a view of the 3-D space. (As we stated in Section 2.1, it is helpful when we probe dependence by rotation to make the axis of the dependent variable the rotation axis.) One example is provided by the stereogram in Figure 9. The independent variables are geographical location. The dependent variable is smoothed values of measurements of coal ash. The raw data, which are reported and analyzed by Cressie [1986], were smoothed as a function of location by locally-weighted multiple regression [Cleveland and Devlin, June 1988] to assist the study of dependence.

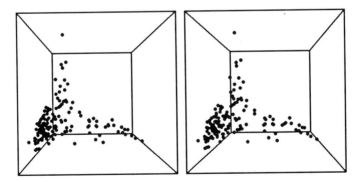

Figure 8. Rotation for Trivariate Structure. Rotation is particularly useful when there are three variables and the goal is to study multivariate structure such as clusters, outliers, and points that lie along one and two dimensional manifolds. In this stereogram there is a cluster with a high density of points and there are two appendages that stick out from the cluster.

(This is discussed in Section 4.6). We can see that the smoothed coal ash values have a sloping valley in the center.

4.3 Three Variables: Dependence of One Variable on Each of Two Variables for the Other One Held Fixed

Suppose the data are three-dimensional and the goal is to study how a dependent variable depends on each independent variable for the other held fixed. This is a common situation and is the one that occurs in most regression studies.

In this scenario, brushing almost always provides far more insight. One reason is that the visual task that we must perform in using rotation to study conditional dependence is too difficult. To see the dependence of y on x for z held fixed to some value, we must visually identify a plane perpendicular to the z axis and study the dependence of y on x for points close to this plane. It is typically quite difficult just to visually detect the right points. Even when detection is possible, it is hard to assess the dependence. In fact, our assessment of dependence in a set of points that lie in or close to a plane is done best when we view perpendicularly to the plane; this is precisely what brushing can do for us, as the next example illustrates.

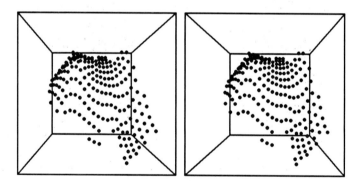

Figure 9. Rotation for General Dependence. Rotation is also particularly useful when there are three variables, the goal is to study the general dependence of y on x_1 and x_2, but there is no need to see how y depends on x_1 for x_2 held fixed or vice versa. In this example x_1 and x_2 are latitude and longitude; the plane of these two variables, geographical location, is the floor of the box in the stereogram. The dependent variable is a smooth function of latitude and longitude and the goal is to see how this function depends on geographical location. Rotation does very nicely in this application.

Figure 10 is a scatterplot matrix of three variables from an experiment of Cleveland, McGill, and McGill [June 1988] in which 16 subjects judged what percent the slope of one line segment was of the slope of another. All line segments had positive slope and the true percents ranged from 50 to 100. The dependent variable is the mean absolute *error*, averaged across subjects and across 12 replications for each of 44 line segment pairs. Let a be the average of the angles of the two line segments and let t be the true percent. One independent variable is d, the distance of a from 45 degrees, and the second is t.

Figures 11 to 14 show brushing with the shadow highlight operation in such a way that we achieve conditioning on t. A long narrow brush is moved from left to right on the (2,3) panel, which conditions on different values of t; at each position we can see the conditional dependence of *error* on d in the (1,2) panel. An important structure is revealed by the brushing. For t held fixed, *error* is nearly linear in d; as t increases, the slope of this linear dependence decreases to zero. But when we condition

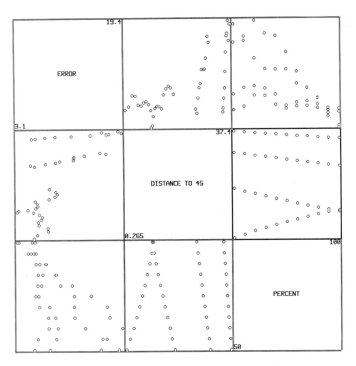

Figure 10. Brushing for Conditional Dependence. Brushing is particularly useful when there are three variables and the goal is to study how y depends on x_1 for x_2 held fixed and vice versa. For the data in this figure the goal is to see how error depends on the other two variables. Figures 11 to 15 show the use of brushing to study the conditional dependence.

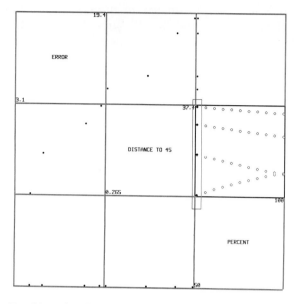

Figure 11. Brushing for Conditional Dependence. The brush is selecting points with percent at low values. The (1,2) panel shows how error depends on distance for percent held fixed to low values. The dependence is linear.

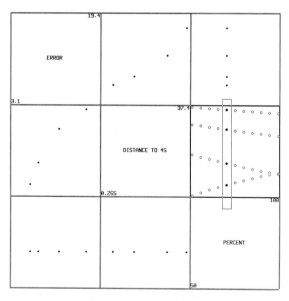

Figure 12. Brushing for Conditional Dependence. Now the brush is selecting points with percent held fixed to middle values. The dependence is still linear but the slope has decreased.

USES OF BRUSHING AND ROTATION IN DATA ANALYSIS 265

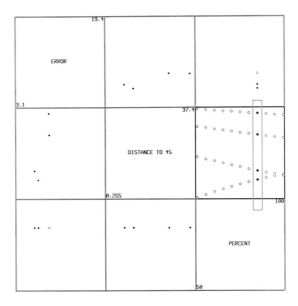

Figure 13. Brushing for Conditional Dependence. The conditioning values for percent have increased and the slope has decreased again.

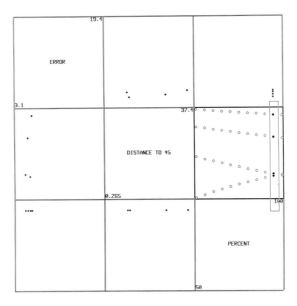

Figure 14. Brushing for Conditional Dependence. Percent is increased and the pattern continues. From this conditioning on a single variable—the brush technique shown in Figures 11 to 14—we see that for percent held fixed, error is linear in distance and the slope decreases as percent decreases.

on d, as shown in Figure 15, the dependence on t is piecewise linear. This simple brushing analysis, which was used in the real-life analysis of these data, helped Cleveland, McGill, and McGill [June 1988] to identify and fit a simple model that explains the dependence exceedingly well. In this model, *error* depends linearly on d for fixed t; for fixed d, *error* has a dependence on t that is linear for t in the range 60 to 100 and is constant for t in the range 50 to 60.

Using rotation, it is difficult to see the above conditional dependence identified by brushing, even when the structure is known. The reader can get some idea of the level of difficulty by trying to see the pattern in Figure 16, which is a stereogram of the data.

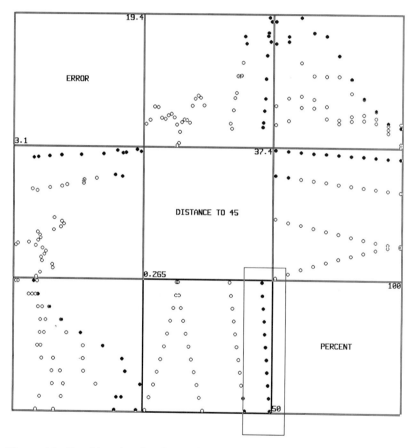

Figure 15. Brushing for Conditional Dependence. Now the brush is conditioning on high values of distance. The (1,3) panel shows how error depends on percent for distance held fixed to high values. The dependence is piecewise linear. As a result of this analysis, a piecewise bilinear model was fit to the data.

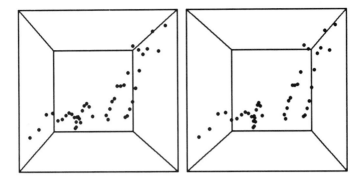

Figure 16. Rotation is Not for Conditional Dependence. It is difficult to carry out a conditional analysis using rotation. This is illustrated by the sterogram of this figure which shows the error, percent, and distance data analyzed in Figures 11 to 15.

4.4 Four or More Variables

Unfortunately, the conclusions about rotation and brushing for the three scenarios in Section 4.1 to 4.3 do not generalize particularly well to four and higher dimensions. The primary reason is that the important property of rotation—the single integrated view of the full multivariate structure—no longer holds. The result is that rotation loses strength relative to brushing. (Of course, as the number of variables or observations increases, visually resolving the individual plotting symbols of the scatterplot matrix becomes a problem.)

The primary reason why brushing gains ground is that the object on which the brush operates—the scatterplot matrix—provides a single integrated view of all the variables. For example, we can perform a simple visual scan along a row of the matrix to see one variable plotted against each of the other variables. We cannot achieve this simple highintegration with rotation even with several clouds rotating simultaneously on the screen.

This tight visual binding of the scatterplots in brushing is important for a variety of data analytic tasks. Consider, as an example, Figure 17. The data are the logarithms of the weights of six body organs for 73 cardiomyopathic hamsters with an inherited heart disease [Ottenweller, Tapp, and Natelson, 1986]. The scatterplot matrix shows readily that five of the organs are pairwise correlated but the sixth, testes, is not correlated with any of the others. There is an outlier in the (3,4) panel. Does the hamster have an unusually large spleen or an unusually small liver? Brushing quickly gives the answer. In Figure 18 the brush in the (3,4) panel is being used to highlight the hamster's measurements on all panels. By scanning the spleen plots in row 4 and then the liver plots in row 3 we can see that the hamster has an enlarged spleen.

We do not mean to imply that rotation is not a useful tool for four and higher dimensions. Rather, our conclusion is that for four or more dimensions we are much more inclined to also invoke brushing for assessing general dependence than we are for three. Actually, the reverse is also true. We would be more inclined to invoke rotation to assess conditional dependence. Part of the reason for this is that even when the

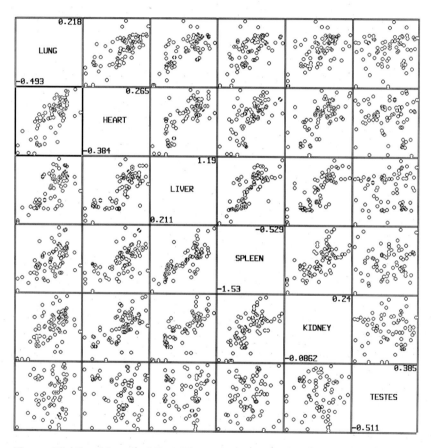

Figure 17. Brushing for More Than Three Variables. Peering into four or more dimensions is a difficult task. One strength of brushing is the integrated view of the data provided by the scatterplot matrix; this view, together with various brush techniques, provides a solution to a number, albeit not all, data exploration tasks. This is illustrated by this figure and Figure 18, which show weights of body organs of 73 hamsters. In the (3,4) panel there is an outlier. Did the hamster have an enlarged spleen or a small liver? See the next figure for the answer.

stated goal is conditional dependence of one variable on the other, our understanding of the dependence can be enhanced by understanding the multivariate structure of the $p-1$ dimensional space of the independent variables, the design space of the study [Chambers et al., 1983].

The usefulness of a combined analysis using brushing and rotation is illustrated by the data in Figure 19, which are 111 daily measurements of ozone, temperature, wind speed, and solar radiation at various sites in the New York City metropolitan region [Bruntz et al., 1974]. The goal of the analysis is to see how ozone depends on the other variables. Solar radiation is certainly an explanatory variable since it is a requisite for the photochemical reactions that produce ozone. The (1,3) panel in Figure 19 shows that ozone and temperature are correlated, but photochemical theory suggests that temperature itself is not responsible for greater ozone generation since the rates of the reactions change negligibly over the

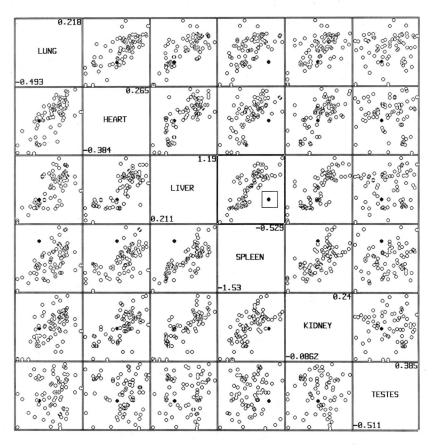

Figure 18. Brushing for More Than Three Variables. The brush is highlighting the outlier and now we can see that the hamster had an enlarged spleen.

range of temperatures in the data set. However, temperature is correlated with the meteorological stagnation that can lead to high ozone; when stagnation is high, temperature tends to be high. For example, in the (3,4) panel of Figure 19 we can see that temperature tends to be high when wind speed is low.

Is temperature a necessary surrogate for stagnation in this data set, or is there little explanatory power beyond solar radiation and wind speed? Figure 20 helps us to see. We have conditioned on both wind

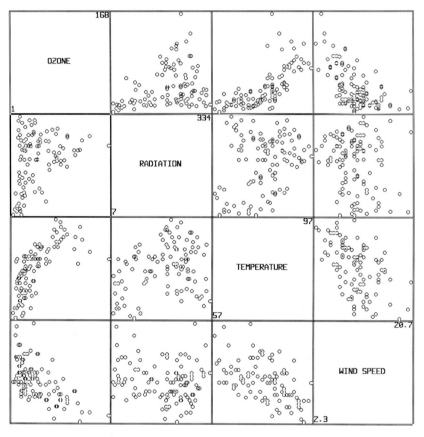

Figure 19. Brushing and Rotation for More Than Three Variables. Often, rotation and brushing will both be useful when there are more than three variables. This happened for these data where the goal was to see how ozone depended on the other variables. In fact, one can combine brushing and rotation by having the result of brushing a scatterplot matrix appear on a rotating point cloud, or by stopping the point cloud and brushing it, with the result also appearing on the matrix.

speed and solar radiation in the (2,4) panel. Panel (1,3) shows a relationship between temperature and ozone, which suggests that temperature is needed. The (1,2) and (1,4) panels show no relationship; this assures us that the temperature variable provides explanatory power beyond what is provided by wind speed and solar radiation. When we move the brush around the (2,4) panel a similar temperature-ozone relationship always appears, but when the brush is in a region of low solar radiation the temperature-ozone slope tends to decrease, which suggests a temperature-radiation interaction. Other insights into the ozone-meteorological data can be gained by conditioning on other pairs of meteorological variables.

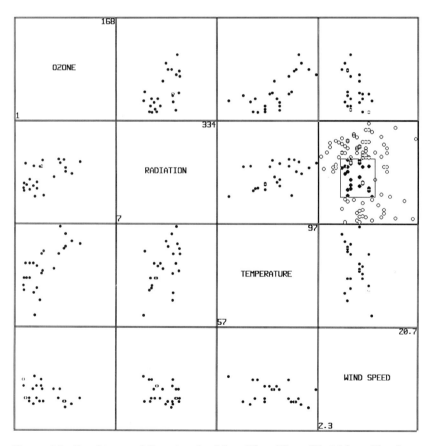

Figure 20. Brushing and Rotation for More Than Three Variables. The data of Figure 19 are being analyzed by conditioning on two variables. The (1,4) panel shows how ozone depends on wind speed for radiation and temperature held fixed to certain ranges.

It is also useful to study the three-dimensional point cloud of the measurements of the independent variables by rotation. In showing us where the measurements of the independent variables lie, rotation lets us see the portion of the three-dimensional space of independent variables for which the conclusions of the analysis are being drawn. Figure 21 is a stereogram showing one 3-D view of the point cloud.

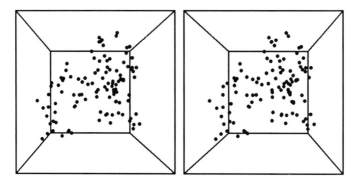

Figure 21. Brushing and Rotation for More Than Three Variables. Rotation is helpful for understanding the data of Figure 19. Using it to study radiation, temperature, and wind speed shows the configuration of the design space of this regression study. This is illustrated by the stereogram of these three variables in this figure.

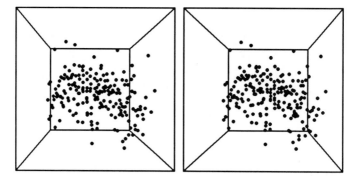

Figure 22. Smoothing. When the goal is to study the dependence of one variable on the others, it is often helpful to smooth the dependent variable and then study the data with brushing and rotation. This can help the signal show through the noise. This figure shows raw data. Figure 9 shows the smoothed data. Here, it is hard to see the pattern revealed by Figure 9.

4.5 Simultaneous Brushing and Rotation

The next logical step, after giving the argument that brushing and rotation can be useful separately for a particular set of data, is to combine the two methods. We have done this in our implementation. The results of brushing the scatterplot matrix can also appear on a rotating point cloud, and a rotating point cloud can have its motion stopped for a brushing whose result also appears on the scatterplot matrix. Stuetzle [1987] has also implemented such a feature. More testing is needed to determine if this combined brushing and rotation will offer a major benefit over the separate use of both tools.

4.6 Smoothing

One theme emphasized by Cleveland and Devlin [June 1988] is that when the goal is to study the dependence of one variable on the others, it often useful, in addition to graphing the raw data, to graph the data in the same way with the values of the dependent variable replaced by a smoothing of the dependent variable as a function of the independent variables using locally-weighted regression. This can greatly facilitate seeing the signal when it is embedded in a moderate amount of noise. For this reason the stereogram of the coal ash data in Figure 9 used smoothed values. The difficulty in seeing the signal when there is no smoothing is shown in Figure 22, a stereogram from the same perspective using the raw data.

5. SUMMARY

Brushing and rotation are two important dynamic graphical methods that aid the data analyst in understanding multidimensional data. We can expect that these methods will be ubiquitous once the requisite hardware and software are available at low cost. In fact, implementation on low-cost hardware has already begun [Donoho, Donoho, and Gasko, 1986; Hulting, 1987; Tierney, 1986; Velleman and Velleman, 1985].

Brushing and rotation have been investigated by implementing them on hardware and with algorithms that allow high performance and a range of important features and using them to study a variety of data sets. The following features of brushing appeared to us to be helpful in the data analyses: (1) A rectangular brush whose size and shape can be easily changed. (2) Instantaneous updating of highlighting on all panels when the rectangle is moved by moving a mouse; (3) Three paint modes:

transient, lasting, and undo; (4) Three brush operations: highlight, label, and delete. The following features of rotation appeared to us to be helpful in the data analyses: (1) Stereo; (2) Monocular perspective; (3) Dynamic control of scaling and symbol size; (4) Rotation about any data axis or any screen axis; (5) Pushing; (6) Rocking.

Brushing and rotation for analyzing measurements of three variables have been explored in three data-analytic settings. (1) Suppose the goal is to examine the trivariate structure. Rotation is particularly useful here for exploring three-dimensional properties such as clusters and collections of points that lie close to one-dimensional and two-dimensional manifolds. Brushing tends to be less useful here, except in special cases. (2) Suppose the goal is to explore the local dependence of one variable as a function of position in the plane defined by the other two variables. Again, rotation is a useful tool and brushing tends to be of less help. (3) Suppose the goal is to explore the conditional dependence of one variable on the other two. Here brushing is particularly appropriate while rotation tends to perform less well.

Once the data are in four or more dimensions the picture becomes less clear. Brushing gains ground on rotation because the scatterplot matrix provides a single integrated view of the data. Even when several rotating clouds are on the screen, the same level of integration is not achieved. Nevertheless, because the data analyst peering into spaces of dimension four or higher is like the blind men attempting to understand the elephant, it is sensible to approach data in these spaces with both tools in hand.

When we study how one variable depends on the others, it is often helpful to smooth the dependent variable as a function of the others and also probe the data with the raw values replaced by the smoothed values.

Combining rotation and brushing by adding a 3-D point cloud to the list of plots that are brushed might achieve still further insight in many applications; more study is needed.

REFERENCES

Andrews, D. F. and Herzberg, A. M. (1985). *Data*. New York: Springer-Verlag.

Becker, R. A. and Cleveland, W. S. (1984). "Brushing a scatterplot matrix: high interaction graphical methods for analyzing multidimensional data," Technical Memorandum. Murray Hill, NJ: AT&T Bell Laboratories.

Becker, R. A. and Cleveland, W. S. (1987). "Brushing scatterplots," *Technometrics*, **29**: 127-142.

Becker, R. A., Cleveland, W. S., and Weil, G. (1987). "An interactive system for multivariate data display," Technical Memorandum. Murray Hill, NJ: AT&T Bell Laboratories.

Becker, R. A., Cleveland, W. S., and Wilks, A. R. (1987). "Dynamic graphics for data analysis," *Statistical Science*, **2**: 355-395.

Bruntz, S. M., Cleveland, W. S., Kleiner, B., and Warner, J. L. (1974). "The dependence of ambient ozone on solar radiation, wind, temperature, and mixing height," *Symposium on Atmospheric Diffusion and Air Pollution*, 125-128. Boston: American Meteorological Society.

Buja, A., Asimov, D., Hurley, C., and McDonald, J. A. (1988). "Elements of a viewing pipeline for data analysis," *Dynamic Graphics for Statistics*, W. S. Cleveland and M. E. McGill, eds., 277-308. Monterey, CA: Wadsworth & Brooks/Cole.

Chambers, J. M., Cleveland, W. S., Kleiner, B., and Tukey, P. A. (1983). *Graphical Methods for Data Analysis*. Monterey, CA: Wadsworth.

Cleveland, W. S. and Devlin, S. J. (June 1988). "Locally-weighted regression: an approach to regression analysis by local fitting," *Journal of the American Statistical Association*, **83**.

Cleveland, W. S., McGill, M. E., and McGill, R. (June 1988). "The shape parameter of a two-variable graph," *Journal of the American Statistical Association*, **83**.

Cressie, N. (1986). "Kriging nonstationary data." *Journal of the American Statistical Association*, **81**: 625-634.

Donoho, A. W., Donoho, D. L., and Gasko, M. (1986). *MACSPIN Graphical Data Analysis Software*, Austin, Texas: D^2 Software, Inc.

Donoho, D. L., Huber, P. J., Ramos, E., and Thoma, H. M. (1986). "The man-machine-graphics interface for statistical data analysis," *Statistical Image Processing and Graphics*, E. J. Wegman and D. J. DePriest, eds., 203-213. New York: Marcel Dekker.

Fisherkeller, M. A., Friedman, J. H., and Tukey, J. W. (1975). "PRIM9: an interactive multidimensional data display and analysis system," *Data: Its Use, Organization, and Management*, 140-145. New York: The Association for Computing Machinery.

Huber, P. J. (1987). "Experiences with three-dimensional scatterplots." *Journal of the American Statistical Association*, **82**: 448-453.

Hulting F. (1987). Unpublished software for the IBM AT, Ames, IA: Iowa State University, Department of Statistics.

Kaufman, L. (1974). *Sight and Mind*. New York: Oxford University Press.

McDonald, J. A. (1982). "Interactive graphics for data analysis," Technical Report Orion 11. Stanford, CA: Stanford University, Department of Statistics.

Ottenweller, J. E. Tapp, W. N., and Natelson, B. H. (1986). "Biological aging in the cardiovascular system of syrian hamsters," Department of Neurosciences Technical Report. Newark, NJ: VA Medical Center, New Jersey Medical School.

Reaven, G. M. and Miller, R. G. (1979). "An attempt to define the nature of chemical diabetes using a multidimensional analysis," *Diabetologia*, **16**: 17-24.

Stuetzle, W. (1987). "Plot windows, *Journal of the American Statistical Association*, **82**: 466-475.

Tierney, L. (1986). Unpublished software for the Apple MacIntosh, Minneapolis, MN: University of Minnesota, School of Statistics.

Tukey, J. W. (1988). "Control and stash philosophy for two-handed, flexible, and intermediate control of a graphic display," *The Collected Works of John W. Tukey, Volume V, Graphics*, W. S. Cleveland, ed., 329-382. Pacific Grove, CA: Wadsworth & Brooks/Cole Advanced Books & Software.

Velleman, P. F. and Velleman, A. Y. (1985). *Data Desk Handbook*. Ithaca, NY: Data Description, Inc.

(a)
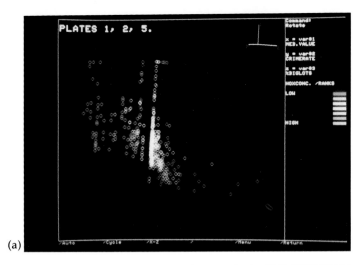

(b)
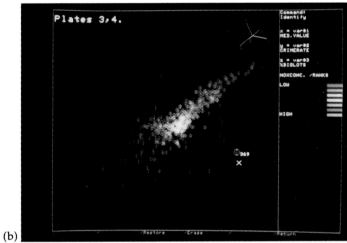

(c)
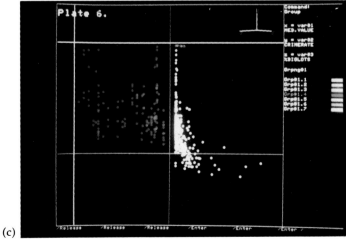

Plate 1

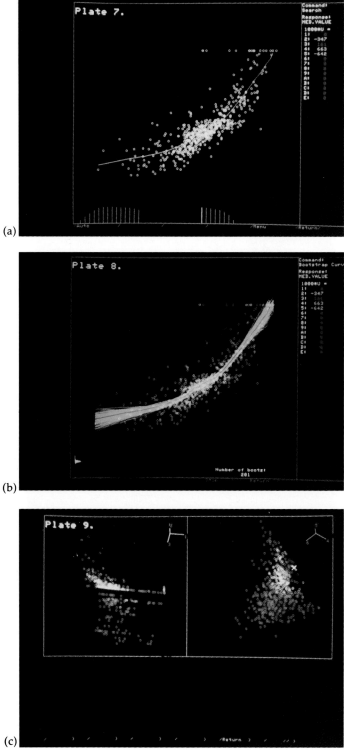

Plate 2

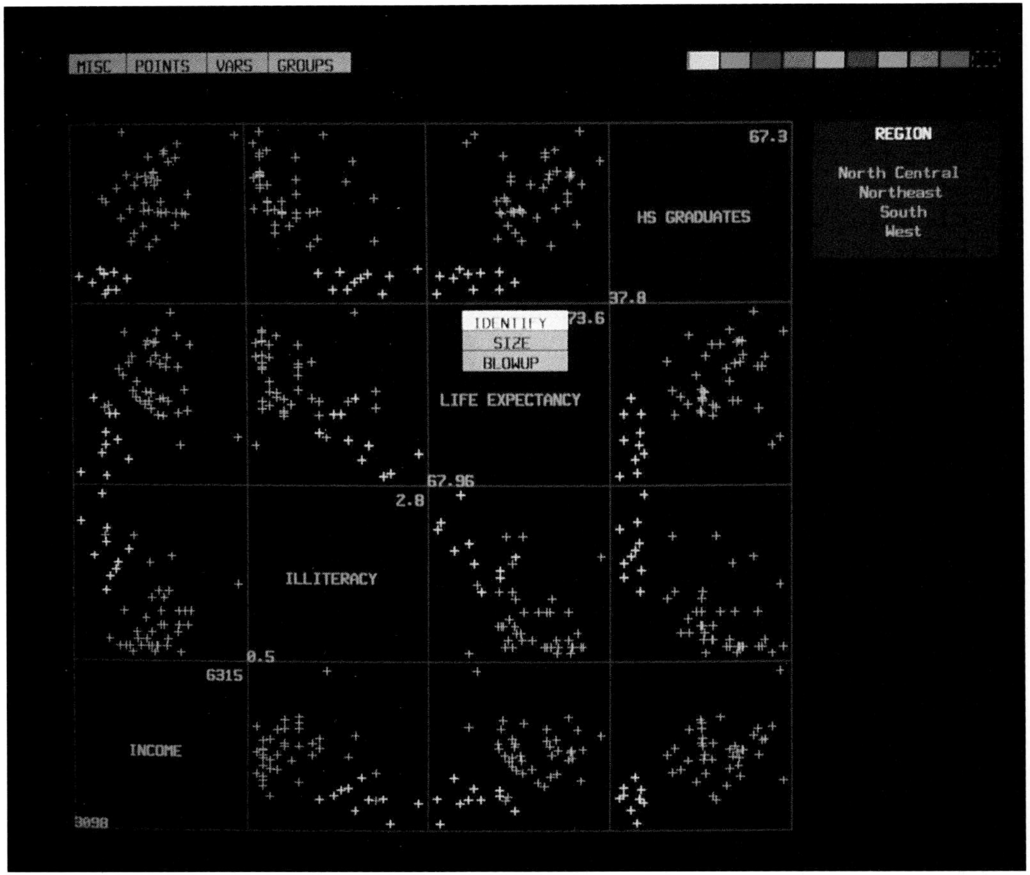

Plate 3. The layout of the screen for brushing a scatterplot matrix. In the upper left corner, a set of pull-down menu items provide high-level control over the operations of the system. The main portion of the screen is taken up by a 4 by 4 matrix of scatter plots, giving all pair-wise combinations of 4 variables. Over the scatterplot matrix is a pop-up menu. The upper-right of the screen contains a set of paint pots, used to provide color to the brush. On the right is a scrolling menu that displays identifiers for the data points, in this case, the regions of the country. The data are four demographic measurements of U.S. states.

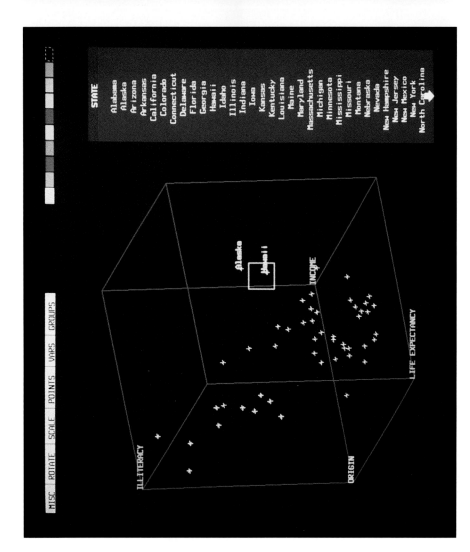

Plate 4. A rotating point cloud display. This figure is much like Plate 3, but the scatterplot matrix has been replaced by a rotating point cloud. The pull-down menus in the upper-left of the screen are somewhat different from those in Plate 3. Notice, too, that the brush is present on the point cloud and is identifying certaining points, and that the scrolling menu now contains the state names.

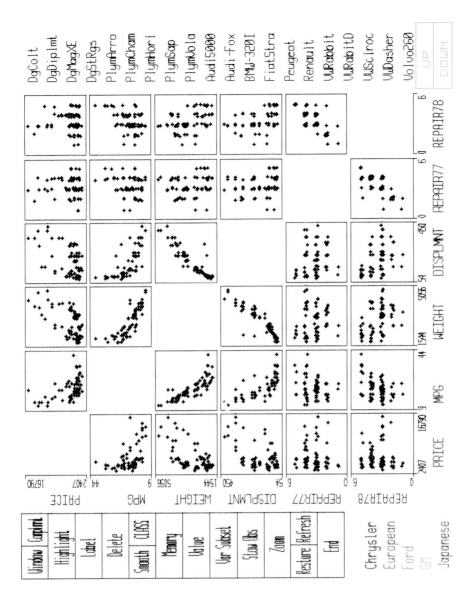

Plate 5. A typical initial display.

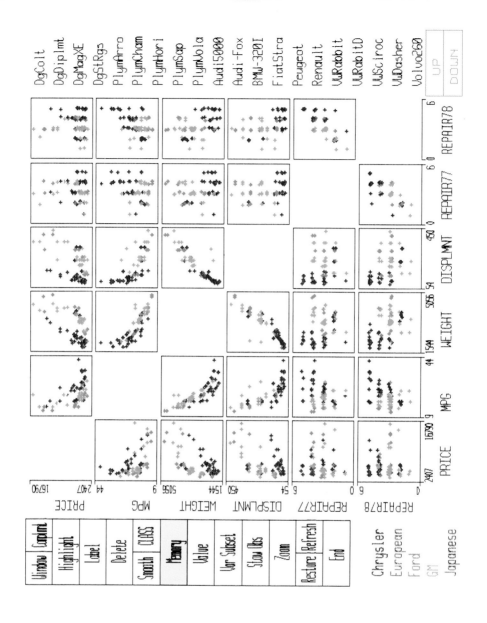

Plate 6. All makes highlighted. Comparing foreign makes (reddish colors) with domestics (bluish colors).

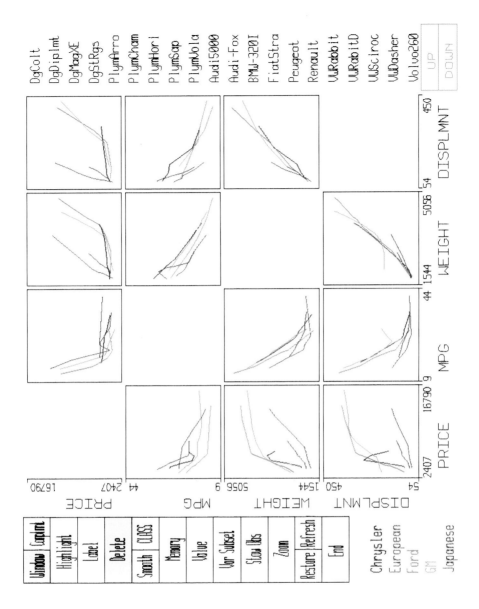

Plate 7. Relationships between variables for each class.

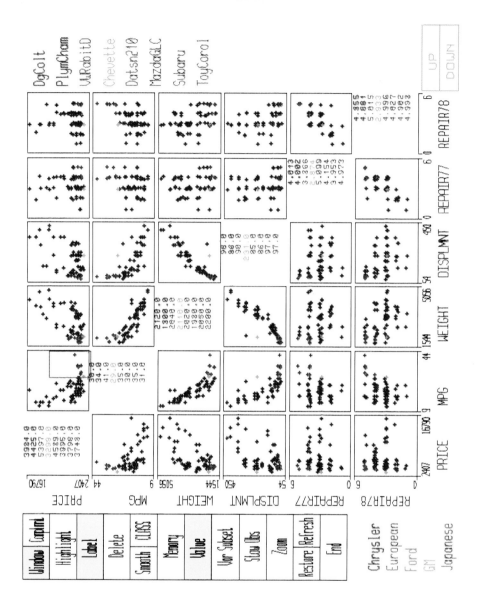

Plate 8. Identifying low price, high mileage cars.

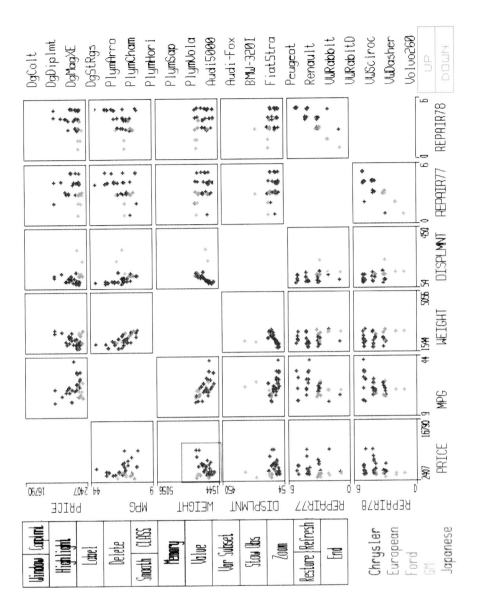

Plate 9. Small cars.

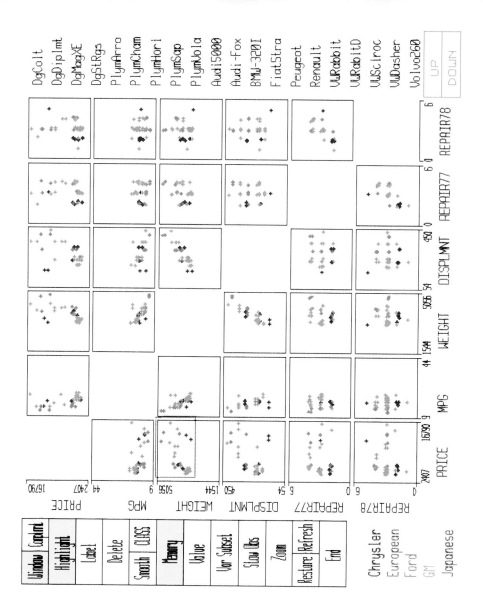

Plate 10. Large cars.

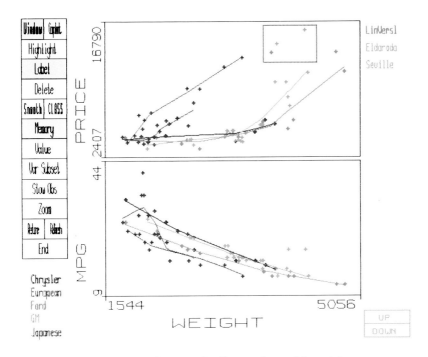

Plate 11, top. Price and mileage plots with weight.

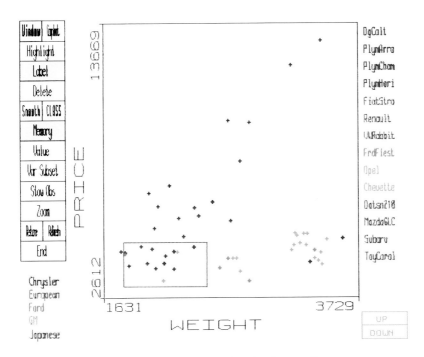

Plate 11, bottom. Price-weight plot for smaller cars.

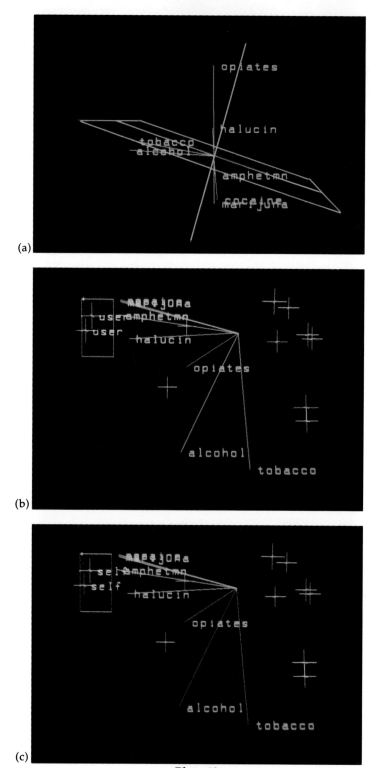

Plate 12

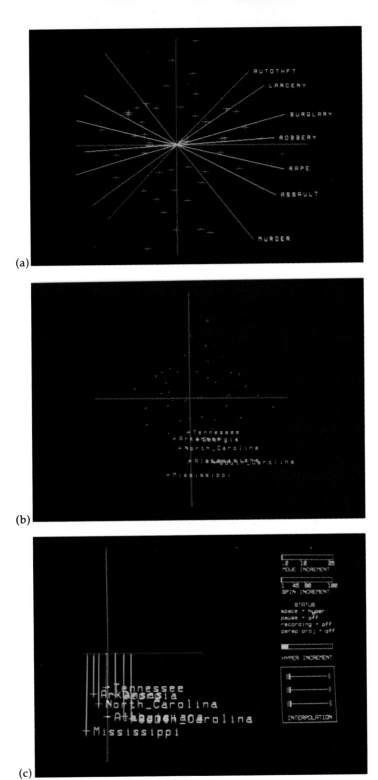

Plate 13

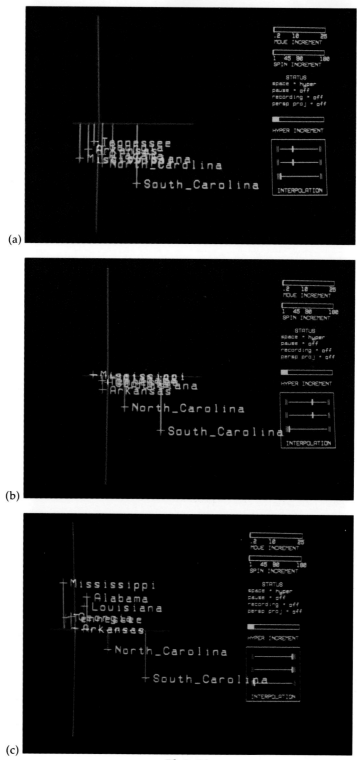

Plate 14

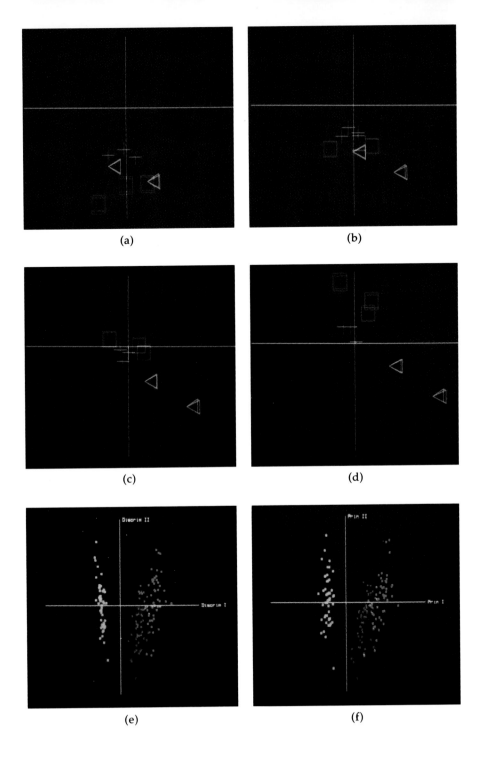

Plate 15

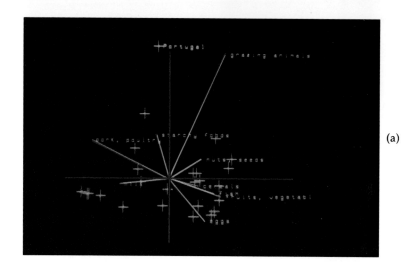

(a)

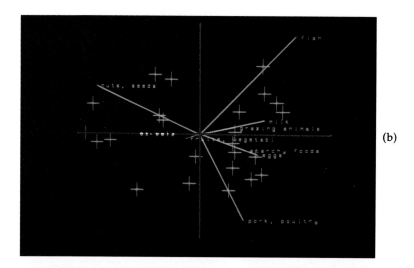

(b)

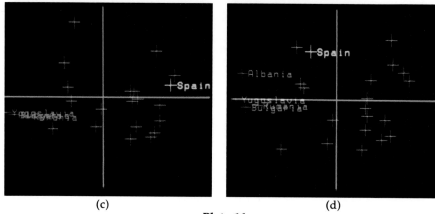

(c) (d)

Plate 16

11

ELEMENTS OF A VIEWING PIPELINE FOR DATA ANALYSIS

Andreas Buja
Bell Communications Research
University of Washington

Daniel Asimov
University of California

Catherine Hurley
University of Waterloo

John A. McDonald
University of Washington

ABSTRACT

We present some new data analytic techniques based on real-time graphics. Traditionally, these techniques have been confined to rotations of three dimensional point clouds, masking, and identification of points. We will be concerned with 1) dynamically scaled and shifted views of data, 2) dynamic projections from arbitrary dimensional data space, and 3) informal graphical inference based on the same randomization schemes as in permutation tests. These operations are most naturally implemented in building blocks which can be organized in what we call a "viewing pipeline".

This research was partially supported by grants DOE, DE-FG06-85-ER25006, and ONR, N00014-81-k-0095 awarded to the University of Washington.

From *Dynamic Graphics for Statistics*, William S. Cleveland and Marylyn E. McGill, eds., copyright © 1988 by Wadsworth, Inc. All rights reserved.

1. INTRODUCTION

In recent years, interest in graphing data has grown under the influence of a few excellent texts on the subject: Bertin [1983], Cleveland [1985], Chambers et al. [1983], Tufte [1983]. Our awareness of the power of good graphics has also increased with the availability of computer graphics hardware and computing environments which are conducive to experimentation. We see efforts to bring complex graphical manipulations closer to the finger tips of users by replacing keyboard interaction with graphical controls, granting a response to an input within fractions of a second. At the extreme end, we find ourselves manipulating plots which change so fast that they appear in motion for all practical purposes. This is the domain of *real-time graphics*: plots are recomputed and redrawn so rapidly that the visual effect of smooth motion is achieved, and at the same time the user is given the possibility of controlling the process at any point in time. This contrasts with *animation*, where sequences of views are precomputed, stored away, and retrieved at the time of viewing. Motion graphics can be generated either way, but non-trivial user control is possible only with real-time graphics. The price we pay is that currently affordable off-the-shelf equipment can handle a real-time approach only on fairly sparse pictures, such as plots of point scatters. Just the same, exploring the present possibilities can be a fascinating endeavor, and this note tries to convey a sense of the usefulness of these techniques.

The best known applications of real-time capabilities are smooth rotations of three dimensional point scatters. They have recently been incorporated in programs running on very small personal computers [Donoho, Donoho, and Gasko, 1986; Parker, 1985; Young, Kent, and Kuhfield, 1986], after over a decade of research on special purpose hardware [Fisherkeller, Friedman, and Tukey, 1975; Stuetzle and Thoma, 1978, Donoho et al., 1982, Friedman et al., 1982, McDonald, 1982]. In another powerful application of fast manipulations of scatter plots, the points are not moved, but the point quality is changed in real-time to interactively highlight subsets of the data across two or more plots by means of color or glyphs; this technique is called "linked scatter plots" [McDonald, 1982] or "brushing paradigm" [Becker and Cleveland, 1984]. We have chosen to concentrate on motion graphics in the present paper.

There are at least two strong reasons why motion graphics is superior to static plots for many cognitive tasks:

— Smooth transitions between images allow us to recognize objects as the same across changing views, i.e., objects retain their identity in our perception; whereas exposure to disconnected, still images presents the task of identifying objects across pictures. This is especially frustrating in a world where all objects look alike, such as in scientific plots featuring mere dots and lines.

— The speed of moving points can convey information in addition to location. Clearly, visual perception is stronger at dealing with location, but speed can still help in finding particular types of structure, such as clusters and exotic points. In projection based motion, one can explicitly state what the speed represents, namely, one or two linear combinations of data variables, depending on whether the motion is 3d-rotation or not.

As real-time manipulation of plots becomes more widespread, the sophistication of users has to grow, as well as the awareness of new possibilities on the side of implementers. This is where we would like to contribute with the present note, which is also a partial documentation of an experimental system, implemented on a workstation[1] by the authors. Our main point is to show how a good number of graphical parameters can be controlled in real-time to one's advantage, and also how certain statistical inferential techniques have graphical analogs, which call for real-time control.

We proceed by outlining elements of what we call a "viewing pipeline", i.e., a series of operations which transform raw data into a scatter plot. The subsequent discussion then describes useful real time controls on these operations or pipeline elements, as well as examples of applications to real and artificial data. The proposed outline follows our currently implemented "Data Viewer" system, but we hasten to state that there is nothing canonical about it, and the operations a data analyst might find useful depends on the data at hand as well as his/her taste and style.

2. COMPOSING A VIEWING PIPELINE

Users of scatter plots are usually aware that some affine transformation of the data has to take place to establish the correspondence between the data scale and the screen scale. This is the only indispensable element of any viewing pipeline for scatter plots; all other elements described below are optional. To obtain a basic scatter plot of two variables from a

---

[1] The workstation we chose is a Symbolics Lisp machine which provides a so-called "integrated programming environment", besides the usual high-resolution bitmapped display and keyboard/mouse for interaction. We found this to be a very advantageous decision because of the support of experimental programming and advanced programming paradigms. In this paper, we do not cover in detail capabilities of our implementation which relate to multiple windows, persistent virtual memory, and large single address spaces which are at the heart of such an environment. For these issues, see the paper by McDonald and Pedersen [1986] which examines the impact of programming environments on data analysis. For some applications, see Stuetzle [1986] on linked windows and Buja, Hurley, and McDonald [1987] on view sharing.

multivariate data set, we need no more than a capability for selecting two variables, and specifying the so-called window-to-viewport transformation (or briefly: viewporting) which affinely maps a rectangle in the two-variable plane to a rectangle in the screen plane [Foley and van Dam, 1982, p. 153ff]. A more detailed discussion of viewporting and its controls follows in Section 3.

Variable selection can be interpreted as a trivial case of projection from the full (say, p-dimensional) data set down to a plane. More general projections are needed as soon as some sort of point cloud rotations in spaces spanned by more than two variables are involved. For the time being, we propose that a general projection mechanism from p-space down to a plane, be accepted as an element of the pipeline, and we defer the discussion of how to control it to Section 4. Although we are interested in 3d-rotations, this is just one particular example of dynamic projections. Our view leads us to a presentation which differs greatly from the one found in computer science texts on 3d-graphics [Foley and van Dam, 1982; Newman and Sproull, 1979]. It is broader in that we consider parallel projections in higher than three dimensional spaces, but narrower in that we do not discuss central projection or perspective, a topic which is relevant if the goal is rendering of 3d-objects. Our projection engine has three different ways of generating dynamic projections, besides static variable plots, and 3d-rotations are just specializations of each (see Section 4).

Allowing parallel projections leads us to a whole set of problems which have to do with orthogonality: projections should be orthogonal with regard to some reasonable metric in p-dimensional data space. In most (but not all) cases, one would like to declare the original raw variables to be orthogonal; then the trivial projections obtained by selecting a pair of variables can be interpreted as orthogonal. In order to specify a scalar product and a notion of orthogonality for arbitrary pairs of vectors, it remains to standardize the variables. The problem is well-known from multivariate analysis, where a similar choice has to be made for principal components analysis. In this situation, the variables are often standardized to zero mean and unit variance, or equivalently, the correlation rather than the covariance matrix is used. This is seldom the scaling of choice for the purpose of graphics, so more desirable options should be included. Some useful choices are: centering at a) a prechosen value, b) the mean, c) the median, d) the mid range, e) no centering, and normalizing with a) a prechosen value, b) the standard deviation, c) the median absolute deviation (MAD), d) the half range, and e) no normalization. Also, one would like to be able to standardize whole sets of variables with common location and scale parameters, in cases where the values of several variables lie on a common scale. Since there are only discrete choices involved, we are not going to say more on this topic, although useful real-time controls for centering and scaling parameters could well be implemented.

Before standardizing the variables, we should be concerned with non-linear variable transformations, which are an indispensable part of any data analyst's tool kit. Interestingly enough, the first real-time graphics system in statistics we are aware of, offered real-time parameter control for non-linear transformations: in E. B. Fowlkes' [1969] system for probability plotting, the parameters c and p of a power transformation $(y+c)^p$ could be manipulated using a knob. Fowlkes' system received much less publicity than PRIM-9 [Fisherkeller, Friedman, and Tukey, 1975], although it predates the latter by several years (for real-time power transformations, see also Becker, Cleveland, and Wilks [1987]). Since we do not have real-time controls for non-linear transformations to offer at present, we are not going to expand on this topic here.

Following these more classical viewing transformations, we introduce a few operations which have to do with statistical inference. An objection made sometimes against extensive data viewing capabilities is that they necessarily lead to the discovery of spurious structure. We disagree with this point of view and, to the opposite, find our present tool kit not powerful enough yet for some purposes. Nevertheless it is instructive to compare actual data with artificial data which are pseudo samples drawn from an assumed model. Flashing through a few dozen plots of artificial data sets can provide striking evidence in favor of (as well as against) a hypothesis. In Section 5, we give examples of hypotheses which can be tested informally using cheap randomization schemes. They can be implemented as a pipeline element with real-time control.

There are several other useful data manipulations which are amenable to real-time control, of which we mention just one without going into further detail: J. A. McDonald's [1986] wrap-around or cyclic plots for finding periodicities in time dependent data; he plots a time series x_i against $(\omega \cdot i) \bmod 2\pi$, with real-time control for the parameter ω.

Our Data Viewer system has a number of additional capabilities such as labeling, identification, brushing, reshapable windows, multiple windows, window management, linked windows, view sharing, and more. However, we would like to focus on

— viewporting,
— dynamic projections,
— informal inference by randomization,

and refer the reader to Becker and Cleveland [1984], Becker, Cleveland, and Wilks [1987], Stuetzle [1986] and Buja, Hurley, McDonald [1986] for some of the other ideas. A flowchart at the end of this paper gives a graphical depiction of the viewing pipeline according to our design.

In the subsequent sections we refer to a good number of plots generated with our Data Viewer system. Therefore it seems appropriate to add a few remarks on the layout of our Data Viewer windows. See Figure 1 for a first example. All windows show some of the following components:

- an upper title for the current major operation;
- a lower title for the current data set;
- a point cloud;
- a set of boxes on the left, each representing one data variable indicated by a variable name, and containing 1) an optional character for the current use of the variable such as "X" or "Y", and 2) an optional vector which is vertical or horizontal if the variable is plotted vertically or horizontally; for a more complete description of these "variable boxes", see Section 4;
- depending on the operation shown in the upper title, the variable boxes may be replaced by a static menu, such as in Figure 2;
- for dynamic projections, there is also a speed bar visible to the right of the variable boxes (Figure 16); the position of the small horizontal dash inside the speed bar is a monotone function of the current speed of the dynamic projection; see Section 4 for more details;

The last three components are the essential tools for interaction: variables are activated and deactivated by sending appropriate mouse clicks to the variable boxes; specialized menu controls are provided for operations like viewporting (Figure 2), informal testing (Figure 38), and brushing; finally, the speed bar is used to control the speed of the dynamic projections via mouse input. Major operations such as viewporting, dynamic projections, informal testing, brushing, data selection, and window manipulations can be chosen by popping up a temporary menu and selecting from the list of mouse sensitive items.

3. VIEWPORT TRANSFORMATIONS

In most graphics software packages, the user is relieved from the task of deciding about the proper "scaling" of the data. What is really involved is the decision on which rectangle in the data plane spanned by the x and y variables should be mapped onto the screen rectangle assigned to the plot. This amounts to specifying an affine transformation from the data scale to the screen scale.

In addition to performing the actual affine mapping of the data, clipping should be done at the boundaries of the assigned rectangle; otherwise, one is confined to plots which always show all the data. In our experience, this restriction proves unbearable: partial blow-ups of plots are often needed in common data analytic situations (see below). Clipping, in the case of point scatters, amounts to a quadruple of trivial checks (one for each boundary: left - right - top - bottom) for each data point. More elaborate computations are required if the plot is to show lines and curves (i.e., connected sequences of short straight lines), but to do this in real-time requires special purpose hardware. Therefore, we assume that all we wish to move in real-time is a point scatter.

There are two ways of expressing viewport operations: either in terms of the viewport, or in terms of the data. Expanding the area covered by the plotted data within a fixed viewport can be achieved by either shrinking the data rectangle mapped onto the viewport, or by expanding the data while maintaining the data rectangle at a fixed size. The latter formulation seems more intuitive for a viewer and should be used in documentation and human interfaces, although the implementation is usually done in terms of viewport manipulations. Our own terminology follows this rule, and viewporting is described according to the apparent effects on the plotted data.

The controls we propose are the following:

1) Simultaneous expanding and shrinking of the data, both horizontally and vertically, with the same factor.
2) Inverse expanding and shrinking of the data, i.e., while the data are shrunk vertically by a factor, they are expanded by the same factor horizontally, and vice versa.
3) Horizontal expanding and shrinking, leaving the vertical axis fixed.
4) Vertical expanding and shrinking, leaving the horizontal axis fixed.
5) Shifting the data horizontally.
6) Shifting the data vertically.
7) Simultaneous horizontal and vertical shifting.

All of these operations involve control of only one parameter each and are computationally very inexpensive. They lend themselves to real-time motion with striking effects, some of which will be mentioned below. In our implementation, motion progresses as long as one of two mouse buttons is depressed while the cursor is located on an item of a temporary menu (see Figures 2 and 3). For example, the top menu item triggers simultaneous shrinking 1) as long as one button is held down, and simultaneous expanding while the other button is depressed. The six top items are used for continuous motion; only the two bottom items show different behavior: they trigger discrete actions on a mouse click.

The inverse expanding and shrinking 2) may appear as the least self evident and most redundant as it may be composed of a sequence of horizontal and vertical scalings 3) and 4). However, in our experience inverse scaling is one of the most often used operations, second only to the obvious simultaneous scaling 1). It is especially convenient for scaling time series to quickly obtain a reasonable aspect ratio. Quite often, time series are very long and cannot be reasonably shown in one plot if a good aspect ratio for discerning the periodicities is required. A natural sequence of steps would be to first expand the time axis in search of a good aspect ratio, regardless of how much of the data is clipped at the right and left boundaries. In a second step, one could then simultaneously scale the horizontal and vertical axis 1) so as not to disturb the aspect ratio any more, while trying to find a better tradeoff between the conflicting goals of showing local details as well as global features. We found, however, that inverse scaling 2) could achieve the desired goal quite often in one single step. A satisfactory scaling can often be found at the cost of clipping a portion of the data at the left and right hand boundaries. Horizontal shifting 5) can then be used to scroll the visible part of the data back and forth on the time axis. (For this, one of two mouse buttons is depressed on the fifth item of the menu, see Figures 2 and 3.)

An additional degree of control can be gained by allowing the viewer to prescribe the center in the viewport at which the shrinking and expanding should take place [Donoho, Donoho, and Gasko, 1986]. On the other hand, if some shifting capability like 5), 6), and 7) is provided, a default center in the middle of the viewport is sufficient in principle.

One way of controlling shifts which we have seen implemented, consists of specifying a point on the screen which becomes the new screen center after the shift [Collier, 1986]. This was done in a context where drawing the plots was computationally too expensive to allow smooth motion, and a discontinuous shift could not be avoided. In our own system, we have chosen to implement simultaneous horizontal and vertical shifting as dragging of the scatter to the desired place, which in some sense is the opposite of prescribing a new center. Remarkably, the MacPaint [1983] program offers this "drag the object" capability as well as a dual which may be called "drag the viewport", using a helicopter view of the viewport as well as the data plane. Whatever controls for 7) one prefers, simultaneous shifting by itself does not make separate horizontal 5) and vertical 6) shifts redundant. In applications like viewing a long time series, where only parts of the data can be shown in a reasonable scaling, we do need strictly horizontal and continuous shifting at a constant speed to examine the series as it flows by from left to right.

Our list of controls is still not entirely satisfactory: it should be supplemented with some options for easy recovery from accidents like losing all of the data from the viewport due to unfortunate shifting and expanding. A means for returning to a default scaling is mandatory. In our implementation, a user clicks on the second item from the bottom to reset all scaling and shifting parameters to defaults (see Figures 2 and 3).

As to the usefulness of scaling control in practice, we would like to give a couple of illustrations. The first concerns the task of identifying points which are near-ties in a plot. In passing, we mention that any reasonable high-interaction system for viewing data should offer some tools for identifying cases in data and points in plots. What we mean is a pair of operations which are in a sense converses of each other:

— Given a point on a scatter plot, we wish to find out which case it represents; this is usually done by interactively moving a cursor near the point in question and provoking a case label to appear, e.g., by clicking the mouse.
— Conversely, given a case in the data, we wish to identify the point which represents it; this typically involves access to the data by some other means, such as a listing.

Figures 2-4 show a series of shifts and expansions of the plot of Figure 1 to unclutter the area of interest and facilitate the identification of points which originally appeared as near-ties.

The second application concerns a genuine scaling problem in plotting time series. Examples of simple structure are trends, periodicities, and exotic points. Unfortunately, there may not exist a single scaling which is guaranteed to reveal these features. To emphasize this point, we constructed a mock time series which is basically a pure sine wave with one single exotic point and a slight linear trend superimposed. Figures 5-8 show the following:

— a plot of all the data, with a scaling which lets us see the periodicities but neither the linear trend nor the the exotic point,
— a blow-up of a subset which shows the exotic point,
— an intermediate plot with the time axis shrunk,
— an unusual scaling with the time axis shrunk even further; this shows the linear trend, but neither the periodicity nor the exotic point.

A real data set which inspired us to make up this artificial series is shown in the next section. Its exotic point could be found with blow-ups, but we chose to use 3d-rotations for this purpose. Another real series is shown in Figures 9-12: the data are monthly flow measurements taken on Willamette river in Oregon. All four plots show the same data, the only difference being the scale (and clipping in Figure 12). Figures 10 and 11 serve as evidence for this claim, but the most striking evidence would be given by watching the transition from Figure 9 to 12 in smooth motion under real-time control, as can be done in our system. In Figure 12 we see a fraction of the data in a reasonable scaling for detecting the yearly cycle, but Figure 9, with the time axis shrunk in an unusual way, suggests a more interesting structure: months of peak river flow seem separated in their intensity from the rest of the year. Whether the effect is real is hard to determine with a sample size of 128, but we have here an example of generating a new hypothesis. Once the separation of the peak months is recognized in Figure 9, we are able to see it in Figure 12 as well, but it does not place the peaks in close enough proximity to suggest the hypothesis.

Cleveland, McGill, and McGill [1986] have investigated the problem of optimal scaling of time series for perception. They are concerned with the visual performance of the human eye regarding slopes. According to their evidence, the eye is at its best for recognizing periodicities when the median absolute slope of adjacent observations is 45 degrees. We can only confirm this result from our subjective experience, by observing that we most often stop rescaling a time series with a strong periodicity when the majority of absolute slopes are approximately 45 degrees. On the other hand, in spite of the beauty of this insight, we would like to point out again that one single scaling is often not enough: partial blow-ups and seemingly unreasonable scaling can prove necessary at times. The relevant slopes are not always adjacent, and in exploration it is seldom a priori known what the important slopes are. Therefore, we need tools for searching among scaled and shifted data. In our opinion, this task is best done interactively, due to the small number of parameters to control. Whether an analog to projection pursuit (which might be called "scale pursuit") is desirable and feasible remains an open question.

4. PROJECTIONS

We mentioned that the notion of projection has to be introduced more formally if one wishes to design a graphics capability which can show, e.g., rotated three dimensional point scatters by two dimensional projections. We make a few remarks on 3d-rotations, but our thrust is on manipulations of more general projections, not confined to 3-space. We introduce some useful higher dimensional techniques as generalizations of 3d-rotations.

Simple 3d-Rotations

The simplest 3d-rotations are those around a coordinate axis: given three variables x,y,z, plot z versus $\cos t \cdot x + \sin t \cdot y$, where t can be incremented in small steps over time, while the plot is erased and redrawn at each step. This should be repeated at least 5-10 times per second to obtain the illusion of smooth motion. Although we perceive the result as motion of a point cloud in space, for analytical purposes it is more fruitful to consider the projection plane as moving. The plane is spanned by the two orthonormal vectors $(0, 0, 1)$ and $(\cos t, \sin t, 0)$, which are in 1-1 correspondence with the vertical and the horizontal screen axes, respectively.

Before we proceed, we would like to show an application of this type of 3d-rotation by way of an example. We demonstrate how an innocuous outlier can be found in a time series consisting of half hourly measurements of tidal levels in a harbor in Alaska. Figure 13 shows a time series plot of these data with the outlier invisible. A search among blow-ups would allow us to find the point as indicated in the previous section, but 3d-rotations will do, too. Figures 14 and 15 show lag plots of order 1 and 2, respectively, i.e., we are plotting x_i versus x_{i-1} and x_{i-2}. The outlier is still invisible. The highly deterministic nature of the data makes for an elliptic double loop pattern (not concentric ellipses as one might guess at first sight). This derives from the presence of two principal frequencies, one being approximately double the other. The many weaker very low frequencies spread out the loop to a band, which conceals the presence of the outlier. Now, we would like to examine the 3-space spanned by the variables x_i, $y_i = x_{i-1}$, and $z_i = x_{i-2}$. Figure 16 shows a rotated projection of the data in this space: x_i is plotted versus $\cos t \cdot x_{i-1} + \sin t \cdot x_{i-2}$, for a specific value of t. The double loop appears to fall in a plane in three dimensional lag space. This implies that the data satisfy some approximate linear relation of the form $x_i \propto \cos t \cdot x_{i-1} + \sin t \cdot x_{i-2}$, i.e., an autoregressive model of order 2. In this plot, we can discern three outliers, a finding which is compatible with the presence of a single outlying value in the time series, due to its appearance in three lag vectors of the form (x_i, x_{i-1}, x_{i-2}). In effect, we replaced an autoregressive fit by a simple visual inspection in a suitably chosen 3-space.

In a second example, we show how the presence of a single product interaction can be detected in regression data. For the purposes of this demonstration we generated error free data where $z = x \cdot y$ and (x, y) are sampled from a standard normal distribution. This interaction surface contains two trivial families of straight lines:

$$\{ (x, y, z) \mid x = x_0, z = x \cdot y \}$$

for each fixed x_0, and same with exchanged roles of x and y. If we look at this surface by rotating it around the x-axis, i.e., by plotting $\cos t \cdot z + \sin t \cdot y$ versus x, we notice that the vertical coordinates

$$\cos t \cdot z + \sin t \cdot y = (\cos t \cdot x + \sin t) \cdot y$$

of the points on the surface vanish when $x = -\tan t$. This line-up along the viewing direction is visible in Figures 17-20 which show rotated views for different rotation angles. As we rotate the points on the surface, i.e., change the angle parameter t, the line-up moves left-right. The same occurs of course if we rotate around the y-axis. These remarks generalize to arbitrary surfaces of the form $z = ax + by + cxy$, and can therefore be exploited for diagnosing the presence of a product interaction in regression data: the presence of two sets of straight lines over $x = \text{const}$ and $y = \text{const}$, respectively, characterizes response surfaces of this form. We were lead to this example by some real data of W. S. Cleveland, who obtained them from an experiment in visual perception [personal communication].

Beyond 3d-Graphics

At the heart of our Data Viewer system is a set of capabilities for manipulating projections onto moving 2-planes in p-space. The technical building blocks are schemes for generating flexible 1-parameter families of 2-planes based on orthogonal transformations [Asimov, 1985; Buja and Asimov, 1986].

The fundamental control we provide is variable selection, for determining the space from which to project. This is done in a high-interaction mode by sending mouse clicks to the variable boxes which play the role of visual representations of the variables. Each box shows three elements: 1) the name of its variable, 2) a letter like "A", "X", or "Y" if the variable is part of the projection space, and 3) a short vector emanating from the center. The vector is the projection of the variable unit vector onto the current projection plane. For example, if we are looking at a simple plot of two variables, the vertical variable has a maximum length vertical vector in its box, and similarly for the horizontal variable (see, e.g., Figures 14 and 15). In case of simple 3d-rotations around a fixed vertical z-axis, the vectors in the boxes of the x- and y-variables are horizontal and proportional in length to the cosine and sine used in the current projection (see Figure 16). The vectors dynamically update themselves to reflect the position of the current projection plane. We prefer this arrangement over the more conventional p-pod, which has the projected variable unit vectors attached to one single point. The unavoidable clutter is only one reason; more importantly, we need some visual representations of the variables for high-interaction manipulations.

Another control is provided for interactive adjustment of motion speed. For this purpose, a rectangular area, the speed bar, is made mouse sensitive: whenever the cursor is within the rectangle and a mouse button is depressed, the vertical mouse position is converted to a speed value. The center area of the rectangle sets the speed to zero, while positions above and below convert to positive and negative speeds, respectively.

We distinguish three types of dynamic projections, each addressing different needs: 1) plot interpolation, 2) correlation tour, and 3) grand tour. Here are comments on each of them:

Plot Interpolation

One of the most frequent uses of simple 3d-rotations is to move back and forth between two variable plots [Andrews, 1981]; in other words, one interpolates an (x,z)-plot and a (y,z)-plot by repeated rotations of 90 degrees. We have expanded on this use by providing interpolation between pairs $(x^{(1)}, y^{(1)})$ and $(x^{(2)}, y^{(2)})$ of plots which do not share a common variable. The algorithm we use is a simple but genuine rotation in 4 dimensions: we plot

$$\cos t \cdot y^{(1)} + \sin t \cdot y^{(2)} \quad \text{versus} \quad \cos t \cdot x^{(1)} + \sin t \cdot x^{(2)}$$

which moves from the $(x^{(1)}, y^{(1)})$-plot to the $(x^{(2)}, y^{(2)})$-plot as t increases from 0 to 90 degrees. The resulting motion graphics seem unusual at first, but we found that their use is straightforward for identifying meaningful subsets across the two plots.

One might wonder why we did not choose convex interpolation, which appears conceptually simpler. There are three reasons:

— Convex interpolation destroys orthonormality of the pair of vectors we project on: in $(x^{(1)}, y^{(1)}, x^{(2)}, y^{(2)})$-space, the vectors $(1-t, 0, t, 0)$ and $(0, 1-t, 0, t)$ are not normalized to unit length. A consequence is the following:

— Under a null-situation such as a multidimensional standard normal point cloud, convex interpolation leads to an artificial shrinkage effect; e.g., the variance of $t \cdot x^{(1)} + (1-t) \cdot x^{(2)}$ at $t = 0.5$ is only half of the variance of either $x^{(1)}$ or $x^{(2)}$, leading to a shrunk plot by a factor of about 0.7.[2]

— Trigonometric interpolation has the advantage that the speed vector of a projected point at $t = 0$ is proportional to the projected vector in the target plot; e.g.,

$$\frac{d}{dt}\bigg|_{t=0} (\cos t \cdot y^{(1)} + \sin t \cdot y^{(2)}) = y^{(2)}$$

We demonstrate this latter effect in Figures 21-24. First (Fig. 21) we show a plot of latitude versus longitude of the 329 largest metropolitan areas of the U.S., and last (Fig. 24) a plot of the ratings of these areas on housing costs versus climate & terrain (see the Rand McNally *Places Rated Almanac*). The two intermediate figures (22 and 23) show two stages of the interpolation from the first into the last plot, with Figure 22 very near the start $t = 0$: we see for example that the cities of Florida fork apart, the Gulf cities moving quickly to the left and the places on the Atlantic moving to the right. In the target plot, left and right mean unfavorable and favorable climate & terrain, respectively. This effect is much less pronounced in convex interpolation where, e.g., the vertical speed component is:

---

[2] If, however, the goal is to interpolate plots in which the pairs $x^{(1)}, x^{(2)}$ and $y^{(1)}, y^{(2)}$ are highly correlated, i.e., the starting and target plots look very much alike, then it may be wiser to use convex interpolation, since otherwise an unpleasant bulging effect may occur: if for example $x^{(1)} = x^{(2)}$, the variance of $\cos t \cdot x^{(1)} + \sin t \cdot x^{(2)}$ is twice the variance of either $x^{(1)}$ or $x^{(2)}$ at $t = 45$ degrees. This is an argument in favor of convex interpolation as used in the VISUALS approach of Young, Kent, and Kuhfield [1986].

$$\frac{d}{dt}\bigg|_{t=0} ((1-t) \cdot y^{(1)} + t \cdot y^{(2)}) = y^{(2)} - y^{(1)}$$

Correlation Tour

Projections of data seem most interpretable if the projection planes are oriented along the variable axes in some way. We provide a capability for selecting two disjoint sets of variables, to be called x-variables and y-variables, respectively, and plot linear combinations of the y-variables against linear combinations of the x-variables. This is very much in the spirit of canonical correlation analysis, the only difference being that we do not optimize our linear combinations, but provide a capability for dynamically searching among all possible linear combinations of x- and y-variables and thus permitting the viewer a less restrictive approach to his/her data. The special data projections resulting from this method have an increased degree of interpretability over general unrestricted projections. They are tailored to searching for dependence between variables. Notice that there will appear two types of vectors in the variable boxes: horizontal ones and vertical ones, depending on whether a box represents an x- or a y-variable (see Figures 29-30). Selection of variables for inclusion or exclusion in the x- or y-set is made by sending their boxes suitable mouse clicks.

There arises then the obvious question of how to *interactively* control the motion of the projections within the constraints defined by the sets of x- and y-variables. Following Asimov [1985], we propose to dodge this problem and instead leave it up to *automatic* algorithms to generate a scanning motion which leads a viewer efficiently through the set of projections. At first sight one might consider this as a step back from real-time graphics towards animation, but this is not a fair assessment since a user can control the process any time by

— selecting or dropping variables,
— stopping, changing and reversing speed,
— storing away a given projection, retrieving it later, and continuing another viewing path from this projection,
— modifying the plot by deleting, highlighting, and coloring points,
— stopping and identifying points.

All this freedom is possible only with real-time graphics, even so, at any given time, the motion path is deterministic.

We notice that simple 3d-rotations are obtained by selecting just one y-variable and two x-variables, or vice versa. Indeed, we have not bothered to implement 3d-rotations as a separate feature but run the correlation tour whenever we wish to use simple 3d-rotations (notice the upper title in Figure 16).

Another special case of the correlation tour can be used for examining regression residuals: choosing the residuals as the only y-variable and some or all (especially more than two) predictors as x-variables, the correlation tour will expose a viewer to a continuous flow of plots of the residuals versus linear combinations of predictors. Finding the residuals "clean", i.e., with no structure recognizable in the tour, drastically increases a data analyst's confidence in his/her regression fit.

Grand Tour

This capability provides scanning of arbitrary projections in variables spaces. As with the correlation tour, the scanning path is determined by an automatic algorithm. The controls are the same with the minor difference that there is no distinction between x- and y-variables; a variable is either selected (and marked "A" in its variable box) or dropped, and nothing else. The vectors in the boxes of the selected variables now point in arbitrary directions (Figures 31-32).

The first implementation and discussion of the grand tour is due to D. Asimov [1985]. Our current implementation uses an algorithm which is briefly described in Buja and Asimov [1986].

The grand tour is most useful in situations where the variables at hand do not have an a priori established meaning, as with principal components coordinates and configurations obtained from multidimensional scaling. In these situations, we can perform interactive factor rotation using the grand tour.

An additional control which we found useful is to confine the projections by an orthogonality constraint. This is the case in repeated measures data where one is often interested in contrasts. By requiring projections to be orthogonal to the $(1,1,...1)$-vector, only contrasts are scanned.

Some Illustrations

In Figures 25-28 we show four pairwise scatter plots of the 4-dimensional particle physics data [Fisherkeller, Friedman, and Tukey, 1975] and for comparison two views obtained from the correlation tour (Figures 29-30) and two from the grand tour (Figures 31-32). In Figures 29 and 31 we recognize easily the 4-fold rod structure consisting of two dense rods forming a right angle, and two sparse rods emanating from the free ends of the dense rods. This structure involves all four variables and cannot be completely seen in either pairwise static scatter plots or dynamic 3d-scatter plots in variable 3-spaces. Figure 30 shows the data in spiral shape, while Figures 31-32 exhibit some faint parallel copies of the obvious rods.

We cannot resist showing also a few examples of artificially created mathematical surfaces just for their beauty: Figure 33 is a projection of points lying on a 2d-torus embedded in 4-space, where there is complete duality between the two sets of circles spanning the torus. Figure 34 shows points on a Moebius strip, also embedded in 4-space. Figure 35 is the hardest to understand: the points lie on a highly symmetric embedding of the Klein bottle in four dimensions. A Klein bottle is a 2-dimensional, connected, closed, non-orientable surface. It is usually drawn as a bottle whose neck bends down and penetrates its own body to close the surface from the inside. No self-penetration is needed for an embedding in four dimensions, however. Figure 36 represents a five dimensional solid cube, or more precisely the set-theoretic product of 2 by 3 by 4 by 5 by 6 points embedded in 5-space; the total number of points shown is therefore $2 \cdot 3 \cdot 4 \cdot 5 \cdot 6 = 720$.

5. INFORMAL STATISTICAL INFERENCE

Powerful viewing capabilities present a problem of over-exploring data and finding spurious structure. In our own experience, this has rarely been a serious issue, perhaps because human vision is a far better instrument for distinguishing between the real and spurious in scatter plots than commonly believed. On the other hand, we tend to use graphical methods for screening data and obtaining rough qualitative insights, while in-depth analysis and subtle quantitative judgements are left to more formal methods, once their applicability is established by the screening process. Whatever the reason for the relative reliability of visual judgements may be, there still arise occasions when one wishes to have tools for sharpening one's perception of random fluctuations in data. It has been suggested that data analysts should gauge their eyes every once in a while on some artificially created pseudo-random data, like multivariate normal point clouds [Diaconis, 1983]. We have followed this advice on occasion, and found it helpful in establishing structure as real, and in realizing that the most frequent types of random structure, such as local clottedness and moderate outliers, are usually not of interest to the data analyst.

Inferential problems are most urgent for small sample sizes. It appears that the comparison of actual data with some extraneous unrelated data is not a powerful enough tool for informal inference, and more careful tailoring of the comparison data to the situation at hand is needed. Bootstrap ideas come to mind, but the high frequency of ties in non-parametric bootstrap samples makes them unsuitable for graphical purposes without additional massaging, such as adding small random noise to each data point to break up the ties. Another approach would use parametric bootstrap: fitting a parametric model and drawing random samples from the fitted model could potentially be very useful, but assuming a parametric model appears somewhat specialized and lacks the universal applicability we have in mind.

We have chosen to implement some very crude non-parametric hypotheses, such as independence between variables, and various types of distributional symmetry. Testing for independence is probably the most important case in an exploratory setting. In order to check whether the variable plotted vertically is independent of the horizontal variable, we randomly permute the y-values against the x-values, and plot the resulting randomized x-y-data. Under the assumption of independence, the permuted plot should not look qualitatively different from the plot of the original data. An example is given in Figures 37-39: the data are the well-known ratings of the 329 largest metropolitan areas in the U.S. (see the Rand McNally *Places Rated Almanac*). Figure 37 shows a plot of health care ratings versus economics ratings. Because of the skewed distribution of the vertical variable, it is not immediately clear whether some dependence is present, but a comparison with the two permuted plots of Figures 38-39 indicates that the dependence is faint, if present at all. We cross-checked our interpretation by exposing audiences to the plots (after covering up menus, titles, and boxes), and asking them to identify the real data; they were unable to do so. We found it a useful mental discipline to ask ourselves whether an unprejudiced person could identify the real data among the permuted ones.

On the other hand, exposing oneself to a large number of comparisons over an extended period of time can lead to a sensitivity of judgement which may allow us to describe very subtle, although unassessable, deviations from independence. As an example, having seen many hundred permuted versions of the plot of health care ratings versus economics ratings, we felt that the actual data show an unusual concentration of places with high health care ratings located over the central portion of the economics ratings. To check this guess, we also plotted normal scores of the two variables against each other; the purpose of this exercise was to normalize the marginal distributions to approximately standard normal and examine whether the resulting scatter plot looks nearly circularly symmetric, as it should in case of independence. Figure 40 shows the normal scores plot for these two variables, indicating that the deviation from circular symmetry is indeed due to a slight pear shape with the narrow side up. This seems to be not unreasonable: neither economically depressed areas nor booming places tend to excel at health care, the former because of lack of resources, the latter because of infrastructure lagging behind business development. The major exception from this rule is Detroit (the isolated point at the top left): depressed as it is, it offers the high health care standards of a big city.

Notice that this data set is non-stochastic, or at least not a random sample in the usual sense. However, the random permutations still seem to make intuitive sense, and their purpose can be explained to lay people.

This is in line with some efforts in giving theoretical justifications to the application of permutation tests to non-stochastic data [Freedman and Lane, 1983] and more generally using randomization schemes for descriptive purposes [Finch, 1979].

Besides the independence test we also implemented various symmetry hypotheses:

— Symmetry of the horizontal x-distribution about the centering value, which is chosen at the variable standardization phase (see Section 2); the median would make most sense in this context. The randomization we use here is to change the sign of the horizontal variable in 50% randomly chosen points.
— Symmetry of the y-distribution: the same, but applied to the vertical variable.
— Simultaneous symmetry of the x- and the y-variable: the previous two randomizations applied simultaneously.
— Symmetry of the joint x-y-distribution: the hypothesis is that the distributions of (x, y) and (y, x) are the same; we call this "45 degree symmetry". The randomization consists of flipping (x, y) to (y, x) in 50% randomly chosen points.

Another hypothesis, which we have not implemented yet, would be rotation symmetry, meaning that the distributions of (x, y) and $(\cos t \cdot x + \sin t \cdot y, -\sin t \cdot x + \cos t \cdot y)$ are the same for all t. The randomization would consist of sampling the angle in the polar coordinate representation of (x, y) from the uniform distribution on $[0, 2\pi]$.

The controls for generating alternative data under the various hypotheses are provided by a temporary menu (see Figure 38). Every mouse click on a menu item generates a new random plot, under the hypothesis represented by the item. Holding the mouse depressed on an item will flash random plots at the viewer at maximum speed. We found this both an amusing and illuminating way of acquiring a sense of randomness under a null hypothesis. For moderate sample sizes, the speed is such that no individual plot can be recognized from the blur formed by consecutive plots. The visual effect comes close to that of a smooth density surface, coded by brightness for high density.

In our current implementation, the randomization schemes act on the projected data, i.e., within the viewing pipeline the inference module is located *after* the projection engine and before the window-to-viewport transformation. Furthermore, randomizations can act only on still plots; i.e., for dynamic projections and scaling-shifting we exit the testing mode and return to the original data. This design decision was made for pure prototyping convenience and should be changed in the near future. The place for the inference module in the viewing pipeline should be after the variable standardization (Section 2) and *before* the projection engine. (The flowchart reflects this final design.) The benefits are at least twofold:

- We can apply dynamic projections to alternative data sets; e.g., we can use the grand tour on a data set where some variables contain permuted values, and gain insight into the effect of purely marginal structure on the views.
- We can define new types of hypotheses; e.g., we may test independence between two pairs of variables (u, v) and (x, y) by permuting the (u, v)-pairs against the (x, y)-pairs, or we may test the hypothesis of complete joint independence of the variables u, v, x, y by permuting the values of all four variables independently of each other.

In a multi-window environment, we can analyze real data in one Data Viewer window simultaneously with a permuted version in another window. This calls for a new capability which forces identical projections in two or more windows. We call this feature "view sharing". One of us [C. Hurley, see Buja, Hurley, and McDonald, 1986] has implemented view sharing in a development version of the Data Viewer.

6. GENERAL REMARKS AND CONCLUSIONS

We outlined an architecture of a viewing pipeline for dynamic scatter plots, and we described some of its elements in greater detail. Especially, we pointed out how

- real-time viewporting,
- dynamic projections, and
- randomization for informal inference

are relevant to data analysis. If we distinguish between

- manipulating data and
- manipulating views of data,

then our emphasis was on the latter. In order to achieve highly interactive manipulations of views, the elements of a viewing pipeline should have graphical representations, at least temporarily. Views are manipulated by poking with a pointing device on the graphical representations. Examples in our Data Viewer are the variable boxes for controlling projections, the speed bar, and the temporary menus for real-time control of the viewporting and randomization elements.

REFERENCES

Andrews, D. F. (1981). "Statistical applications of real-time interactive graphics, *Interpreting Multivariate Data*, V. Barnett, ed., 175-185. Chichester: Wiley.

Asimov, D. (1985). "The grand tour: a tool for viewing multidimensional data," *SIAM Journal on Scientific and Statistical Computing*, **6**: 128-143.

Becker, R. A. and Cleveland, W. S. (1984). "Brushing a scatter plot matrix: high interaction graphical methods for analyzing multidimensional data," Technical Memorandum. Murray Hill, NJ: AT&T Bell Laboratories.

Becker, R. A., Cleveland, W. S., and Wilks, A. R. (1987). "Dynamic graphics for data analysis," *Statistical Science*, **2**: 355-395.

Bertin, J. (1983). *Semiology of graphics*. Madison, WI: The University of Wisconsin Press.

Buja, A. and Asimov, D. (1986). "Grand tour methods: an outline," *Computer Science and Statistics: Proceedings of the 17th Symposium on the Interface*, 63-67. Amsterdam: Elsevier.

Buja, A., Hurley, C., and McDonald, J. A. (1986). "A data viewer for multivariate data," *Computer Science and Statistics: Proceedings of the 18th Symposium on the Interface*, 171-174. Washington, DC: American Statistical Association.

Chambers, J. M., Cleveland, W. S., Kleiner, B., and Tukey, P. A. (1983). *Graphical Methods for Data Analysis*. Monterey, CA: Wadsworth.

Cleveland, W. S. (1985). *The Elements of Graphing Data*. Monterey, CA: Wadsworth.

Cleveland, W. S., McGill, M. E., and McGill, R. (1986). "The aspect ratio of a two variable graph," Technical memorandum. Murray Hill, NJ: AT&T Bell Laboratories.

Collier, G. (1986). "Toth II: a hyper text browser," Technical Memorandum. Morristown, NJ: Bell Communications Research.

Diaconis, P. (1983). "Theories of data analysis: from magical thinking through classical statistics," Technical Report No. 206. Stanford, CA: Stanford University, Department of Statistics.

Donoho, A. W., Donoho, D. L., and Gasko, M. (1986). *MACSPIN Graphical Data Analysis Software*, Austin, TX: D^2 Software, Inc.

Donoho, D. L., Huber, P. J., Ramos, E., and Thoma, H. M. (1982). "Kinematic display of multivariate data," *Proceedings of the Third Annual Conference and Exposition of the National Computer Graphics Association*, **1**: 393-398. Fairfax, VA: National Computer Graphics Association.

Finch, P. D. (1979). "Description and analogy in the practice of statistics," *Biometrica*, **66**: 195-208.

Fisherkeller, M. A., Friedman, J. H., and Tukey, J. W. (1975). "PRIM9: an interactive multidimensional data display and analysis system," *Data: Its Use, Organization, and Management*, 140-145. New York: The Association for Computing Machinery.

Foley, J. D. and van Dam, A. (1982). *Fundamentals of Interactive Computer Graphics*. Reading, MA: Addison-Wesley.

Fowlkes, E. B. (1969). "User's manual for a system for interactive probability plotting on graphic-2," Technical Memorandum. Murray Hill, NJ: AT&T Bell Telephone Laboratories.

Freedman, D. A. and Lane, D. (1983). "Significance testing a nonstochastic setting," *A Festschrift for Erich L. Lehmann*, P. J. Bickel, K. A. Doksum, and J. L. Hodges, Jr., eds., 185-208. Monterey, CA: Wadsworth.

Friedman, J. H., McDonald, J. A., and Stuetzle, W. (1982). "An introduction to real time graphics for analyzing multivariate data," *Proceedings of the Third Annual Conference and Exposition of the National Computer Graphics Association*, **1**: 421-427. Fairfax, VA: National Computer Graphics Association.

Huber, P. J. (1987). "Experiences with three-dimensional scatter plots." *Journal of the American Statistical Association.* **82**: 448-453.

"MacPaint," (1983). California: Apple Computer Inc.

McDonald, J. A. (1982). "Interactive graphics for data analysis," PhD thesis. Stanford, CA: Stanford University.

McDonald, J. A. (1986). "Periodic smoothing of time series," *SIAM Journal on Scientific and Statistical Computing,* **7**: 665-688.

McDonald, J. A. and Pedersen, J. (1986). "Programming environments for data analysis, part III," Technical report, Seattle, WA: University of Washington, Statistics Department.

Newman, W. M. and Sproull, R. F. (1979). *Principles of Interactive Computer Graphics.* New York: McGraw-Hill.

Parker, David (1985). "Software running on an IBM PC jr.," Cambridge, MA: Massachusetts Institute of Technology.

Stuetzle, W. (1986). "Plot windows," Technical Report. Seattle, WA: University of Washington, Statistics Department.

Stuetzle, W. and Thoma, M. (1978). "PRIMS-ETH: A program for interactive graphical data analysis," Research Report 19, Fachgruppe fuer Statistik. Zuerich: ETH.

Tufte, E. R. (1983). *The Visual Display of Quantitative Information.* Cheshire, CT: Graphics Press.

Young, F. W., Kent, D. P., and Kuhfeld, W. F. (1986). "Visual exploratory data analysis and the VISUALS system for dynamic hyper-dimensional graphics," *Annual meeting of the Psychometric Society.*

298 ELEMENTS OF A VIEWING PIPELINE FOR DATA ANALYSIS

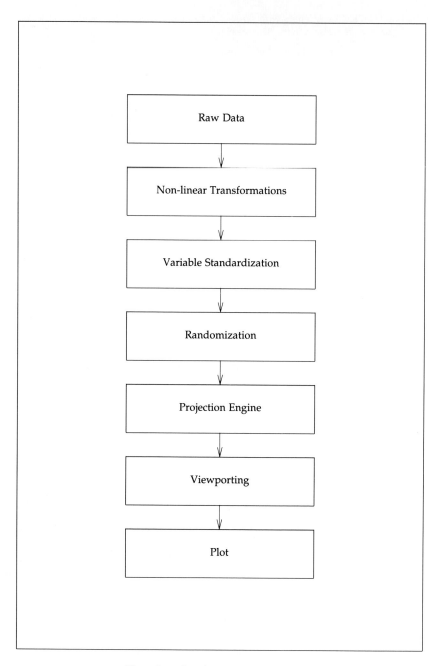

Flowchart for the Viewing Pipeline

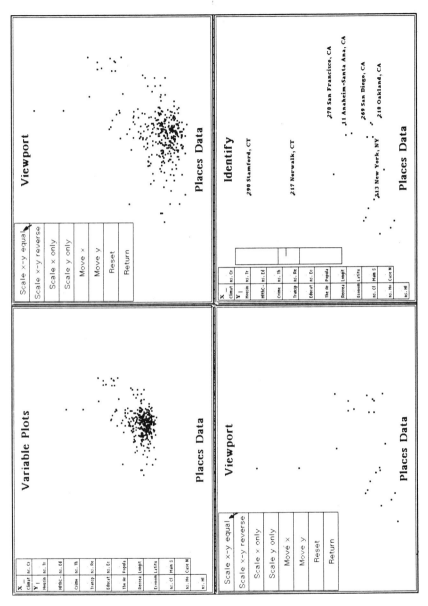

Figures 1-4

300 ELEMENTS OF A VIEWING PIPELINE FOR DATA ANALYSIS

Figures 5-8

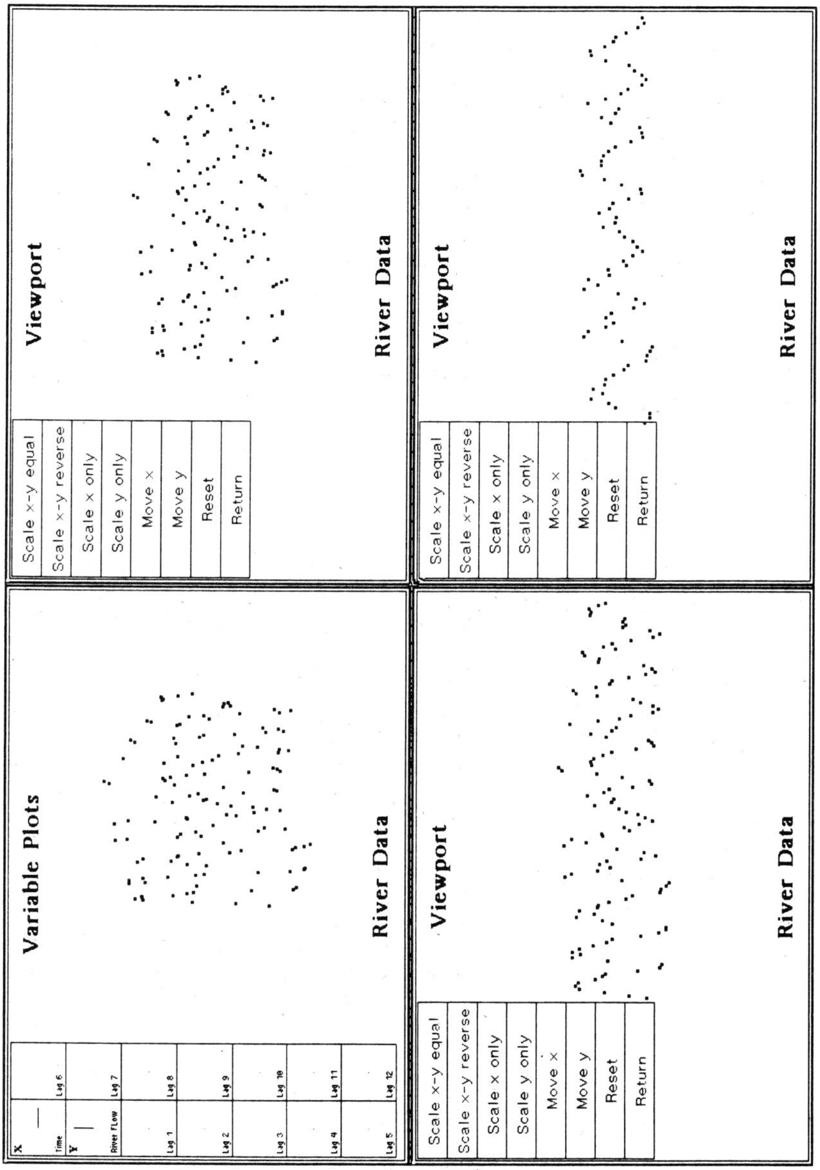
Variable Plots
X
Time
Y
River Flow
Lag 1
Lag 2
Lag 3
Lag 4
Lag 5
Lag 6
Lag 7
Lag 8
Lag 9
Lag 10
Lag 11
Lag 12
River Data
Viewport
Scale x-y equal
Scale x-y reverse
Scale x only
Scale y only
Move x
Move y
Reset
Return
River Data
Viewport
Scale x-y equal
Scale x-y reverse
Scale x only
Scale y only
Move x
Move y
Reset
Return
River Data
Viewport
Scale x-y equal
Scale x-y reverse
Scale x only
Scale y only
Move x
Move y
Reset
Return
River Data

Figures 9-12

302 ELEMENTS OF A VIEWING PIPELINE FOR DATA ANALYSIS

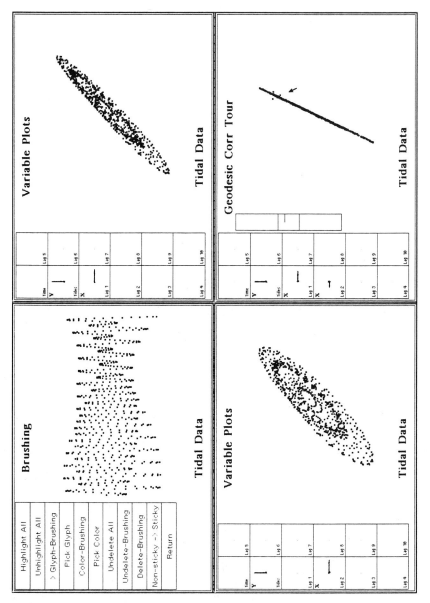

Figures 13-16

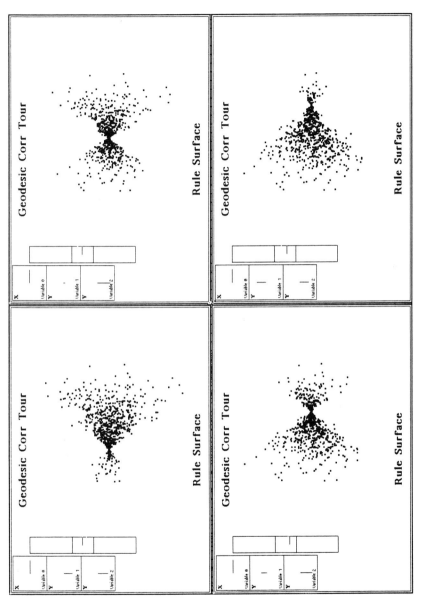

Figures 17-20

304 ELEMENTS OF A VIEWING PIPELINE FOR DATA ANALYSIS

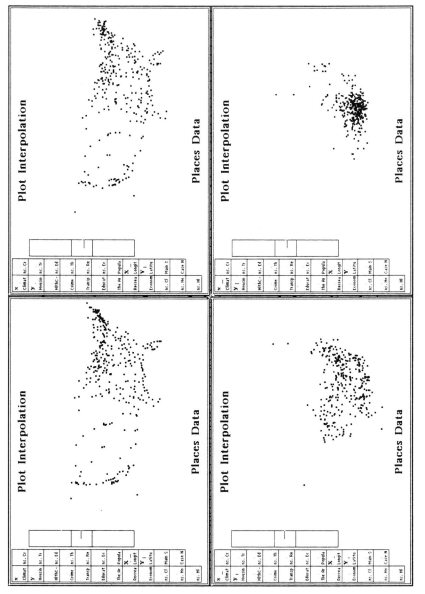

Figures 21-24

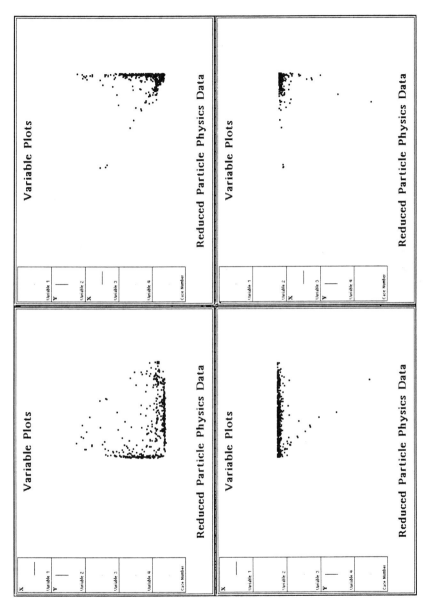

Figures 25-28

306 ELEMENTS OF A VIEWING PIPELINE FOR DATA ANALYSIS

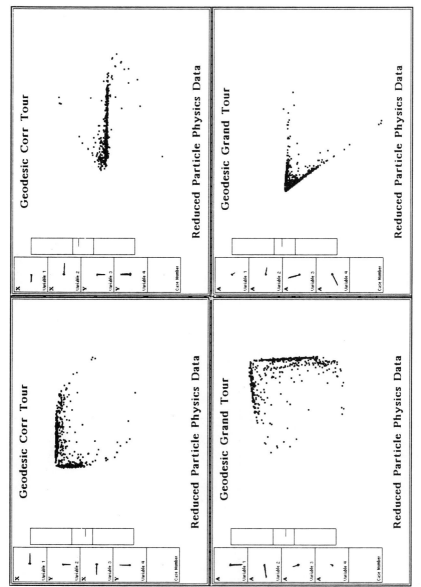

Figures 29-32

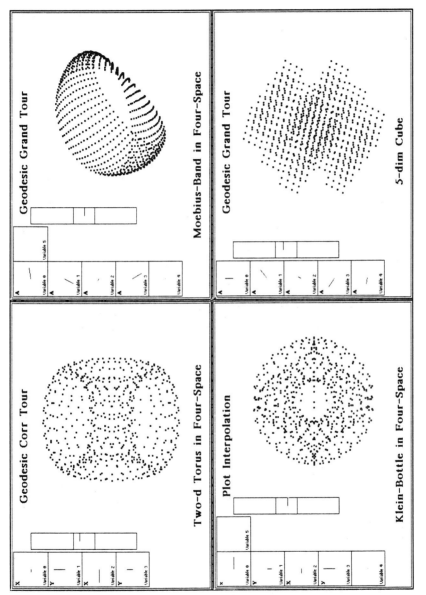

Figures 33-36

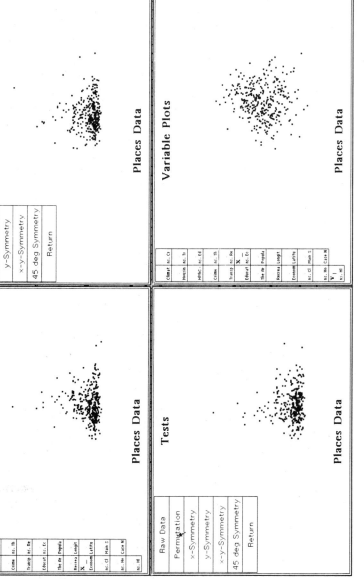

Figures 37-40

12

EXPLOR4: A PROGRAM FOR EXPLORING FOUR-DIMENSIONAL DATA USING STEREO-RAY GLYPHS, DIMENSIONAL CONSTRAINTS, ROTATION, AND MASKING

D. B. Carr
W. L. Nicholson

Battelle Pacific Northwest Laboratories

ABSTRACT

This paper presents an approach to the visual exploration of four-dimensional data based on the representation of each data point using a single symbol, the stereo-ray glyph. The result is a three-dimensional scatter plot with ray angle showing the fourth coordinate. With this representation and tools for changing the view, local lower dimensionality and density anomalies can be identified. The tools have been integrated in a computer program called EXPLOR4. The tools include tracker-ball controlled three-dimensional rotation in a selected subspace and real-time mouse-controlled point masking. Masking criteria include point density, distance from a hyperplane and distance from a point. Point and subspace constraints required for the operations are provided through interactive graphical specification of four-dimensional point sets.

From *Dynamic Graphics for Statistics*, William S. Cleveland and Marylyn E. McGill, eds., copyright © 1988 by Wadsworth, Inc. All rights reserved.

1. INTRODUCTION

In this paper we discuss display and enhancement procedures for four-dimensional (4-D) data. Our experience with stereo-cued 3-D scatter plots has motivated us to find a workable representation for 4-D data. Stereo-cued 3-D scatter plots often show structure that is not evident in 2-D marginal plots. How much are we missing by not representing variables?

Numerous methods have been used to represent points with four or more coordinates. Methods that use multiple coded symbols to represent multiple dimensions include Andrew's curves [Andrews, 1972], Chernoff faces [Chernoff, 1973], trees and castles [Kleiner and Hartigan, 1981], and stars [Chambers, et al. 1983]. For these methods the spatial position of the symbols does not directly represent any of the individual coordinates and the transition from the visual representation back to the data is awkward. With just four dimensions we prefer methods that are in some way enhancements of the 2-D scatter plot. Such methods include slicing and casement displays [Tukey and Tukey, 1983], the grand tour [Buja and Asimov, 1986], brushing and polygon subset selection for scatter plot matrices [Becker and Cleveland, 1987; Carr, et al. 1987], and linked kinematic displays [Donoho, et al. 1982]. However, such methods result in multiple representations for each data point and the gestalt impression generated by a single symbol or glyph is lost. The compromise discussed in this paper is the scatter plot with third and fourth variables being represented by stereo depth cues and by ray angles, respectively (see Carr, et al. (1986) for an anaglyph stereo ray glyph example).

Our justification of the stereo ray glyph (see Figure 1) in static displays is as follows. When perceptual accuracy of extraction [Cleveland and McGill, 1984] is used as a criterion for picking a visual representation, position along a common scale is the best. Thus, for showing two coordinates (x and y) the scatter plot is the best method. For showing a third coordinate (z) a stereo depth cue provides limited resolution because our binocular depth perception is of limited accuracy. However, most of us live in a three dimensional world and understand spatial coordinates. For people with binocular depth perception, this understanding overshadows any advantages for other static representations. (We also use motion parallax cues from rotation to facilitate our depth perception, but primarily use rotation to find good static views to study.) We use a ray, for the fourth coordinate (w). In the perceptual ordering of display variables, line length ranks slightly ahead of angle. However, due to the potential for ambiguous stereo fusion and misinterpretation caused by overplotting we prefer to use angle. We claim the stereo ray glyph is a workable representation for 4-D data.

A workable representation does not mean that we suddenly become able to understand the complexities of four dimensions. Rather the stereo ray glyph provides a reasonable starting place for a very difficult problem. Before we can begin to understand complicated data patterns, we have to calibrate our vision to understand simple things like distance. (A test concerning the evaluation of distance might be used to compare multivariate representations.) In our 4-D display, the x, y and z variables are each allocated eight inches of plotting space (the apparent depth is a function of the viewing distance so a particular viewing distance is required for z). The potential range for angle is 180° (ambiguity results when a greater range for angles is allowed). Thus 45° corresponds to about 2 inches. With this relatively simple correspondence, rough intuitions concerning distance can be learned. Besides learning about distance, we have to gain experience and vocabulary to describe the simple patterns that register on our eye-brain systems. While our understanding and intuitions in four dimensions will undoubtedly remained limited, some tasks are not as difficult as might be supposed.

This paper restricts attention to two relatively simple visual tasks: 1) recognizing the existence of local lower dimensional structure, and 2) recognizing density anomalies such as clusters, holes, and density based outliers. Most of the paper is devoted to motivating (Section 2) and describing (Section 3) software tools that help in accomplishing these tasks. Section 4 indicates directions of future work.

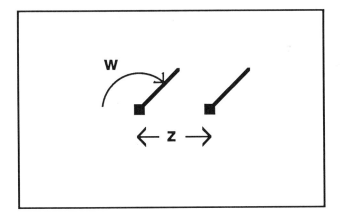

Figure 1. The Stereo Ray Glyph. The midpoint of the stereo dots represents two coordinates, x and y. The stereo separation represents a third coordinate, z. The ray angle represents the fourth, w. For the ray angle we use a speedometer analogy. The angles are restricted from 0° to 180° and for large values the ray points to the right.

The software tools to be described have been integrated in an computer program called EXPLOR4. EXPLOR4 is an evolving research tool that is currently implemented on a VAX 11/780 with a Ramtek for the display. The code is only a few hundred lines of FORTRAN, with a substantial portion devoted to the user interface (pull down menus, surface emulation, tracker ball, data tablet communication, etc.). The manipulation routines such as rotation and masking are quite simple. A good programmer with some user interface tools should be able to put together a similar program for other environments with modest effort.

2. CONSTRAINTS AND GEOMETRIC EXAMPLES

In this paper we consider two visual tasks, identifying the existence of lower-dimensional structure and spotting density features. Since looking for visual evidence of lower dimensions is not commonly discussed as part of the data analysis process, we devote this section to it.

2.1 Dimensionality

A mathematical description helps in a discussion of dimensionality. Table 1 reminds us that dimensionality corresponds to the number of parameters required in a parametric description of a structure. For example, if values of u and v generate points of a structure, $(x1 = f1(u, v), x2 = f2(u, v), x3 = f3(u, v))$, then that structure is two-dimensional and is embedded in 3-D space. Our primary use of parametric representations is to generate test data for subjective evaluation of display methods. We select parametric functions, generate values for the parameters either systematically or with pseudo random numbers and then compute the "data" values. The range of the parameters, the density of the sample in the range, and the wildness of the functional forms all influence how much we can see, but that is getting ahead of the story.

TABLE 1. Parametric Representations and Dimensionality in K Dimensional Space

Representation			Dimension	Name
$X1 = f1(u)$	$X2 = f2(u)$	...$Xk = fk(u)$	1	Curve
$X1 = f1(u, v)$	$X2 = f2(u\ v)$	...$Xk = fk(u, v)$	2	Surface
$X1 = f1(u, v, w)$	$X2 = f2(u, v, w)$	...$Xk = fk(u, v, w)$	3	Hypersurface

For interpretation purposes, we think of dimensionality in terms of constraints rather than in terms of parametric representations. Under favorable conditions, a parametric representation translates into constraints on the display variables. Table 2 suggests that a single constraint equation lowers the dimensionality by one.

CONSTRAINTS AND GEOMETRIC EXAMPLES 313

TABLE 2. Single Constraints and Dimensionality

Number of Coordinates	Constraint	Structure Dimensionality
2	$f(x, y) = 0$	1
3	$f(x, y, z) = 0$	2
4	$f(x, y, z, w) = 0$	3

As an example consider the constraint

$$f(x, y) = y - x^{**}2 = 0.$$

Given a local region for x, the variation in y is severely constrained. This notion generalizes in several ways. In four dimensions we can, for example, look at local regions for (x, y, z) and decide if w is constrained or varies independently of (x, y, z). Mathematically, consider fixing three variables (x, y, z) of the four variables in the constraint equation. The result is a new equation in just one variable. If the equation has real roots that vary smoothly as a function of (x, y, z), then the analyst has a good chance of spotting local lower dimensionality. In practice, measurement error complicates the identification of lower dimensionality. Nonetheless, 4-D stereo ray glyph scatter plots will often suggest lower dimensionality. We would certainly hope so when simple linear or quadratic models are being entertained.

2.2 Finding Constraints

Which tools help in spotting constraints? Consider two lower dimensional examples. Figure 2 shows three petals generated by

$$x = \cos(u)*\cos(3u) + \text{noise}$$
$$y = \sin(u)*\cos(3u) + \text{noise}.$$

We see the structure at a glance. However with four-dimensional data we are restricted to looking at lower dimensional scatter plots plus cues for the remaining variables. What happens in our 2-D case when we look at single variables? The overplotting in 1-D marginal plots based on Figure 2 makes those plots remarkably uninformative. Consequently, we have shown densities in the margin plots. (While overplotting is typically less severe for higher dimensional marginal plots, density representations should always be considered.) The holes in the data help convince us that the densities are not very informative. About the only impressions we get from these 1-D density plots and those based on other rotations concern notions of symmetry. Selecting the 1-D views based on principal components or projection pursuit helps little. Whatever projections we use, unique identification of the structure is highly unlikely. That does not deny the utility of projection methods. For linear structures, projection methods can be very informative. But with curvilinear structures, the shadows cast by projection can leave much structure hidden.

So what works? Slicing works! Imagine two parallel lines separated by a quarter of an inch. Move them slowly from the bottom to the top of the plot. Locally in the y coordinate, the x coordinates are obviously restricted. Not only does slicing suggest lower dimensionality, it helps in understanding the data density since it shows the structure of the petals. Since the petals can be represented in two dimensions, it is possible to develop a mental image of petals from the slices.

In higher dimensions where overplotting problems are not so severe, a second method also works. This method represents the additional coordinate using a ray. Figure 3a shows a scatter plot matrix of a three variable mystery data set (see also Carr and Nicholson, 1985). The

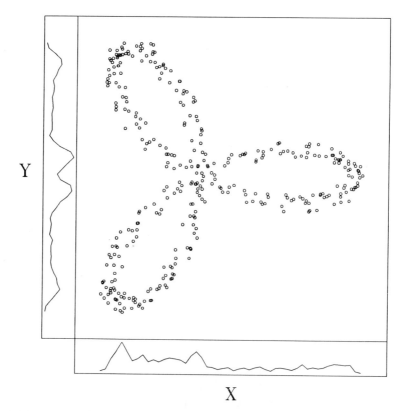

Figure 2. Petals and Marginal Density Plots. Much structure remains obscure in the marginal views no matter what rotation is chosen.

nice edges suggest that the data is constrained. If we add the ray glyph as in Figure 3b, it becomes apparent that the data is constrained in the inside of the cloud as well. On the left side of the plot, the ray angles are all the same. Hence in this region the data is at most 2-D. As we follow the data up and to the right (or down and to the right), notice how smoothly the rays change towards large (or small) values. At the right side of the plot the y coordinate varies and the rays chatter (show large and small values) while the x values are restricted. Thus the data is again 2-D. From this plot some people can identify the data as points sampled from a mobius strip. Figure 3c is a side-by-side stereo view of the mystery data set as a scatter plot of y versus x with z the stereo-cued depth variable. With such a stereo view or, as an alternative, with dynamic rotation views, the points are easily identified as being locally constrained to two dimensions and globally lying on a mobius strip. The point of the example is to illustrate that the ray glyph is a useful symbol for displaying 3-D data in a 2-D viewing space.

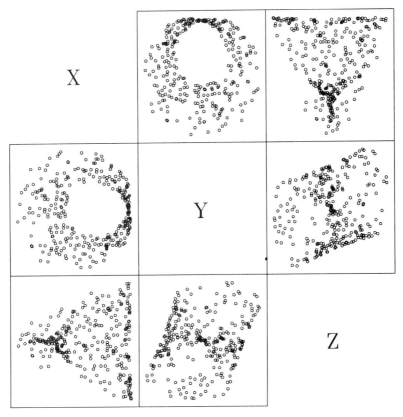

Figure 3a. A Scatter Plot Matrix. What is the 2-D structure embedded in 3-D space.

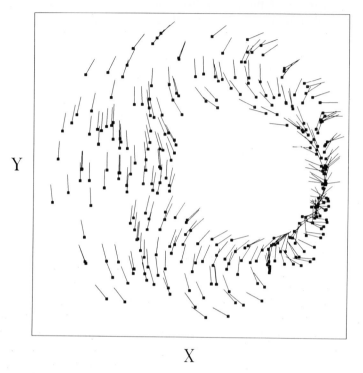

Figure 3b. A Ray Glyph Plot. The ray angle represents the third coordinate. The local constraints on the ray angle suggest local 2-D structure.

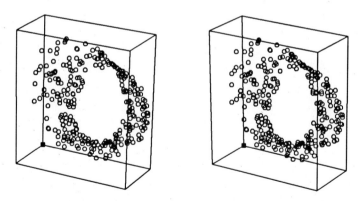

Figure 3c. A Stereo View. When using the viewer, make sure the tops of the two images are aligned. The mobius strip structure is obvious. This and subsequent stereo views associate data variables with display variables so that a point whose first three coordinates are large will plot at the top right (y and x) and above the page (z). The 3-D frame orientation indicates the rotation and a square dot indicates in the most distant corner. Here the rotations are 20° in the x–z direction and 20° in the y–z direction. Cube-shaped frames generally indicate separate coordinate scaling into the same range. Other shapes indicate global scaling for the spatial coordinates.

In a similar manner, the ray glyph can be used to display 4-D data in a stereo-cued 3-D viewing space. Before describing the EXPLOR4 analysis tools we illustrate this 4-D display medium used by EXPLOR4 with an intuitive example of points systematically arranged on the edges of a 4-D cube. Specifically, the data set consists of the 16 corner points plus four equally spaced interior points on each of the 32 edges, a total of 144 4-D points. Not all stereo ray glyph displays of 4-D data are equally informative. Rotation is a good tool for investigating which displays are most informative. Figure 4a is a side-by-side stereo view of the above described four dimensional cube data with three coordinates displayed spatially, and the fourth coordinate displayed as the angle of the stereo ray glyph plotting symbol. There is a good deal of overplotting, as indicated by the ray multiplicity. Each corner point is six overplotted points, and each interior edge point is two overplotted points. Thus the outlined 4-D cube appears as two superimposed outlined 3-D cubes with each corner point being an end-on view of six points along a common edge.

Any rotation that reduces the overplotting must involve the nonspatially displayed coordinate. Figure 4b, as the result of such a rotation, is more interesting. In EXPLOR4 rotations are controlled with a tracker ball, which demands predefinition of a 3-D rotation space. This is done by specifying the orthogonal 1-D fixed subspace. Here the fixed subspace is the line through the center of the cube with direction (1,1,1,1). A 30° horizontal rotation of the tracker ball followed by a 30° vertical rotation of the tracker ball displays more of the 4-D character of the data. Now all 32 edges are apparent as outlined 1-D structures and four edges emanate from each corner point. What do we really discern from this display? We do not see four spatial dimensions. What we do see is two interlinked stereo-cued 3-D cubes. The ray glyph symbol shows that the nonspatial coordinate varies in a smooth constrained manner along each of the 3-D edges. Further, the data appear to be partially ordered points on a

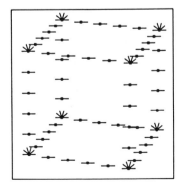 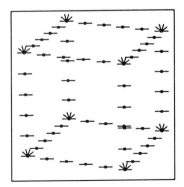

Figure 4a. A 4-D Cube. Each edge contains 6 points. The ray angle in the plot show overplotting. For rotation to reduce this overplotting, it must involve the 4th coordinate.

4-D network. Thus, as far as local dimensionality is concerned, the display is very successful. Clearly the structure locally is either a line segment or line segments with common intersection. Whether or not one interprets these line segments and intersections as the edges and corners of a 4-D cube depends upon one's mental set, but the local dimensionality is obvious!

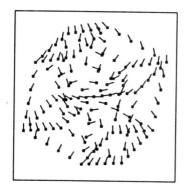 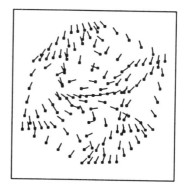

Figure 4b. A Rotated 4-D Cube. The rotation subspace is orthogonal to the vector (1,1,1,1). The rotation is 30° in the x–z direction and 30° in the y–z direction.

2.3 A Lower-Dimensional Example

Friedman and Tukey (1974) use data from a high-energy particle physics scattering experiment to illustrate a projection pursuit algorithm for identifying clustering structure in high dimensions. In the experiment, positively charged pi-mesons bombarded a stationary target of protons contained in hydrogen nuclei. Final reaction products of the experiment were a proton, two positively charged pi-mesons, and a negatively charged pi-meson. These products can be completely described by seven independent quantities, each functions of particle energies and momenta. Our 4-D data set is a partial description of 500 repetitions of the experiment. The as-recorded data were viewed with EXPLOR4 using global scaling. Six outlier points were discarded to improve the resolution. Figure 5 shows a view rotated 20° about the horizontal axis, 20° about the new vertical axis and rescaled to maximize ray angle resolution. While the overplotting is extreme, the view clearly shows that most of the experimental repetitions cluster along one of two planes, z and angle constant or x and angle constant. Further, that for z constant, the clustering is along one of two linear structures. Even the outliers tend to lie along linear structures, particularly the half dozen points extending up on the far right of the view. Two outliers, extreme in ray angle, appear in the top center left. Rotation and low-density masking help discern additional low dimensional structure.

Thus, a quick view with EXPLOR4 shows that these four characteristics of the scattering experiment repetitions group into several clusters each defined by two or three linear constraints. Further, the few ungrouped repetitions may belong to less frequently occuring clusters also defined by linear constraints.

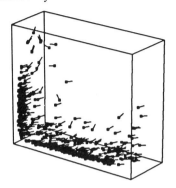
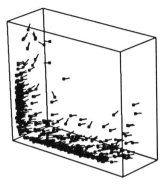

Figure 5. Particle Physics Data. Much structure in the plot appears one dimensional. Six points with extreme coordinates have been deleted to increase resolution. Poor choice of plotting scales can induce false impressions of lower dimensionality. Other views of this data set continue to show much lower-dimensional structure.

3. ANALYSIS TOOLS AND THE USER INTERFACE

EXPLOR4 tools for analysis and display of 4-D data are stereo ray glyphs, slicing, other types of masking that help focus attention on portions of the data, and rotation. Slicing masks out part of the data so that attention is focused on the remainder. We use several other masking options, most notably point density masking. The process of estimating the data density at each point and showing only the data above a certain density has been called sharpening by Tukey and Tukey (1981). Here the concept is the same except that masking hides points either inside or outside a dynamically specified density interval. This is described further in 3.5. The final tool is tracker ball controlled rotation. As suggested above, projection techniques like rotation help little in seeing the interior of the data cloud. However the "cloud surfaces" are a different story. Rotation can bring "cloud surfaces" to x and y extreme coordinates where overplotting is less and the structure is more easily observed.

The user has access to EXPLOR4 tools through the user interface. Figure 6 shows the five elements of the screen layout, pull down menus across the top, a 4-D display at the center left, a variable name and linear combination display at the top right, a masking bar at the center right, and four 1-D density plots at the bottom right. Rather than describing all of the pull down menu items in detail, we sketch the basic process of entering data, discuss the scaling and display strategy and then go into more detail concerning the control and use of rotation and masking.

3.1 Entering the Data

The user starts by creating an edited data file that includes variable names, transformations, and other data modifications. After invoking EXPLOR4 from a standard terminal, the display appears on the Ramtek. Mousing on the appropriate file menu item prompts the user for the file to be read. The file name is entered on the terminal. Next, the variables are selected by mousing in the variables menu. The selected variable names appear in the linear combinations window. The first selected variable is initially represented as the x coordinate in the display. The next three variables are initially represented as the y, z and w coordinates in the display. The user can alter the variable order and substitute other variables at will.

3.2 The Scaling and Display Strategy

The 4-D display and univariate density estimates appear (or are updated) as each variable is selected. For interpretation purposes it can be important to know the translation and scaling that takes place. Initially the midrange for each variable is translated to 0 and is equated with the center of the display. For the x, y, and z display variables the center of the display is the center of an 8 inch cube. For the ray symbol, the display center corresponds to a ray pointing straight up. Two types of scaling are provided. The first simply multiples all the coordinates by a constant, so that no rigid body rotation will produce a value outside the display space. This preserves the relative distances when the user has prescaled the data using spherizing or other methods and obviates the need for clipping algorithms. However, if the user has not carefully scaled the extremes, this may lead to poor resolution for some variables. The default method scales each variable into the interval [–1 1] before applying the global scaling. Since the one time global scaling covers all variable reordering (of 4 variables) and rotation, resolution is lost in individual displays. In Figures 4a and 4b the ray variable has been scaled separately from the spatial variables to increase resolution. Future versions of EXPLOR4 may have scaling options optimized for rotations in a particular 3-D subspace or for individual views.

In EXPLOR4 the display variables are the frame of reference and not their precursors. Interpretation and action corresponds to what is shown. The univariate density estimates correspond to the display variables x, y, z and w. Similarly, the user defines 4-D points in the display space and not in the initial data space.

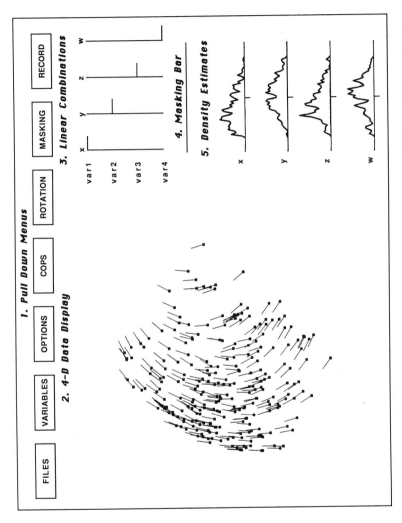

Figure 6. The Screen Layout for EXPLOR4. Text has been added to indicate the five components. The 3-D representation under the 4-D display label is the left eye view of the 3-D structure embedded in 4-D space.

3.3 Constraint Point Sets

When working with 4-D data it is helpful to use 4-D points as benchmarks added to the data and as a means of controlling operations such as projection, rotation, and masking. We name sets of such points and refer to them generically as "cops" or constraint point sets. The edit cops menu allows us to define new cops. When the define option is invoked, a new point (with a larger dot) appears at the center of the 4-D display and also as a dot on each univariate axis in the density estimate window. In the current implementation the user edits the 4-D point by mousing for each coordinate on the respective density axis. The mouse changes the point in real time in both the 4-D display and in the density display. The software draws the points (and erases them) in a different surface than the data, with all surfaces visible. Thus the data is not touched. Our mouse is actually a data tablet with four buttons located to form a diamond. The bottom button terminates the editing for the current point and a new point for editing appears at the display center. After we have defined and named our constraint point set, we are free to add the set to the data matrix as benchmarks using an option under edit cops. The large dot points are shown along with the data and are treated similarly except they can go out of range so are subject to clipping.

In practice the main use of cops is to control the action in 3-D rotation, hyperplane masking, box masking and distance masking. Both 3-D rotation and hyperplane masking require the definition of a 3-D hyperplane. To define a hyperplane two points suffice, a point in the hyperplane and a second point that gives the opposite end of a vector normal to the hyperplane. For box masking two points, (min x, min y, min z, min w) and (max x, max y, max z, max w), that define opposite vertices also suffice. For distance masking, just a single point is needed. When a display action is chosen, the algorithm selects the most recently used cops with the appropriate number of points. The current stack of cops is maintained under the cops menu and can be reordered by mousing. The name of the cops being used is given for each display action.

3.4 Three-Dimensional Rotation

For 3-D rotation the constraint point set determines the center of rotation (first point) and a 1-D fixed subspace (the vector as indicated above). As 3-D rotation embedded within a 4-D display is complicated, we always represent the fixed subspace with the ray and the rotation subspace with the spatial variables. Unless the fixed subspace vector is in the w direction, this gives up continuity with previous displays. We are willing to pay this price to avoid wiggling rays.

We control the rotation in three space using a tracker ball. The tracker ball motion corresponds to motion on the screen. Rotation occurs in the x–z direction, the y–z direction, and in mixtures involving z. (Our tracker ball does not rotate in the x–y direction, but all views can be obtained). Tracker ball buttons allow the motion to be restricted to either the x–z or the y–z direction. Our tracker ball does not allow as precise control as might be thought so the restriction helps in obtaining the desired display. The tracker ball is convenient but certainly control as provided through packages like MacSpin is adequate.

The display under rotation updates as fast as it can. This is a function of the number of points in the display and the number of users on the system. Under adverse conditions updating takes longer than real time (fraction of a second). Tracker ball control is awkward but is not a serious problem since the display maintains depth via the stereo representation. Each updated display is drawn in a hidden surface and popped into view when complete. This involves tricks with the color table and on our Ramtek is very fast.

For rotation the linear combinations are shown in the linear combination window as in Figure 6. This general idea was borrowed from Grand Tour software [Buja and Asimov, 1986]. In EXPLOR4 the vertical line under x is the zero line. For each data variable, the endpoints of a horizontal segment connected to the vertical line show the weight from –1 to 1. For our 3-D rotation, the weights for the w variable remain the same, since it corresponds to the fixed subspace. The linear combinations are updated for each tracker ball query cycle. In contrast the three active density plots (which are drawn in a different surface) are only updated when the tracker ball is inactive. This compromise was made to improve tracker ball response.

3.5 Masking

Once we have found an interesting view via rotation, we often proceed with data masking. We have implemented a form of masking that involves the computation of the masking criterion only at the beginning of the masking operation. For each data point a criterion is computed. Our current criteria include point density (no cops required), distance from a point (a one point "cops" required), and directed distance to a hyperplane (a two point "cops" required.) Masking then consists of comparing the criterion for each data point to a dynamically specified acceptance interval. If the criterion value is in the interval the corresponding point is shown. Since only a range comparison is required before redisplay, masking works faster that rotation.

To interactively specify the acceptance interval for which the data is to be shown, the criterion range is equated to the length of a masking bar and the data tablet mouse is used. Mousing above the masking bar defines the acceptance interval for which data is displayed. The left button controls the smallest value for the interval and the right button controls the largest value. By holding down either left or right button and moving the mouse, one side of the acceptance interval is altered and the display updated in real time. Both lower and upper values can be altered simultaneously by moving the mouse with the top button held down. The bottom button terminates masking.

A line above the masking bar from the last left button position to the last right position indicates criterion values for which points are displayed. The positions are initialized at the criterion extremes so all data points are shown at the beginning. If the left button value exceeds that of the right button value, the values are switched before the comparison is made. In this case, points whose criterion values fall between the switched values are masked. The visible line segments continue to indicate which points are displayed. Thus data points with extreme criterion values can be shown while those with intermediate values are hidden.

The masking routines only cause the 4-D display to be updated. As with rotation the new view is drawn in a hidden surface and substituted when complete. The univariate densities are updated only when no mouse button is depressed. Again simultaneous updating is desirable.

Masking remains in effect for subsequent operations unless otherwise explicitly reset. The last masking operation can be turned off and all masking can be turned off.

The most useful masking operations are hyperplane masking and density masking. Hyperplane masking can show holes in the data cloud. We typically mask in the z direction since our binocular vision is poor, the w direction because we want to see the spatial relationships for a fixed angle range, and in the (1, 1, 1, 1) direction when we want the sum of the variables constrained. Density masking is helpful in finding modes

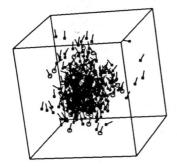 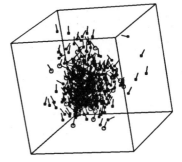

Figure 7a. NHANES Data. The data variables serum cholesterol (x), TIBC (y), albumin (z), and height (w) are thought to be associated with increased cancer risk. Octagons distinguish the cancers cases. Note the heavy overplotting.

and low density regions. Distance from a point masking is useful for calibrating our distance intuition or for removing data in a local region. We haven't used box masking. Distance and box masking would be more useful if masking region size were fixed and the location could be altered in real time with the mouse.

3.6 A Density Masking Example

The last example illustrates density masking and uses data from the National Health and Nutrition Examination Survey I (NHANES I). In the survey, a probability sample of the U.S. population was given a medical exam, dietary questionnaire, and blood tests during the period 1971-1974. These 14,407 adults aged 25-79 have been followed for disease occurrence. Figure 7a shows the four variables, serum cholesterol (mg/dl), serum total iron binding capacity (TIBC: ug/dl), serum albumin (mg/dl) and height (cm) for non-smoking men over age 55 as x, y, z, and w respectively. The cases known to have cancer are distinguished with an octagon rather than a square dot. With 419 cases the overplotting is substantial.

Figure 7b shows 54 low density points based on 4-D density masking. At least three subsets are of interest. Figure 7c, the left eye view without the 3-D frame, shows 2-D outlines of the three subsets. The first subset contains three points with extreme ray angles. These low density points appear where there was heavy overplotting in Figure 7a. The unusual heights (angles) give the points low density even though their other coordinates are common. This potential for overplotting of low density points makes density masking more important for 4-D data than for 2-D data.

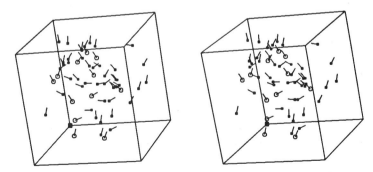

Figure 7b. Density Masking. Fifty-four points with the lowest 4-D densities are shown. This plot reveals a previously overplotted lower-dimensional structure.

The other two subsets show lower dimensional patterns. One subset is a tight 4-D cluster of 7 points with similar ray angles. With little variation in any of the coordinates this subset is described by saying it has dimension zero. Such a cluster would not appear in a low density plot if the smoothing parameter in the density estimate were much smaller. The remaining subset contains 9 points and lies along a gentle curve in the flattened view in Figure 7c. With the exception of 1 point, the angles show little variability. Since the points lie close to a gentle curve and vary in depth, the subset is roughly two dimensional and can be thought of as lying in a 2-D plane.

The eye-brain system is so powerful at finding patterns that hypotheses generated need to be tested by independent data. The observed patterns may not stand the test of replication. A large data set such as the NHANES can be partitioned so that hypotheses generated in one subset can be rigorously tested in another.

Another assessment based on Figure 7b compares the 11 observed cancer cases to the expected number from a random sample. Here 56 of the 419 men had cancer. The expected number in a random sample of size 54 is 7.2 and 93% of random samples will contain an observed count in the range 3 through 11. Looking for samples with excess cancer cases is a natural activity. Epidemiologic evidence indicates that high cancer risk is associated with low serum cholesterol, low iron binding capacity, low albumin and tall stature. This suggests a normal vector for hyperplane masking. The low density view in Figure 7b is not expected to have an extreme number of cancers cases because it contains cases from both tails of the univariate distributions.

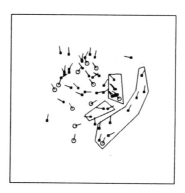

Figure 7c. The Left Eye View. The outlines in the plot surround three subsets discussed in the text.

4. FUTURE WORK

EXPLOR4 is being moved to a Silicon Graphics Iris workstation. In this section we discuss our plans for EXPLOR4 and our vision for its application.

EXPLOR4 was research software designed to investigate the use of specific tools for exploratory analysis of 4-D data. For general use, some standard tools need to be added. These include easy access to a rich data transformation language such as S [Becker and Chambers, 1984]; a feature for interactive point labeling and the ability to plot more than simple stereo ray symbols. Modifications will exploit developments in windowing technology the facilitate multiple views of the data and corresponding density representations. With the multiple views, linkage will be provided between graphically selected data subsets as discussed by McDonald [1982]; Becker and Cleveland [1987]; Carr, Littlefield, Nicholson, and Littlefield [1987]; and Stuetzle [1987]. The current color analyph stereo views will be replaced by polarized stereo views. Thus color will become available for distinguishing subsets, and viewing will be more comfortable.

In addition to the above proven graphical tools, we plan to include some speculative features. With large data sets we would like to replace portions of the data with skeletal structures (in lower dimensions see Fabbri [1984]). This should reduce overplotting and speed rotation. We would like to enhance plots by applying higher dimensional versions of techniques such as principal surface smoothing [Hastie 1984] to data subsets. We plan to add previously demonstrated 4-D rigid body motion in the form of rocking [Nicholson and Littlefield, 1982]. Finally, we intended to flesh out EXPLOR4 bookkeeping capabilities. Frequently we want to show past dynamic graphical sequences to others. However we usually have failed to save a good starting point and only have our memory as a record of the subsequent process. We will experiment with editable bookkeeping for dynamic graphical displays [Tukey, 1988].

Future work will demonstrate the application of EXPLOR4 concepts to a variety of data sets. Seismic monitoring where x and y show location, z shows depth and w shows magnitude seems a natural application. A similar example uses the spatial display variables to represent the nodes of 3-D PDE solution sets and ray angle to represent a scalar value. Even if none of the data is stochastic, a 4-D view can help provide understanding. Other candidate data sets include isotopic data, biological landmark data and compositional data where four components are subject to constraints. The constraints should be evident as lower dimensionality.

Eventually EXPLOR4 will be pushed into representing five interval or ratio variables. Ray length is suitable for showing one more dimension so a viable 5-D symbol is available. We have little hope of finding a viable symbol for 6-D data. We have looked at 6-D plots of PDE solution set velocity vectors. The base of each vector was located in 3-D space. The relative coordinates of the tip of each 3-D vector represented three additional variables. This stereo ray did not work well because the depth angle is poorly represented by small stereo separation differences and the angle complicates interpretation of ray length. For the near term the task of interpreting 4-D graphics provides sufficient challenge.

5. ACKNOWLEDGEMENTS

The authors would like to thank Rik Littlefield, Mike Kosorok, and David Thurman for their work in implementing EXPLOR4. This work was supported by Basic Energy Sciences, U.S. Department of Energy under Contract DE-AC06-76RLO 1830.

REFERENCES

Andrews, D. F. (1972). "Plots of high-dimensional data," *Biometrics*, **28**: 125-136.

Becker, R. A. and Chambers, J. M. (1984). *S: An Interactive Environment*. Belmont, CA: Wadsworth.

Becker, R. A. and Cleveland, W. S. (1987). "Brushing scatterplots," *Technometrics*, **29**: 127-142.

Buja, A. and Asimov, D. (1986). "Grand tour methods: an outline," *Computer Science and Statistics: Proceedings of the 17th Symposium on the Interface*, 63-67. Amsterdam: Elsevier.

Carr, D. B. and Nicholson, W. L. (1985). "Evaluation of graphical techniques for data in dimensions 3 to 5: Scatterplot matrix, glyph, and stereo examples," *American Statistical Association 1985 Proceedings of the Statistical Computing Section*, 229-235. Washington, DC: American Statistical Association.

Carr, D. B., Nicholson, W. L., Littlefield, R. J. and Hall, D. L. (1986). "Interactive color display methods for multivariate data," *Statistical Image Processing and Graphics*, E. J. Wegman and D. J. DePriest, eds., 215-250. New York: Marcel Dekker.

Carr, D. B., Littlefield, R. J., Nicholson, W. L. and Littlefield, J. S. (1987). "Scatterplot matrix techniques for large N," *Journal of the American Statistical Association*, **82**: 424-436.

Chambers, J. M., Cleveland, W. S., Kleiner, B. and Tukey, P. A. (1983). *Graphical Methods for Data Analysis*. Monterey, CA: Wadsworth.

Chernoff, H. (1973). "The use of faces to represent points in k-dimensional space graphically," *Journal of the American Statistical Association*, **68**: 361-368.

Cleveland, W. S. and McGill, R. (1984). "Graphical perception: theory, experimentation, and application to the development of graphical methods," *Journal of the American Statistical Association*, **79**: 531-554.

REFERENCES

Donoho, D. L., Huber, P. J., Ramos, E. and Thomas, H. M. (1982). "Kinematic display of multivariate data," *Proceedings of the Third Annual Conference and Exposition of the National Computer Graphics Association*, **1**: 393-398. Fairfax, VA: National Computer Graphics Association.

Fabbri, A. G. (1984). *Image Processing of Geological Data*. New York: Van Nostrand Reinhold. pp. 61-63.

Friedman, J. H. and Tukey, J. W. (1974). "Projection pursuit algorithm for exploratory data analysis," *IEEE Transactions on Computers*, **C-23**: 881-890.

Hastie, T. (1984). "Principal curves and surfaces," Technical Report Orion 24. Stanford, CA: Stanford University, Department of Statistics.

Kleiner, B. and Hartigan, J. A. (1981). "Representing points in many dimensions by trees and castles (with discussion)," *Journal of the American Statistical Association*, **76**: 260-276.

McDonald, J. A. (1982). "Interactive graphics for data analysis," Technical Report Orion 11. Stanford, CA: Stanford University, Department of Statistics.

Nicholson, W. L. and Littlefield, R. J. (1982). "The use of color and motion to display higher-dimensional data," *Proceedings of the Third Annual Conference and Exposition of the National Computer Graphics Association*, **1**, Fairfax, VA: National Computer Graphics Association.

Stuetzle, W. (1987). "Plot windows," *Journal of the American Statistical Association*, **82**: 466-475.

Tukey, J. W. (1988). "Control and stash philosophy for two-handed, flexible and intermediate control of a graphical display," *Collected Works of John W. Tukey, 5 - Graphics*, W. S. Cleveland, ed., 329-382. Pacific Grove, CA: Wadsworth & Brooks/Cole.

Tukey, P. A. and Tukey, J. W. (1981). "Graphical display of data sets in 3 or more dimensions," *Interpreting Multivariate Data*. V. Barnett, ed., 189-275. Chichester: Wiley.

Tukey J. W. and Tukey, P. A. (1983). "Some graphics for studying four-dimensional data," *Computer Science and Statistics: Proceedings of the 14th Symposium on the Interface*, K. W. Heiner, R. S. Sacher and J. W. Wilkinson, eds., 60-66. New York: Springer-Verlag.

13

MACSPIN: DYNAMIC GRAPHICS ON A DESKTOP COMPUTER

Andrew W. Donoho
D^2 *Software, Austin, Texas*

David L. Donoho
Miriam Gasko
University of California, Berkeley

1. INTRODUCTION

MacSpin is a program for dynamic display of multivariate data. It uses rotation to display 3-dimensional scatter plots, and also offers dynamic graphics primitives such as animation, identification, and highlighting. Using these features, the user can visually find trends, clusters, outliers, and other patterns in multivariate data.

MacSpin has a broad range of data manipulation and calculation features. These allow the user to interactively transform, edit, split, and categorize his data as patterns in the display indicate. MacSpin also provides many display options and a number of statistical summaries. All these features are available to the user at the click of a mouse button. Thus the program speeds up the detective work required to understand the visual clues provided by the graphical display.

Dynamic graphics systems have been developed before at various research institutions; these systems ran on large, expensive, and/or exotic

---

From *Dynamic Graphics for Statistics*, William S. Cleveland and Marylyn E. McGill, eds., copyright © 1988 by Wadsworth, Inc. All rights reserved.

machines and had limited use. MacSpin runs on the Apple Macintosh, a personal computer. Like other Macintosh software, it places a great deal of emphasis on completeness of display options, on quality of the user interface and on ease of use. It can fairly be said that MacSpin offers a wider array of features than earlier systems; that where it overlaps with previous systems, MacSpin offers a considerably more complete implementation; and that all these features are packaged together intelligently, so that users who do not need certain features will not be bothered by them. MacSpin makes dynamic graphics perform up to the potential sketched out by earlier systems, and it makes dynamic graphics accessible to a broad audience of scientists and engineers.

In this paper, we begin by telling a few stories that show how the program is used on real data; we discuss some history of dynamic graphics; we discuss the environment and user interface of MacSpin; and finally, we discuss some system design issues.

2. SOME STORIES

In our experience, the best way to explain what dynamic graphics can do is to use it on real data and illustrate the features as one goes along. The MacSpin documentation does this for examples ranging from data on Galaxies and earthquakes to Urban Planning. Here we give brief examples for the CARS and DIABETES datasets.

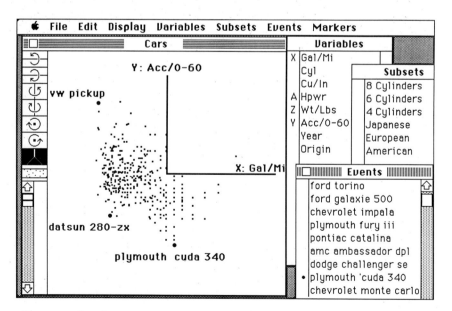

Figure 1. Gasoline consumption and acceleration time for 418 cars reviewed by *Consumer Reports* between 1971 and 1983.

2.1 Cars Data

This dataset consists of data on all the cars road-tested by *Consumer Reports* magazine between 1971 and 1983. The data will help us see how the auto industry changed over the last decade or so. The names of the 418 cars are listed in the events window on the MacSpin display (Figure 1 lower right): the portion we see includes the **Plymouth Barracuda** and the **Plymouth Fury III,** cars from the early 1970's. The **Variables** window (partially obscured by the **Events** window) shows the variables we have measured for each car; things involving performance (Gallons per Mile, Seconds to reach 60 MPH from a full stop), size (Horsepower, Weight, ...), and miscellaneous (Model Year, Continent of Origin).

X-Y Plots. The view in the plot window (Figure 1) shows all the cars (American, European, and Japanese) in an x-y plot, with x = **Gal/Mi** (i.e., fuel usage per mile) versus y = **slowness** (Secs. 0-60). The points represent individual cars. By moving the cursor to a point and clicking, we can find out its identity. The point at the upper left (slow but economical) is a **VW pickup;** the point in the lower right (fast gas-guzzler) is a **Plymouth Barracuda.** By holding down the control and option keys as we identify, the full data record pops up (Figure 2). This shows us that the fast, efficient car in the lower left is a 1981 **Datsun 280 ZX.**

X-Y-Z Plots. The x-y plot shows the general trend of the auto industry -- what combinations of speed and economy are available. By rotating the plot, we can get an extra dimension into the display. We point at a rotation icon on the far left (Figure 1), and hold down the mouse button. The 2-d plot becomes a rotating 3-d plot, the previously hidden z-variable coming into play (Figure 3). When we do this for the Cars data, and bring z = **Weight** into the display, we see a cloud of points rotating smoothly in space. The cloud is shaped like a sausage and shows the combinations of economy, speed, and weight being built during the 1971-1983 period.

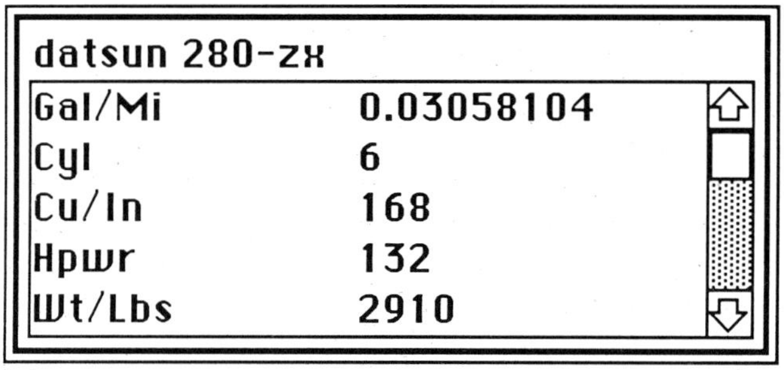

datsun 280-zx	
Gal/Mi	0.03058104
Cyl	6
Cu/In	168
Hpwr	132
Wt/Lbs	2910

Figure 2. Info Pop-up for **Datsun 280 ZX.**

As we rotate the plot, we notice a few interesting things. First, one point turns out to be an outlier (Figure 3). We stop the rotation and identify it; it is an International Harvester truck. Somehow a truck has slipped in to a database on cars! When we scroll to the truck's name in the events window, we see that *Consumer Reports* road-tested a few other trucks, too. By pointing at their names on the list, we can highlight them in the plot window. They are also outliers. By choosing **Exclude** from the events window, we can (temporarily) remove them from the display (Figure 3). The rotation has helped us identify and remove outliers.

Highlighting Subsets. Further rotation shows that the data consist of three clusters. Seeking for an explanation, we bring the **Subsets** window (partially hidden in Figure 1) to the front. This shows some subsets of the data predefined (by us) as being interesting to look at. By pointing at the name of any subset, we can highlight its members on the display. When we do this, we see that the 3 clusters consist of 8, 6, and 4 cylinder cars, respectively (Figure 4). We could also highlight **American, European,** and **Japanese** subsets in turn, and find out where they are on the display.

Focus on Subsets. We can focus the display on a specific subset of the data, say on **American** cars. We select first American cars from the **Subsets** window and then select **Focus** from the events menu. MacSpin now temporarily excludes imported cars from the display.

Animation permits us to study the effect of a fourth variable, for example time, on a display. Suppose we are interested in how the American auto industry has changed over time. We then drag the **Year** variable to the scroll bar in the lower left. This will let us scroll through the data model-year by model-year. We begin at 1971. The cars made then are concentrated in the lower left of the display: fast, heavy, gas-guzzling cars (Figure 5). As we scroll smoothly forward, we see that the data drift systematically with time towards the upper right -- toward slower, lighter, more economical machines of 1983.

Transformations. The researcher can also transform existing variables to create new ones. Features like this make MacSpin useful not just for displaying data but also for manipulating it to get the right display. We just saw that cars became more economical over the period 1973-1981. Did they just become lighter and smaller, or was there an actual increase in mechanical efficiency? Dividing **Gal/Mi** by **Weight** gives us a standardized measure of fuel efficiency in which the effects of weight are taken out. Looking at plots with this new variable shows that American cars got more efficient and not just smaller over this period. Variable transformations are all included in a special transformations window (Figure 6), and executed by pointing and clicking with the mouse.

Markers. MacSpin also makes it easy to get a hard copy of the data plots. (Since you can mark subsets with special symbols, you can use these to convey some of what the dynamic exploration showed you.)

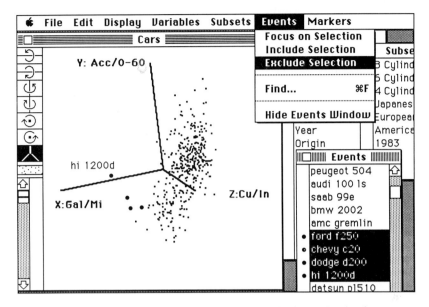

Figure 3. Highlighting and excluding trucks from the display.

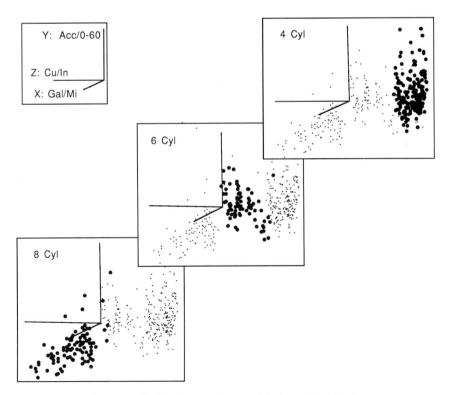

Figure 4. Highlighting subsets of 4, 6, and 8 cylinder cars.

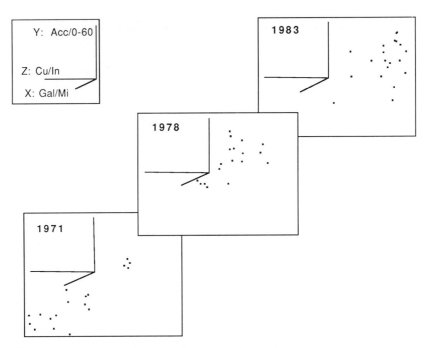

Figure 5. Animation showing changes in the performance of American cars over time: the years 1971, 1978, and 1983 are shown.

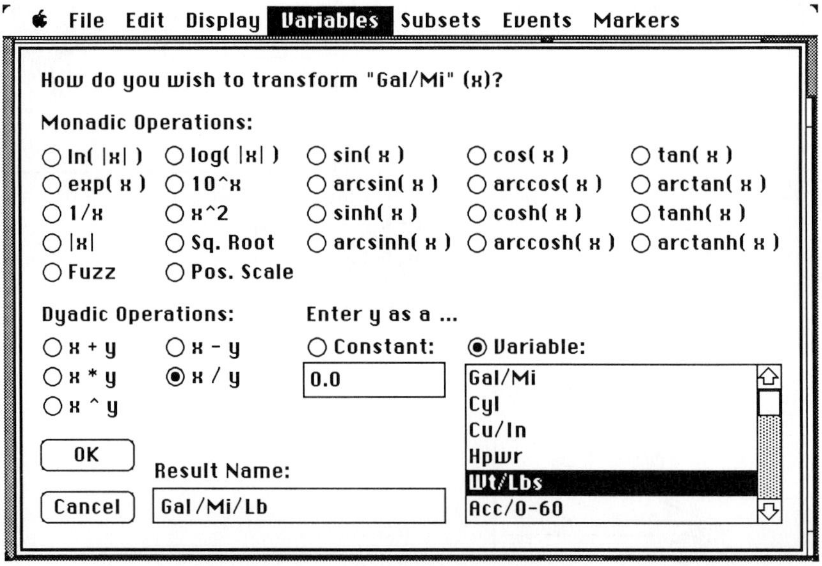

Figure 6. Transformations window.

"Screen dumps" are generated using the Command-Shift-3 sequence. Figure 7 is derived from a screen dump. 1971 model cars are marked with a box, and 1983 model cars with an asterisk. The resulting image was cropped, and captions were drawn in, using MacPaint. Figure 7 conveys the interesting message that the *most* efficient 1971 model cars are less efficient than the *least* efficient 1983 models.

2.2 Diabetes Data

The **Diabetes** data were provided by Reaven and Miller of Stanford University. The original graphic analysis of it was done on the PRIM-9, and reported in *Diabetologica* in 1979. The data describe 145 nonobese and nonketotic patients who agreed to participate in a medical experiment. The purpose was to assess the relationships among various measures of plasma glucose and insulin in order to illuminate the etiology of "Chemical" and "Overt" diabetes. Each patient underwent a glucose tolerance test, and the following quantities were measured: **Age, Relative Weight, Fasting Plasma Glucose, Test Plasma Glucose** (a measure of insulin intolerance), **Steady State Plasma Glucose,** Plasma Insulin during the test. In addition we have the doctors' classification of these patients as overt diabetic, chemical diabetic, or normal. Age and Relative Weight turned out to be unimportant hence are excluded from our analysis (as they were by Reaven and Miller).

The opening view of our demonstration has the following variables assigned to the three axes: Fasting Glucose on the X-axis, Test Glucose on the Y-axis and Test Insulin on the Z-axis (Figure 8).

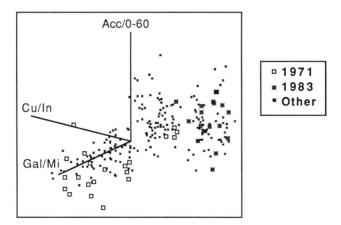

Figure 7. American cars, with special markers given to model years 1971 and 1983.

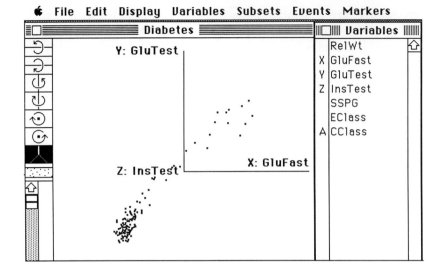

Figure 8. Data for 145 patients who underwent a glucose tolerance test.

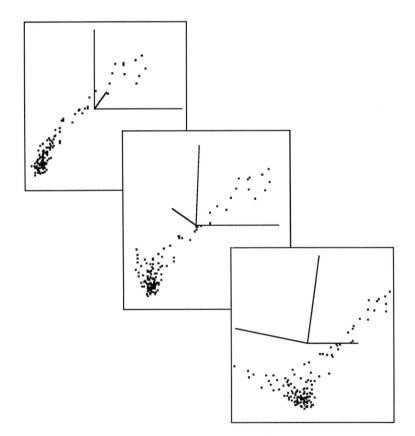

Figure 9. Rotation showing the boomerang aspect of the data.

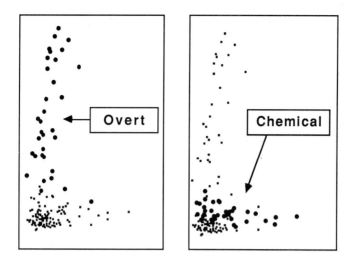

Figure 10. Chemical and overt diabetics shown occupying the two arms of the boomerang.

This view, showing the data distributed in a sausage-shaped cloud, supports the interpretation that there is but one direction in which abnormality develops, as we progress from normal patients to Chemical to Overt diabetics. However, as soon as we start to rotate the data around the X-axis, and tilt it a bit to better show the third dimension, Z, we can see that the pointcloud has, in fact, the shape of a boomerang (Figure 9). We can no longer accept that there is just one direction of disease development.

The most natural question at this point is: What makes the two arms of the boomerang? As the doctors have classified the diabetic patients as either Chemical or Overt, we can highlight each subset separately. As Figure 10 shows, each arm corresponds to one of the groups.

Our conclusion is that Chemical and Overt diabetes are two different syndromes, not just one manifested at different levels of intensity.

3. SOME HISTORY

MacSpin, of course, is not the first system for dynamic graphics. Over the last decade, dynamic graphics systems have been developed at a few research institutions. These were by and large *one-of-a-kind installations* and required expensive and/or exotic equipment. The first system of this kind, PRIM-9, developed by Friedman, Fisherkeller, and Tukey at the Stanford Linear Accelerator Center in the Early 1970's, ran on an IBM mainframe with a $400,000 display unit, and cost several hundred dollars an hour to use. Other systems were built, for somewhat less cost, at Harvard (PRIM-H), at Stanford, (Orion) and at Bell Labs (Scatmat). These systems served to show that dynamic graphics could be used to analyze data in many different fields... so that a lot of potential was there. For example, PRIM-9 was used to analyze data from particle physics experiments, ... but it was also used to study the different diabetes syndromes. The PRIM-H system was used to inspect data on circadian rhythms of monkeys, on demographics of developing nations, and on clustering of remote galaxies.

Unfortunately, such systems were hidden away in labs and only a few people ever had access to them. Also, because they were used by so few people, issues like ease of use and capacity for working with real data on an everyday basis were secondary. The designers of these systems created them to make movies and videotapes illustrating the tremendous potential of dynamic graphics, but they did not use their systems for everyday work.

One difference between MacSpin and earlier systems makes this clear. Earlier systems were all demonstrated at conferences by projecting a videotape or movie that had been shot at the research institution and carefully edited. It would have been impractical to install a VAX and appropriate graphics devices on site for a live demonstration. At the 1986 Annual Meeting of the American Statistical Association, MacSpin was demonstrated live, using a Macintosh one of the authors carried into the lecture hall and plugged in, and a small projector which took the video signal direct from the back of the Macintosh. Dynamic graphics has become as portable as the 22-pound Macintosh.

(Of course, one can still use big computers in conjunction with dynamic graphics. For example, we like very much to use the S statistical system developed at AT&T Bell Laboratories, running on a VAX at Berkeley. We use the Macintosh as a graphics terminal to the VAX, and can rapidly transfer to MacSpin any data we have in S which we would like to view with MacSpin. In this way can use the VAX from home over a phone line and do dynamic graphics at home, which the earlier systems could never have allowed us to do).

4. AN ENVIRONMENT AND A USER INTERFACE FOR DYNAMIC GRAPHICS

A key feature of MacSpin, not present in earlier systems, is the supporting environment that MacSpin offers. This environment includes facilities for data input/output, data editing, graphical hardcopy, data transformation, and variations in the display options -- we will give details below.

It is hard to overstate the importance of this supporting environment. In our experience, successful use of dynamic graphics involves not just looking at a display of raw data, but a mixture of massaging the data (transforming and editing it), modifying the display in order to make the display more informative, and saving documentation of what has been seen. Indeed, exploration of data is inherently a process of discovering structure, and then removing it so that further structure becomes apparent. Thus, in principle, there ought to be a very tight loop in which the user alternately sees and then acts based on what he sees. The stories above give examples of the way this process happens in analyzing real data.

When the environment is not rich enough to support this tight loop, users become frustrated; they want to follow up on what they see, but cannot. In a few minutes of interacting with their data, many questions come to mind: "What happens if we look at gallons/mile instead of miles/gallon?", "Are all these outliers 4 cylinder cars?", "How well can we separate Chemical Diabetics from Overts using just two variables". The inability to immediately follow up on such questions imposes a burden on the user: he must remember the questions, and then improvise methods for answering them using, say, a statistical package or database management system. At the very least, this puts a burden on the user's memory; if the user has to use some other system or package to answer his questions it means that he must make constant context switches, and preserve his memory and sense of purpose across switches. The end result can easily be exhaustion, frustration, and loss of effectiveness.

So to encourage true high-interaction between user, data, and display requires a rich environment; but it also requires a clean *user interface* to that environment. It must be easy for the user to access the tools of the environment, or he will still get exhausted by the effort of using the system, and again lose the benefits of high interaction. Similarly, the visual design of the user interface must be clean and uncluttered so that the user's eye is not too much distracted by too much "screen junk". And the sequence of visual displays encountered in using the system must not be too varied or discontinuous, or else fatigue will arise from constant context switching.

We believe that MacSpin offers an excellent combination of richness of environment with quality of user interface. We now describe both elements of the combination in more detail.

4.1 The Environment

The facilities that MacSpin offers can be classified as follows:

→ Display Options. Changing how data get displayed.

→ *Information Pop-ups.* Providing additional information about the data.

→ *Subset Operations.* Defining, combining, removing subsets of data.

→ *Transformations.* Defining and transforming variables.

→ *Input/Output.* Data import/export, Data editing, Graphical Hardcopy.

→ *Documentation.* Manual with worked-out examples on real datasets.

We discuss these facilities in turn.

Display Options

These range from very simple ones:

Background color -- the ability to display data as white points on a black background or as black ones on a white background;

Axis Visibility -- the ability to have the tripod visible or invisible as rotation happens;

Big Pixels -- the ability to display points using either big or regular pixel widths;

Save Orientation -- the ability to "save" a certain 3-d viewpoint, and return to it later;

to more sophisticated ones:

Depth Cueing -- the ability to display points with size of marker indicating depth;

Set Origin -- the ability to set the origin of the rotation coordinate system to any point in the dataset, or to the centroid of any subset of points;

Rotate N Degrees -- the ability to numerically "navigate" the orientation so that the plot is seen from any numerically specified viewpoint in 3-space;

Spatial Scaling -- the ability to force the scaling of the plot axes so that proportions seen in the display reflect the actual numerical proportions present in the dataset (that is, in stretching or shrinking the plot to fit on screen, the same stretch or shrink is applied to all three axes).

Automatic Pilot -- the ability to have the plot rotate continuously, without need for continual pressing of the mouse button. The rotation can be either in one direction, or else rocking back and forth a few degrees about a fixed viewpoint.

Other features control display of individual data points. Thus, when data points are highlighted because they have been selected, the user has available two different intensities. And, as we have seen, -- points can be displayed with markers -- articulated symbols rather than just dots can be used to differentiate between subsets of data. the markers can also be temporarily suppressed to go back to a plain display.

In our experience, for each dataset, some subset of these display options will be essential. For example, spatial scaling is needed in looking at spatial data, markers are needed for data with categorical variables, and a white background is useful when preparing artwork. All these features, as discussed later, are available through simple point-and-click operations with the Macintosh mouse.

Information Options

MacSpin can provide the user with additional information beyond what normally is showing in the display. Much of this is accomplished through "pop-up windows" which the user calls up by holding the option and or command keys and pressing the mouse button. The resulting information is about either MacSpin or about the user's data.

> For information about MacSpin itself, there are pop-ups which: tell the user the current viewpoint (in latitude, longitude, etc.); tell him the current settings of the animation controls; and tell of the settings of the automatic rotation and rocking features.
>
> For information about the dataset, we have already seen the pop-ups that give the full data record on an individual; there are also pop-ups which give a range of summary statistics on any variable in the dataset, pop-ups which give a list of the individuals in the dataset, sorted by a variable, and pop-ups which count and list the members of a subset.

These pop-ups allow the interested user to quickly get answers to many of the basic questions that come up during exploration, without cluttering the screen and forcing the uninterested user to see the information.

Other features which could be listed as information-providing include the ability to inspect a spreadsheet-like view of the entire dataset, which the user can scroll through and edit.

Subset Operations

MacSpin allows the user to define and combine subsets of data. The definition can be based on selection from the plot window, using the mouse, or based on the label of a point matching a certain pattern (e.g. to make a subset "Ford" containing all individuals whose labels contain the string "Ford"), or based on a numerical criterion (e.g. to make a subset "Overt Diabetics" containing all individuals with GluFast > 120). The subsets can be combined via intersection, union, and difference. Once subsets are defined, they can be highlighted, specially marked, focused on or hidden from the display; and statistics of that subset alone can be computed.

As we have seen, the subset manipulations are the key to much exploratory work with real data. They allow one to focus attention on American cars, to isolate outliers in a subset and browse through them for clues as to where they came from, to compare different time periods in the dataset. We believe that MacSpin is unique in providing this broad range of subset operations. We have found these operations so frequently useful that the prospect of having to use a system without them would depress us greatly.

Transformations

As we have seen, MacSpin allows the user to transform his data, by any of the typical operations (add two variables, take logs of a third, ...). The user may also create linear combinations of several variables; this is useful interpreting linear regression and principal component analyses done elsewhere. MacSpin also allows the user to create new variables containing random numbers, and to create variables containing arithmetic progressions. These "artificial" variables are frequently useful. Thus, conditioning on a random variable allows one to select a random subset of the data; focusing on such a subset can be useful when the dataset consists of so many thousands of individuals, so that the display has gotten saturated with points. Progressions are useful, for example, so that one can select the last 418 points in the dataset, or if one wants to deal with time series.

Input-Output

MacSpin reads in Ascii files with white space (blanks or tabs) separating the entries. It also accepts data from the Macintosh clipboard. This means that the process of transferring data from, say, a VAX running S, or from another Macintosh program is quick and painless. If there are problems reading the dataset, a box appears showing the offending line in an edit box. The user can at that point correct the error, discard the line, or abort the read. MacSpin also can save data in the form of text files or on the clipboard. Some users have reported that they use MacSpin as a data manipulation tool, because of its subset and transformation features, when they are dealing with spreadsheets. They take the data from their spreadsheet program into MacSpin, manipulate it, and take it back into the spreadsheet program.

MacSpin allows three forms of graphical hardcopy. First, there is the "screen dump" which every Macintosh application offers. Second, there is a print command which allows the user to add captions and other embellishments. Third, the user can paste a figure onto the clipboard, and from there into MacDraw or another drawing program, where the figure can be edited, captions added, borders added, etc.

All of these abilities are easily used by anyone with a little Macintosh experience.

Documentation

MacSpin comes with a 189-page manual and 32-page supplement. The documentation explains not only use of the features, but how the features fit into the process of data analysis. Seven datasets are analyzed in detail, and stories similar to the Cars and Diabetes stories are given for each. The datasets cover fields ranging from geophysics to botany. The program itself now comes with 15 datasets spanning an even wider range than is covered in the manual.

We believe that this documentation represents the first real attempt to explain what dynamic graphics is and what it can do outside of research articles. Judging from the response we have received, we believe that it does get some basic points about data analysis and graphics across to a large audience.

4.2 User Interface

The many features we have described and/or hinted at in the last subsection would be overwhelming to the user without a good user interface. That interface needs to make the easy things easy, to hide unnecessary details from the person who doesn't need them, but to provide lots of power so that users don't get the sense of frustration we talked about earlier.

Luckily, by following the Macintosh User Interface Standard, we have been ahead of the game from the start. As the Section 5 points out, the Interface Standard has taught us a lot about designing user interfaces using the mouse and windows, to get a visually and intellectually satisfying result.

As a crude example of how the Interface Standard helps one avoid clutter and yet offer both simplicity and power, consider the device common to all Macintosh programs, the device of having a menu bar at the top of the screen with pull-down menus. In MacSpin, this gives a pleasing classification of the tasks available to the user: *File, Edit, Display, Variables, Subsets, Events, Markers*. The user thus knows that commands associated with the display are on the *Display* menu; he can get them any time, but doesn't have to look at them when he doesn't need them.

As another example, consider the pop-up windows, that bring information to the sophisticated user, then vanish when the user is finished with them. The novice does not need to know that they are available; the expert calls for them in an instant by holding down the option key as he clicks the mouse button. Functionality is provided without intellectual or visual clutter.

Nevertheless, it must be admitted that MacSpin offers an awful lot of features. To offer so many in another environment would perhaps be foolish; no one could remember them all, along with all the other things he has to remember, simply to work in that environment. Fortunately, all Macintosh applications work alike in certain ways, and this keeps all the complexity manageable. In learning a new program, the user only has to learn about the task unique to the program. All the other things about program operation -- starting up, opening files, editing data, moving and resizing windows, transferring data, printing -- work the same as in any other program. The user already knows them. In contrast, a user of UNIX may have to remember the 8 different ways of quitting a program that his 8 most commonly used programs employ, and 8 different sets of option flags, and different file formats, and so on. In the Macintosh world, there is much less of the "removable complexity" so omnipresent in less unified systems such as UNIX. Therefore individual applications can present a greater level of functionality without fear that users will be too confused or exhausted to make use of it.

4.3 Nonstandard Applications

Of course, the proof of the pudding is in the eating. If MacSpin's user interface and full-featured environment are so worthwhile, then there should be things that users have come up with in the course of analyzing their data that illustrate this. We can give three examples.

Mercedes Benz Plot

The Fortune 250 data supplied with MacSpin give information about the top 250 companies in the Fortune 500 listing for 1985; the information includes the rankings of the company in 1980, 1975, 1970, and 1965. Suppose we make a plot in MacSpin with axes giving the 1985, 1980, and 1975 rankings. However, we look at it from an unconventional viewpoint: so that each pair of axes makes a 60-degree angle (the Mercedes-Benz symbol). We position the axes in this way by using the "navigation". When we do this, we get the plot shown in Figure 11. When viewed in this way, a point is in the upper right if it has been shrinking (rank in 1985 > rank in 1980 > rank in 1975; recall that a higher rank means smaller sales (Exxon's 1985 rank is 1)). A point is in the lower left if it has been growing (rank in 1985 < rank in 1980 < rank in 1975). By going through the subsets of this dataset (Chemicals, Drug, Oil, Computers, ...) and highlighting each in turn, one rapidly sees which industries have been growing and which have been shrinking. This example shows a

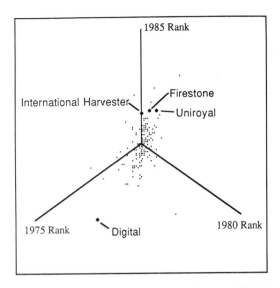

Figure 11. Mercedes-Benz plot of the Fortune 250 data.

nice, simple application of the navigation and subset highlighting features.

An Earthquake Movie

The Quakes dataset supplied with MacSpin gives the location of 1000 earthquakes occuring in the South-West Pacific between 1964 and 1980. The events are labelled simply by their chronological order. If the user selects the high-intensity highlighting option, then when he points at a label in the Events window, the position of that earthquake is highlighted on the display with a big, fat, dot -- practically, a 'ball'. Now suppose he brushes the mouse through the list of labels; then the quakes get highlighted one after the other, in a smooth fashion. The user sees the sequence of earthquakes "happening" bam-bam-bam, all over the South-West Pacific. It looks as if a ball is bouncing all over the screen randomly. Occasionally, many earthquakes occur in a small region of space and time; such an aftershock sequence shows up vividly in the movie, as the "bouncing ball" stops hopping all over the screen and stays confined to a relatively small region. The aftershock sequences practically leap out at the viewer.

Flooding Oklahoma

Figure 12 shows a digital terrain map of an area in Oklahoma. The map consists of 4096 points, with X-Y coordinates in a regular 64 by 64 grid. The Z variable is altitude. By animating on a fourth variable, one can see easily how it correlates with topography. Even on an ordinary Macintosh, the speed of animation on this large dataset is fast enough

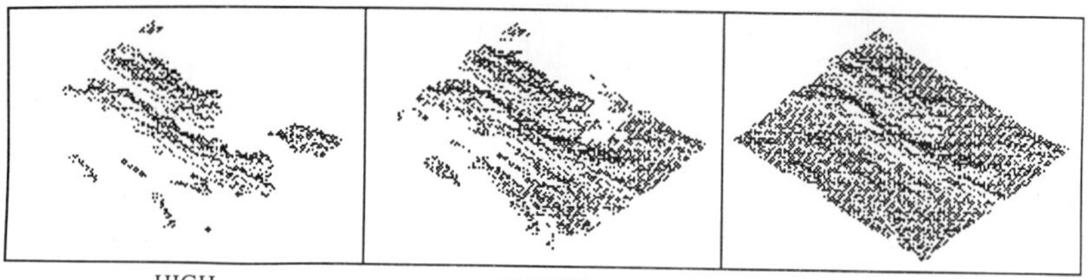

| HIGH | MEDIUM | LOW |

Figure 12. Stages in the flooding of a digital terrain map.

that by animating back and forth, one gets a very good feel for anomalies and correlations. The example would not be possible except for the spatial scaling feature in MacSpin, which allows physically faithful scaling of plots.

These examples serve to illustrate the point that a rich environment allows one to see many different things, in many different types of data.

5. DEVELOPING FOR THE MACINTOSH

Of course, the difference between MacSpin and earlier systems is far more than just the availability on a personal computer. As we have indicated, the choice of the Macintosh has had far-reaching effects on the functionality of the program.

As you probably know, the Apple Macintosh is a personal computer with a high-resolution bitmap display and a Motorola 68000 microprocessor. Like other personal computers, it is available inexpensively -- especially under some of the discount plans that exist for universities. There are now 1 million Macintoshes in use.

MacSpin follows the Macintosh User Interface Standard. This standard was created by Apple Computer to ensure that all software developed for the Macintosh follows a uniform protocol for issuing commands, displaying information, opening files, and so on. In our opinion, this standard is an important innovation in the computer world. Such a standard does not exist on any other computer -- whether it be an IBM PC, a SUN, a VAX, or an IBM mainframe. The effects of this standard are: Every Macintosh user already knows how to open files, edit them, print out results,... for *every* program that he might consider using on the Macintosh. It is no exaggeration to say that most Macintosh applications do not need user manuals. In a very real sense, most Macintosh users can profitably employ programs like music synthesizers, paint programs, page-layout programs -- even without being an expert in the fields addressed by those programs, and without being a computer programmer. Every Macintosh developer is taught, by the standard, how to produce an application with a clean, smooth user interface and a sharp appearance. Since most programmers, and most people, don't have any skills in graphical design or in user interfaces, this means that the ease of use and appearance of Macintosh software are an order of magnitude better than those of any other computer.

The Macintosh User Interface Standard is derived from years of experience with graphics user interfaces on the SMALLTALK-80 system developed at Xerox. For MacSpin in particular, following the the User Interface standard meant a prolonged period of apprenticeship for us, learning the standard and its rationale. For example, part of our task was to take standard User Interface notions which seemed designed for word processing and similar applications, and *discover* what they meant in a data analysis application. Eventually, we discovered things that were natural for data analysis that were not considered in the standard -- so we had to create new user interface notions in several places (our information and hyper information pop-ups are examples of this). In retrospect, we believe that only through such an apprenticeship can one develop the knowledge necessary to produce a polished user interface. In particular, we don't think that a statistician could produce a very good user interface just using his subject matter knowledge and just experimenting with the computer on his own.

Once the design is done, implementing a good user interface is hard work. This can be judged from the fact that MacSpin is 36,000 lines of Pascal code (not counting comments). Thus MacSpin is as large as, or larger than, most statistics packages for microcomputers -- and MacSpin is a display program only. Again, we doubt that we would have invested this sort of effort in MacSpin, except for the social pressures brought to bear by the Macintosh community to do work meeting a certain standard of polish and ease of use.

6. EXPERIENCES

MacSpin has expanded by a few orders of magnitude the number of people who have a chance to use dynamic graphics in their work. Many users of MacSpin are, surprisingly, not scientists, or professional statisticians; we were somewhat skeptical of this trend at the beginning, but have found that the statistically uninitiated can find many uses for this program. In fact, some of the more interesting applications have been found in areas that only a novice would have the freshness of mind to think up. MacSpin has been used by amateur astronomers to look for patterns in local star clusters and by a homemaker to compare nutritional content of different recipes.

Our own experiences, and feedback from users, have indicated a few broad conclusions.

[1] While rotation of point clouds is eye-catching, this is only a small part of what a good dynamic graphics program should do. Abilities such as subset highlighting and animation on a fourth variable play a large role in helping a user understand the data he is looking at. These allow the user to dig into the reasons that certain patterns appear on the point cloud display. In our experience, the MacSpin user spends about 20% of his time watching rotations, and the remainder of the time using other dynamic, interactive features of the program. Point cloud rotation without all the other features of MacSpin is just a gimmick -- it's eye catching, but it doesn't help answer any questions that come up during the analysis.

[2] MacSpin involves the user fully in the analysis of his data. While using MacSpin, the user is constantly asking questions -- "Who is that outlier?", "Where are the American cars" -- and getting instant, graphical feedback. MacSpin establishes a very tight loop between the user and the display; this is only possible because of the graphics/mouse user interface. Moreover, using graphics and a mouse is not enough: very subtle defects in the user interface would be enough to sabotage the interaction loop entirely.

[3] After building MacSpin, when we look at other people's work in dynamic graphics, we notice two common problems in their designs which would, we believe, sabotage the interaction process. One of these is constant *context switches*. By this we mean that the user is not presented with one basic display format and one uniform style of interaction, but instead, with frequent changes: a scatter plot is present; it goes away, and is replaced by a menu; the menu goes away, and is replaced by the scatter plot, etc. While the menu is present, the user cannot see the scatter plot, and vice versa. This means that the user constantly has to adjust to -- parse -- his changing visual environment, rather than focusing on his data. Also, the user is forced to remember things he saw in one view -- e.g. the list of variable names -- so that he can use the other view effectively. This means that the user's short term memory is occupied with incidentals rather than with the significant issues of analysis. In MacSpin we try, as much as possible, to leave the basic plot on the screen, and pop-up small windows and menus on the screen which leave everything else visible. Thus, to the user, it seems that there is one basic visual scene, with minor variations. Also, things like the list of existing variable names and subset names are still visible while a dialog box asking for the name of a new subset or variable is present -- so if the user forgets one of the names, he can refer back to the list.

A second common problem is *clutter* -- presenting too much information and too many options to the user at once. These command items -- often rarely used -- are a constant drag on the visual system. In contrast, the standard MacSpin display presents almost no command items; only things having to do with the dataset are visible -- the scatter plot window, the variables list, the subsets list, and the event names list. Commands pull down or pop up and are conveniently organized for easy recovery. One could say that there is a lot of power hidden behind a chaste, attractive facade.

7. A FINAL WORD

We should emphasize that the real value of dynamic graphics is in the *process* of human-data interaction. This interaction between the user and his data is very helpful in most investigations; it helps the user choose a strategy for further analysis and rapidly find out about the flaws and peculiarities of his dataset. Dynamic graphics does not replace other types of analysis, such as those one might want to do with S or SAS. However, it is an important addition to the analyst's toolkit.

SUGGESTIONS FOR FURTHER READING

Becker, R. A. and Cleveland, W. S. (1984). "Brushing a scatter plot matrix: high interaction graphical methods for analyzing multidimensional data," Technical Memorandum. Murray Hill, NJ: AT&T Bell Laboratories.

Donoho, D., Huber, P. J., and Thoma, H. (1981). "The use of kinematic displays to represent high dimensional data," *Computer Science and Statistics: Proceedings of the 13th Symposium on the Interface*, 274-278. New York: Springer-Verlag.

Fisherkeller, M. A., Friedman, J. H., and Tukey, J. W. (1974). "PRIM-9: An interactive multidimensional data display and analysis system," SLAC-PUB-1408. Stanford, CA: Stanford Linear Accelerator Center.

Friedman, J. H. and Tukey, J. W. (1974). "A Projection Pursuit Algorithm for Exploratory Data Analysis," *IEEE Transactions on Computers*, **C-23**: 881-890.

McDonald, J. A. (1982). "Interactive graphics for data analysis," PhD thesis. Stanford, CA: Stanford University.

Reaven, G. M. and Miller, R. G. (1979). "An attempt to define the nature of chemical diabetes using a multidimensional analysis," *Diabetologia*, **16**: 17-24.

14

THE APPLICATION OF DEPTH SEPARATION TO THE DISPLAY OF LARGE DATA SETS

Thomas V. Papathomas
Bela Julesz

AT&T Bell Laboratories

ABSTRACT

The importance of stereo and animation in displaying large data sets, in which the spatiotemporal variation of one or more variables is given quantitatively, is examined. Several techniques are described for depicting univariable meteorological episodes with a high degree of realism. In addition, a new method is presented which employs depth separation for the simultaneous display of multiple variables, each of which depends on two spatial variables and, possibly, time. Examples are given from the application of this method to practical problems in the areas of VLSI modelling and computer-aided tomography.

1. INTRODUCTION

The ever increasing speed and memory capabilities of computers has resulted in the numerical solution of models for large-scale problems. The solution, in many cases, is in numerical form. This is certainly the case when modelling phenomena whose variables have a spatiotemporal

---

Reprinted in part, by permission, from: T. V. Papathomas, *et al.* (1987). "Stereo animation for very large data bases: Case study-meterology," *IEEE Computer Graphics and Applications*, 7(9): 18-27. © 1987 IEEE.

From *Dynamic Graphics for Statistics*, William S. Cleveland and Marylyn E. McGill, eds., copyright © 1988 by Wadsworth, Inc. All rights reserved.

distribution, such as meteorological episodes, seismological events, semiconductor behavior, heat transfer, etc. When large spatial and temporal resolutions are used in the model, the size of the resulting numerical output dataset can be enormous.

The companion problem, of almost equal importance, is how to present these results so that they are easily comprehended and assimilated. The problem is a difficult one, especially when the original model involves more than one variables whose spatiotemporal distribution is numerically approximated. Numerous techniques have been developed for the visual display of such simulation data, as well as for depicting statistical information. The present volume contains a wide selection of papers on this issue. Four excellent books on the subject are by J. Tukey [1977], Tufte [1983], Chambers et al. [1983], and Cleveland [1985].

In this paper, the role of stereo and animation in depicting voluminous multidimensional data points is examined from a dual point of view: that of a visual psychophysicist on one hand, and a display specialist on the other. We have found in psychophysical experiments that stereo and motion cues enhance the perception of complex visual stimuli and aid substantially in discerning details that would go unnoticed in these cues' absence. We have applied this methodology in developing a very efficient system for displaying meteorological data of up to five dimensions, and we report briefly on our methods which resulted in very realistic renderings of the weather episodes.

We also present a novel technique that we developed for displaying multivariable time-varying phenomena. This technique can best be used to portray *simultaneously* two variables f_1 and f_2 which are functions of two spatial parameters x and y and time t. The key idea is to display $f_1(x,y,t)$ and $f_2(x,y,t)$ on two different depth planes p_1 and p_2 by utilizing appropriate binocular disparities for the left-eye and right-eye images, and to combine the two images as if the frontal plane p_1 were *translucent*. This allows the two events associated with f_1 and f_2 to be perceived simultaneously with a higher degree of success than multi-color or side-by-side displays. We have experimented with this technique with fractal images and then applied it to display data from computer-aided tomography (CT) and Very Large Scale Integration (VLSI) modeling.

2. STEREO DISPLAY OF MULTIDIMENSIONAL DATA

In this section we briefly review some key ideas from psychophysical experiments in visual perception, particularly in the areas of stereo and motion perception. After a few general remarks on the efficient coupling of images to the human visual system, we present some results from our research on the interaction between stereo and motion perception, on which our novel display techniques are based.

2.1 Man-Machine Symbiosis

Human visual perception is performed by the most complex structure of the known universe, the visual cortex, that contains at least 10^{10} neurons, where each neuron in average contains 10^4 synapses (gates). This enigmatic processing network can perform prodigious feats when properly coupled to the visual stimuli. For instance, as one of us showed [Julesz, 1960], if one presents to each eye an array of random dots, provided the correlation between the two arrays consists of horizontally shifted subregions of random dots (even without any monocularly perceivable gap or boundary) and the shift stays within some critical limit (called Panum's fusional area), any normal human observer will be able to establish the binocular correlations in about 60 msec and obtain a fused image in vivid depth [Julesz, 1971], even with large arrays of dots, in excess of 10^6 pixels.

Of course, this information has to be matched to human heuristics. For instance, if the correlation between identical but shifted areas were vertically displaced (instead of horizontally) no stereoscopic fusion could be obtained. Similarly, if a horizontally displaced area would be displaced by more than Panum's fusional limit, then in the absence of monocular cues, such as contrast, velocity, etc. gradients, fusion would become difficult, or impossible. Obviously for stereoblind people (2% of the population), even horizontally displaced areas within Panum's fusional area cannot be fused.

We have shown years ago [Julesz and Oswald, 1978] that if one adds to a random-dot stereogram some monocular cue, for instance, contrast gradients, latency of stereopsis is greatly reduced. The same applies for introducing in dynamic random-dot stereograms some monocular velocity cues, in the form of some moving contours, that are correlated in the two eyes' views. This additional information enhances stereopsis (stereoscopic depth perception) and disambiguates depth localization in cases where two transparent depth planes are stacked above each other, and it can be ambiguous whether a certain dot (pixel) belongs to one or the other depth plane. The next subsection discusses this problem in detail.

Here we tried to capitalize on additional human abilities beyond stereopsis. It is a common experience, that one can look through a window and perceive simultaneously events that appear through the window and also events that are reflections from the window. As a matter of fact, if the two events are very similar — although portrayed in different depths — humans are able to detect this sudden similarity most effectively. In Section 4 of this article we describe some of our efforts to portray events (in x, y, and t) in two transparent depth planes stacked above each other.

2.2 Perception Enhancement through Stereo and Motion

Recently, we conducted a series of experiments which vividly demonstrated the value of motion cues in stereopsis [Julesz and Papathomas, 1985]. Our stimuli were random-dot stereograms containing N families of planar surfaces with distinct binocular disparities. Family f_k consisted of several vertical bars at a fixed disparity, rendered in random-dot stereogram format, for $k = 2, \ldots, N$, with f_1 reserved for the background, which was shown at zero disparity. Initially, the kth family f_k was portrayed on rows $k, k+N, k+2N, \ldots$ for $k = 1, 2, \ldots, N$. Thus f_1 covered the entire length of rows $1, 1+N, 1+2N, \ldots$; it also covered those parts of the rest of the rows that were not occupied by any vertical bar.

When these stimuli were viewed statically, it was extremely difficult to discern the various surfaces in depth, even for $N = 2$; for $N > 2$ it was impossible to achieve stereoscopic fusion. In the absence of motion, the bars were perceived as fuzzy, non-rigid objects and their dot patterns were nearly impossible to be seen. On the contrary, when the $N-1$ families of vertical bars were moved laterally at various speeds or directions from each other, the constituent bars were easily segregated and were perceived as rigid translucent surfaces at different depths. It became significantly easier to spot them and even to study the two-dimensional random-dot patterns that defined them. Even when two or more bars belonging to different families were occupying the same horizontal position, the percept was that of translucent bands, sliding over each other.

In the first phase of the experiment, each bar had the same dot pattern from one frame to the next. This yielded the strongest depth segregation, although such segregation was also observed when the lifetime of each dot was limited to a few successive frames in the second phase of the experiment. In the limit when the lifetime was only one frame, the percept was as weak as the static one. In the third phase, we rendered each bar not by assigning to it r/N rows (where r is the total number of rows), but by randomly speckling the area A of the bar with A/N dots at the desired disparity. The results were very similar with this last type of stimuli.

In the absence of both stereo and motion cues, i.e. when a single image was viewed statically and monocularly, obviously there are no discernible bars and the percept is that of a random-dot background. When stereo alone was utilized statically, the bars appeared, albeit in a fuzzy fashion. When motion alone was present, i.e. when the animation sequence was viewed monocularly, the bars were segregated but there was no depth information. The segregation was weakened considerably when one bar passed over another. It was when both stereo and motion were utilized that the percept became exceedingly clear, demonstrating the significance of both in comprehending complex visual stimuli.

3. DYNAMIC DISPLAY OF UNIVARIABLE METEOROLOGICAL DATA

It was evident from our earlier experiments that our techniques could be tried effectively in the display of large data sets whose elements varied in the spatial and temporal dimensions. Indeed, in collaboration with meteorologists, we have developed a system for displaying stereo animated sequences of the distribution of a single meteorological variable in space and time [Schiavone et al., 1986]. In this section we shall briefly describe the form of the data, the graphics system and the techniques we developed. The problem was to take the numerical output data of a weather simulation program and to display them in a format which would couple them effectively to the human visual system.

3.1 The Problem

The meteorology community has always pushed for novel ways of displaying large sets of statistical and simulation data, but this demand will increase in the near future. A major restructuring of the USA's National Weather Service will take place over the next five years to incorporate the latest technologies for observing and forecasting the weather with at least an order of magnitude increase in spatial and temporal resolution. The goal is to better forecast the smallest, but almost always the most damaging, meteorological features. This forecasting revolution will place additional information assimilation burdens on forecasters from a dual perspective: from the increase in the data set size that they must examine and from the reduced lead times that they will be afforded to forecast these extremely fast-breaking meteorological events.

A number of innovative methods have been developed for depicting weather episodes, most of which are based on the power of animation [Grotjahn and Chevrin, 1984; Hausman, Kroehl and Kamide, 1985; Hibbard, Krauss, and Young, 1985; Hasler et al., 1985; Schlatter, Schultz, and Brown, 1985]. Anaglyph stereo display [Hasler, des Jardins, and Negri,

1981] and perspective projections [Hasler et al., 1985; Klemp and Rotunno, 1985; Vonder Haar et al., 1986] have also been explored. Finally, a combination of animation and stereo is being investigated by several groups [Hibbard, Krauss, and Young, 1985; Hasler et al., 1985; Vonder Haar et al., 1986].

We use meteorological data in our case study but our approach integrates animation and three-dimensional display techniques in a way which is significantly different from the above-cited techniques. The meteorological data set is the numerical output of LAMPS, a simulation program which models meteorological phenomena. LAMPS (Limited Area and Mesoscale Prediction System) was developed at Drexel University [Kreitzberg and Perkey, 1976] to predict rainfall accumulations caused by intense thunderstorm systems. A collaboration was established among meteorologists, graphics experts and visual perception scientists to create and evaluate novel display methods for the voluminous numerical data output of the program. Our effort culminated in the development of a display method to aid the meteorologist in visualizing time-varying three-dimensional meteorological fields and it demonstrated that anaglyphic stereo animation is a valuable technique for the portrayal of the data [Schiavone et al., 1986].

The heart of the LAMPS software system is an atmospheric dynamics model for simulating severe weather episodes, such as thunderstorm convection systems. Its numerical output data are the values of various parameters for each (x,y,z) coordinate over a time interval. The size of the data grows rapidly as the spatial and temporal resolutions are increased (one of these runs resulted in an output of 14.7 MBytes). The problem is to find techniques for displaying these data effectively. Before we discuss the techniques, we will briefly describe the model and its output.

3.2 Data Structure-Display Format

LAMPS simulates the evolution of precipitating weather systems which have horizontal resolution on the order of 50 km. This spatial scale is known among meteorologists as the "mesoscale." It produces a four-dimensional array of values $v(x_i, y_j, z_k, t_m)$, where v is the parameter of interest, x_i, y_j and z_k are the discrete spatial coordinates of longitude, latitude and altitude, and t_m is the time interval, with

$$1 \leq i \leq n_x, \quad 1 \leq j \leq n_y, \quad 1 \leq k \leq n_z, \quad 1 \leq m \leq n_t \tag{1}$$

Successive points along the x-, y- and t-axes are equidistant, i.e.

$$x_i = x_0 + i\Delta x, \quad y_j = y_0 + j\Delta y, \quad t_m = t_o + m\Delta t \tag{2}$$

where Δx, Δy and Δt are the corresponding intervals, and (x_o, y_o) locate the geographical origin of the studied episode.

For the z-axis, there is a monotonic function [Kreitzberg and Perkey, 1976] that maps the subscript k to an altitude z_k such that $z_1 = 0$ and

$$z_m > z_k \quad \text{for} \quad m > k \tag{3}$$

Thus the volume of interest is the closed parallelepiped bounded by the longitudinal planes $x = x_1$ and $x = x_{n_x}$, the latitudinal planes $y = y_1$ and $y = y_{n_y}$, and the altitudinal planes $z = 0$ and $z = z_{n_z}$. The values for n_x, n_y, n_z, and n_t were 113, 113, 15, and 25, respectively.

Since it was not possible to process the data at video rates, it was decided to use one of the best alternatives available, i.e. to use a frame buffer with adequate memory to store the entire animation sequence in semiconductor memory (RAM). The system that we used was a VAX 11/750 computer connected via a DMA link to an ADAGE RDS3000 Raster Display System. The latter possesses a memory of 1024×1024 pixels each 32 bitplanes deep, a bit-slice processor for animation at video rates, a frame buffer controller for zoom, pan and timing control, and a cross-bar switch for mapping different combinations of bitplanes onto the input of a color look-up table, whose output feeds the red, green and blue beams of the display device.

Thus the output data of the LAMPS program, which was run on the CRAY 1 supercomputer, were transferred over the ETHERNET network to our VAX 11/750 minicomputer to be processed locally for graphical display. The next issue that had to be addressed was the format to be used for depicting these data.

To make the discussion concrete, let us assume that $v(x_i, y_j, z_k, t_m)$, the meteorological variable of interest, is the *cloud water concentration* of the air parcel centered at (x_i, y_j, z_k) at time t_m. Thus points for which v is greater than a threshold value v_T can be thought of as forming a cloud. We selected the "particle system" format [Reeves, 1983; Magnenat-Thalmann and Thalmann, 1985] in which every point in the volume of the cloud is illuminated.

Some of the psychophysical considerations that we alluded to in the introduction also led us in the same direction: 1) Julesz's pioneering work on stereo and movement perception, summarized in his book [Julesz, 1971] proved conclusively, through the use of random-dot stereograms, that stereopsis is possible even without monocular cues [Julesz, 1960]. The presence of monocular cues in our case, due to the shapes of clouds and to the dithering rendering (see next subsection), makes the stereo perception even easier. 2) More than a million points were used in some random-dot stereograms [Julesz, 1971; Julesz and Johnson, 1968], well above the maximum number of approximately 192,000 (113×113×15) points in our model. 3) Our findings on the interaction between movement and stereo perception, outlined in subsection 2.2, indicated that the movement aided in the perception of depth, and this was also observed in the present case.

The main advantages of selecting the particle-system format are: 1) *Computational efficiency*: All points are treated similarly and there is no post-processing of the graphical output to identify edges, hidden surfaces, etc. 2) *Adaptability for parallel computers*: As a result of item 1 above, this method is easily extended to parallel-processor architectures with minimum effort and considerable gain in processing time. 3) *Scene realism*: Particle systems are particularly suited for depicting fuzzy objects [Reeves, 1983] to which clouds certainly belong. 4) *Motion realism*: We were able to verify a claim made elsewhere [Reeves, 1983] that this format is able to represent dynamic changes of form that are difficult to achieve with classical surface- or line-based renderings.

3.3 Novel Stereo Animation Techniques

In this section we shall briefly describe the data transformations and the computers and peripherals used in the project. We shall also discuss the display formats that were used and the techniques that resulted in noticeable improvements. Finally, a few display variants are presented, which added flexibility to our system.

The data specifying the initial conditions, which were used by the meteorological model, were obtained from radiosonde measurements and were provided to us by Kreitzberg and Perkey of Drexel University. The model was run on a CRAY 1 supercomputer and its output numerical data were transfered to our VAX 11/750 via an ETHERNET link. These data were used as the input to the graphics algorithm, which produced one 512×512-pixel frame for each time step. Each image consists of two subimages, one for each eye for stereo rendering, and each subimage is stored on a single bitplane, i.e. in binary fashion. As each image is generated, it is assigned to and stored in consecutive bitplane pairs; the subimages for the left and right eyes are stored in odd and even bitplanes, respectively, of the frame buffer in the RDS3000. Appropriate horizontal disparities, depending on z, are introduced between the left- and right-subimages to produce the perception of depth.

After all the data have been processed and all the frames have been created, the contents of the frame buffer memory are saved on disk for future retrieval. A bit-slice processor in the RDS3000 system is able to display the frames in sequence and produce animation at video rates (30 frames/second) by exercising the frame buffer controller, the cross-bar switch, and the look-up table. For each frame, the right and left subimages are directed to the red and green beams, respectively, of a large projection TV system to produce stereo anaglyphs, which are viewed with red-green glasses to produce scenes of vivid depth. The large memory of the RDS3000 allows animation intervals of up to 6.4 seconds, which are more than adequate for depicting quite complex meteorological episodes [Schiavone et al., 1985].

One frame is generated for each time slot t_m, $m = 1, 2, 3, \ldots, n_t$. The particle system representation is used in each frame: each point (x_i, y_j, z_k) belonging to the cloud is treated as a light source and the pixel (x_P, y_P), corresponding to its perspective projection point, is illuminated. Although this rendering does not address the issue of shadows cast by clouds onto other clouds [Reeves, 1983], the images obtained were satisfactorily realistic. An example of a frame with its two subimages is shown in Figure 1.

The next issue is how the point (x_i, y_j, z_k) was projected. The left eye (EL) and right eye (ER) were assumed to be at the same altitude and latitude, and separated by an interocular distance e:

$$z_{EL} = z_{ER} = z_E; \quad y_{EL} = y_{ER} = y_E; \quad x_{ER} - x_{EL} = e \qquad (4)$$

This assumption doesn't impose serious restrictions on the viewing angles, but reduces the computation time considerably, as explained below.

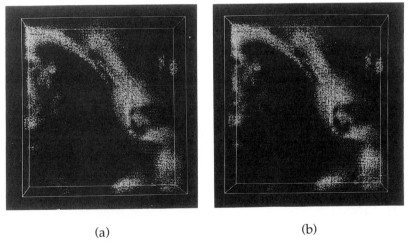

(a)　　　　　　　　　　　　(b)

Figure 1. The left (a) and right (b) subimages for a particular time frame in the animation of a cloud formation. The particle system scheme is used here with jittering and dithering (see text for details).

The perspective projection equations, which produce the projection point (x_P, y_P) from the object coordinates (x_i, y_j, z_k) and the eye coordinates (x_E, y_E, z_E), are

$$x_P = \frac{x_E z_k - x_i z_E}{z_k - z_E}; \quad y_P = \frac{y_E z_k - y_j z_E}{z_k - z_E} \qquad (5)$$

The disparity d is defined as $x_{PL} - x_{PR}$, the difference in x_P between the left and right eyes. For simplicity in comparing different algorithms, we let $n_x = n_y = n_z = n$, although this is not the case in the model's parameters. This results in n^3 points in the volume of interest, for which we need $6n^3$ additions/subtractions (+/−), and $12n^3$ multiplications/divisions (*/÷) for the two eyes. However, we can reduce these requirements from $O(n^3)$ to $O(n^2)$, based on the following observations from Equation (5): 1) Two distinct points with $z_{k1} = z_{k2}$ will have the same disparity $d = f(z_{k1})$. 2) Two points with $z_{k1} = z_{k2}$ and $y_{j1} = y_{j2}$ have the same y projection y_P. 3) Since $y_{EL} = y_{ER}$, points with the properties of item 2 above have the same y projection for both eyes. 4) Similar to 3, two points with $z_{k1} = z_{k2}$ and $x_{i1} = x_{i2}$ have the same x projection x_P. The algorithm we developed scans the volume first along "z-slices," thus computing d only once for each z value. We then scan each z-slice along "y-lines," thus computing y_P only once per line, taking advantage of observation 2. Finally, x_P has to be computed only once (for the first y-line), according to observation 4 (this value is stored); corresponding points on subsequent y-lines have the same x_P, which is retrieved from memory, thus avoiding a lengthy computation. Overall, this results in a reduction of the computational complexity from $O(n^3)$ to $O(n^2)$. For more details, consult Schiavone et al., 1986.

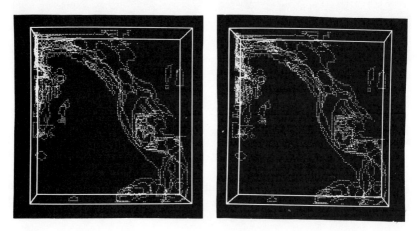

Figure 2. Same as Figure 1, but utilizing the outline scheme on the same set of data.

In addition to the particle system format, which we termed "*full-body*" scheme because it displayed every point with $v(x_i, y_j, z_k) > v_T$, we also used the "*outline*" scheme. In the latter, the volume was scanned along z planes (slices); within $z = z_k$, only those points lying along the boundary of the cloud were illuminated, thus enhancing the outline of the cloud (see Figure 2). This permitted some transparency through upper cloud slices, which was useful for the meteorologists in certain cases. The following techniques offered noticeable improvements in the quality of the graphics.

1) *RANDOM JITTER*: In the original phase, each point was projected on the (x,y) viewing plane. However, since the data points occupied lattice positions specified by (1), each projected z slice resulted in a very regular equidistant lattice distribution, as if depicting a crystal structure (see Figure 3(a)). Worse yet, superimposed z slices created undesirable Moiré patterns, which interfered with the intended perception of the particle system as a cloud. This was solved by first obtaining the pixel (x_P, y_P) for the projected point and then assigning randomly one of the nine immediate neighboring pixels (including (x_P, y_P)) to that point, using the uniform random distribution. The result, a sample of which is shown in Figure 3(b), produced a dramatic improvement in the realism of the depiction. A much more realistic example is shown in Figure 1.

2) "*DITHERING*" It would be nice to indicate, along with the four standard dimensions (x,y,z,t), a measure of the magnitude of the parameter of interest, v, at each point as a fifth dimension. Unfortunately, this is impossible to achieve with modulating the intensity of the point, since only one bit per pixel is alloted to each subimage of every frame (dedicating more bits would increase the storage requirements per image to a point where we would not have an adequate number of frames for a smooth animation). However, since the display array is 512×512 pixels while the particle system's z slices are 113×113 points, there are more than nine neighbors to each point (including itself). Thus, we can use a 3×3 dithering matrix [Foley and van Dam, 1982] to show ten ranges of values for v (see Figure 3(c)). A sample of an ideally dithered z slice is shown in Figure 3(d), with a more realistic example appearing in Figure 1.

3) *EDGES OF METEOROLOGICAL DOMAIN:* The twelve edges of the parallelepiped faces that enclose the volume of interest were shown as solid lines to define the model's domain. This was a simple idea whose implementation established a visual reference frame in which the episodes could be perceived with added clarity. This feature is applicable to both, the particle system and the outline display schemes, whereas the other two (jittering and dithering) are only applicable when the particle system format is utilized.

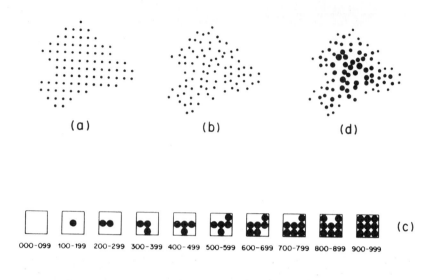

Figure 3. (a) Cloud points on a z slice. (b) The same set of points with "jittering" applied to them. (c) A set of dithering patterns. (d) The same set of points as in (b), including one form of dithering.

We initially developed the efficient display algorithm that we just described, whose input is the direct numerical output of LAMPS, and whose graphical output we termed "spatial" display mode. We next looked into the possibility of preprocessing the LAMPS output data before submitting it to the display algorithm. This resulted in the following display modes, in addition to the standard one:

1) "MAGNITUDE" MODE: This mode is used when the magnitude of v at a given altitude, say $z = z_3$, is the variable of interest. In this case, rather than using dithering to portray the magnitude of $v(x_i, y_j, z_3, t_m)$, we map the point to a different one (x_i, y_j, z', t_m) whose altitude z' is proportional to the magnitude of v; in this sense, we create a surface as a function of (x, y) whose height, i.e. z value, varies with v.

2) "TRAJECTORY" MODE: This mode provides visual information on the deformation and divergence characteristics of the atmospheric fluid as it is acted upon by the longitudinal and latitudinal wind velocities. The idea is to start with a well-defined surface or volume at $t = t_1$ and observe its deformations as it is subjected to the wind velocity field during the course of the episode. Only the points of the surface or volume are assigned values greater than the threshold v_T, so that they are the only points to be illuminated. The position of a point in frame t_{m+1} is obtained from that in frame t_m by displacing it according to the local

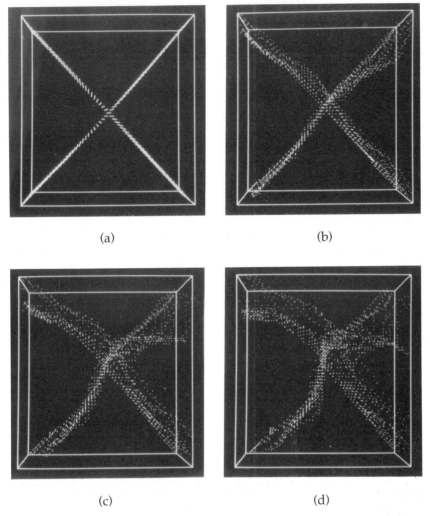

Figure 4. (a) The initial surfaces and subsequent frames showing their deformations at $t = t_m$ for $m = 3$, (c) $m = 6$ and (d) $m = 9$.

wind velocity in frame t_m. A few typical frames from such an animation sequence are shown in Figure 4, where the original surface consisted of the intersecting planes $x = y$ and $x + y =$ constant.

3.4 Results — Extensions

The animation sequences generated by our system proved to be a very valuable tool for meteorologists who want to study the performance of weather simulation models. By providing an easy method of comprehending the output data, it helps them form the intuitions necessary for discovering new ideas. One of the possible uses of the system is to first display the entire episode (which can cover a geographical area larger than the U.S.) and to view it for the purpose of locating in it a particularly interesting sub-domain, which can be either spatial or temporal or a combination of the two; the next step would be to run LAMPS over the new domain with increased resolution and view the new animation sequence. This ability to zoom in space and/or time is a feature that is made possible by the visual feedback afforded by the transformation of the numerical data into moving images.

It takes about 50 minutes of processing time to produce an animation sequence with $n_x = n_y = 113$, $n_z = 15$ and $n_t = 25$, on the VAX 11/750 with the modified display algorithm; the same task would require tens of hours if the complexity of the algorithm was not reduced from $O(n^3)$ to $O(n^2)$ by taking advantage of the special geometry of the situation. This turnaround time of about an hour, although long by absolute standards, is reasonably short for providing visual feedback and guiding the meteorologist in searching interesting subdomains.

To limit the processing time, we used a simple variant of the particle system mode, [Reeves, 1983] in favor of other computer graphic techniques of displaying clouds [Gardner, 1985; Blinn, 1982]. The relatively short time, about one hour, required by our implementation to process complex output data from the LAMPS model for animation illustrates the method's potential for weather forecasting operations.

The program was tried successfully with data from meteorological simulations, utilizing both the outline and the full body schemes, as well as the standard ("spatial"), magnitude and trajectory display formats. In addition to cloud water concentration, the spatial and temporal distributions of other key variables, such as pressure, temperature, and humidity, were depicted on our system. The resulting animation sequences gave the meteorologists an accurate perception of the time evolution of weather episodes. Other potential fluid dynamics applications for our method, beyond the mesoscale weather forecasting application, include the study of electromagnetic (microwave and optical) propagation in the atmosphere and the study of convection in crystal melts.

One mode in which our system could be, but has not yet been, used is to provide the developer of the model with useful visual feedback for comparing the performance of the prediction model against the actual episode that ensued starting from the same initial conditions. Major errors between the predicted and actual behavior of the atmospheric fluid could be isolated by the powerful human visual system; these deviations could be subsequently analyzed and an attempt could be made to correct their source, thus resulting in ever improving models.

4. NOVEL STEREO TECHNIQUE FOR MULTIVARIABLE PHENOMENA

In this section we shall present a technique for portraying two two-dimensional images simultaneously on the same visual field in a manner that allows us to attend to one of them in detail while still following the other image "with the corner of the eye," without much difficulty. If each image changes with time, this technique can be extended to display two two-dimensional episodes simultaneously or, more importantly, to delay one with respect to the other in order to discover correlations that may involve a time lag. In the special case where the images are highly correlated, such as successive computer-aided tomographic scans, more than two images can be displayed, as explained in subsection 4.4.

To make the presentation more concrete, let $f_1, f_2, \ldots, f_N$ be the image intensity functions, with each f_k being a function of the spatial variables x, y and (possibly) of time t. The problem is to display these functions simultaneously either statically or, in the case of temporal variations, as animation sequences.

The problem of simultaneously portraying multivariable image data has been addressed by graphics experts and visual perception scientists. An evaluation of methods for the simple case where each f_k depicts a single-curve graph of the form $y = g_k(x)$ was performed quite early [Shutz, 1961]. Experiments in which two visual events were combined on a mirror or a screen showed that human subjects could attend to one of them, keeping the other event in the background [Kolers, 1969, 1972; Neisser and Becklen, 1975]. Although color in itself is not a good separator, one technique [Robertson and O'Callaghan, 1985] separates the two variables by representing one of them by the topography of a surface and the other one by the color of the same surface. In what follows, we present our technique, which uses depth as a separator, and several applications in the areas of fractals, VLSI modelling, and CT-scanned images.

4.1 The Concept

Our technique is easily understood by referring to Figure 5a which shows $f_1, f_2, \ldots, f_N$, placed on distinct parallel depth planes $p_1, p_2, \ldots, p_N$, respectively. The inter-plane distance Δz is assumed constant. For our purposes, we make these planes normal to the line which bisects the angle formed by the left and right lines of sight. To portray all the f_k simultaneously, we consider that each plane p_k has some degree of translucency r_k. The values of r_k determine how the individual functions f_k contribute to the overall image. To quantify matters, refer to Figure 5b, the top view of a geometrical model for the situation of Figure 5a, in which it is assumed for simplicity that the eyes are at an infinite distance from the scene so that the projection lines are parallel

(parallel, not perspective, projection). The combined images for the left and right eyes, f_L and f_R, respectively, are formed onto the projection planes p_L and p_R, which are parallel to the p_i and form an angle θ with the normal to the projection lines.

In Figure 5b, the planes $p_1, p_2, \ldots, p_N$ have a common origin $(x=0, y=0)$. The functions f_k have a dynamic range of grey levels from 0 (black) to I_{max} (white); this is the same as the range of the graphics system. Furthermore, $f_k(x,y)$ is defined for integer values of its arguments with (x,y) denoting the pixel at column x and row y. Let us assume that each image is K rows by M columns. Our goal is to obtain f_L and f_R by selecting a plane, say p_2, to be the plane of fixation for which the horizontal disparity, or parallax, is zero. As shown in Figure 5b, this means that the x coordinates x_L and x_R for the left and right images, respectively, are identical to those of f_2. However, the x coordinates for all the other images $(f_1, f_3, f_4, \ldots)$ must be displaced horizontally by appropriate disparities to create the percept of depth. From Figure 5b, we derive the equation which introduces the disparities:

$$f_L(x_L, y) = \frac{1}{N} \sum_{k=1}^{N} f_k(x + (k-2)\Delta h, y) \qquad (6)$$

where

$$\Delta h = \Delta z \cdot \tan\theta. \qquad (7)$$

Δh is the inter-plane parallax. The equation for f_R differs from Equation (6) only in that the plus sign becomes a minus sign. Here we have assumed that all the planes have the same translucency. However, there are cases in which non-uniform degrees of translucency are desired. Equation (6) can be easily generalized to cover such cases:

$$f_L(x_L, y) = \sum_{k=1}^{N} w_k f_k(x + (k-2)\Delta h, y) \qquad (8a)$$

where

$$\sum_{k=1}^{N} w_k = 1 \qquad (8b)$$

Here each w_k is nonnegative and it is inversely related to the translucency of plane p_k. If $w_k = 0$, it means that it does not contribute anything to the overall image, i.e. it means that p_k is completely translucent; on the other hand, $w_k = 1$ means that all other w's are zero, hence f_k is the only image present (p_k opaque). Equations (8a) and (8b) guarantee that $0 \le f_L(x,y) \le I_{max}$.

NOVEL STEREO TECHNIQUE FOR MULTIVARIABLE PHENOMENA 369

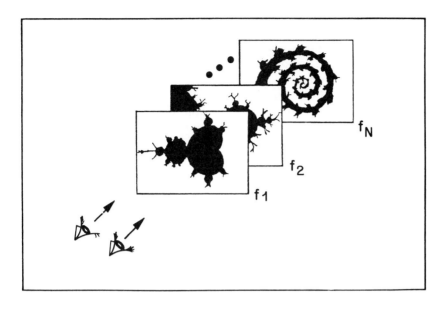

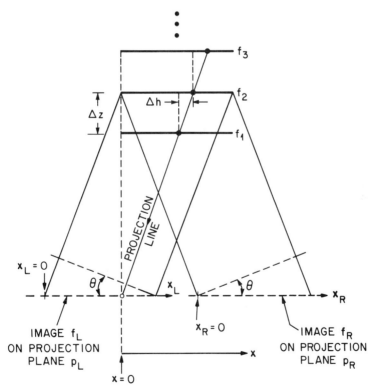

Figure 5. (a) Displaying multiple images in depth. (b) Top view of a geometrical model for Figure 5a. The y axis pierces the page coming toward the reader.

The form of Equations (6) and (8a) indicates that the arguments x_L and x_R may become negative or exceed M (see also Figure 5b). Thus one must first compute the extrema for x_L and x_R and then render the images for f_L and f_R on the screen. In the case of side-by-side stereograms, f_L and f_R are displayed next to each other with an appropriate margin between them, as in Figures 6, 7, 8, and 9. In the case of red-green anaglyphs, the images for f_L and f_R are first assigned green and red intensity levels, respectively, and are then spatially superimposed so that the origins of x_L and x_R coincide. Their color values are combined additively in the superposition process.

4.2 Experiments with Fractal Forms

In order to try the method of Section 4.1, we experimented with images resulting from deterministic fractal functions. There are two advantages in using fractal forms instead of images with conventional geometrical shapes: first, the former provide much more complex and natural-looking pictures [Mandelbrot, 1983; Norton, 1982]; with minor modifications, they have also given rise to non-trivial animation sequences [Papathomas and Julesz, 1987]. Second, the programming effort and complexity is much smaller for fractals, in which the same simple iteration loop produces all the visual stimuli [Dewdney, 1985]; by contrast, to render complex scenes with geometrical shapes, one requires rather elaborate routines for curve fitting, polygon fill, clipping, hidden surface removal, etc.

The complexity of the fractal forms limited the number of planes that could be viewed simultaneously to $N = 2$. The Mandelbrot set, generated by the iteration formula

$$z_{k+1} = z_k^m + c \qquad (9)$$

with $m = 2$ and $z_0 = 0$, where z and c are complex, was used in the first stage of experimentation. Animation was created by translating and expanding or contracting the region of the complex c-plane which produced the image. Thus f_1 and f_2 (see Section 4.1) were two different regions of the Mandelbrot set (see Figure 6) and animation resulted by independently translating and expanding/contracting these regions.

However, the affine transformations that were employed in the first stage did not result in complex metamorphoses of the basic figures. To achieve this, we developed a variation on the Mandelbrot set by allowing the power m in the iteration formula of Equation (9) to vary smoothly within a range of values. This produced very striking results under animation, resembling biological growth and evolution [Papathomas and Julesz, 1987]. Figure 7 shows a typical frame of an animation sequence.

In this frame the frontal and background figures are the modified Mandelbrot sets with $m = 2.06$ and $m = 1.94$, respectively.

The experiments with fractal images demonstrated that the technique of Section 4.1 was indeed valuable in portraying two two-dimensional events simultaneously. The next step was to use this technique in practical applications with numerical data from various sources. The following two subsections describe such applications.

Figure 6. Two regions of the standard Mandelbrot set separated by depth.

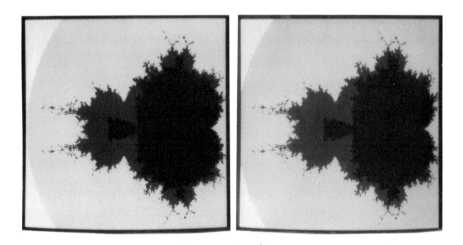

Figure 7. The two modified Mandelbrot sets are shown in two depth planes: in the frontal and rear planes are shown the sets resulting from the iteration equation (9) with $m = 2.06$ and $m = 1.94$, respectively.

4.3 Experiments with VLSI Data

The data for this experiment were produced by a simulation program [Bank et al., 1985], which models the transient behavior of silicon devices by using a combination of multistep integration, Newton-iterative methods, and sparse matrix techniques to solve the coupled system of partial differential equations characterizing the devices. In particular, the animation data were obtained by modeling an undesirable phenomenon known as "latchup" in CMOS devices, which is due to a parasitic bipolar thyristor structure. At each time step, the two-dimensional distribution of three key variables is computed over a cross section of the device. These variables are the electrostatic potential $u(x,y,t)$, the electron density $J_n(x,y,t)$, and the hole density $J_p(x,y,t)$.

The animation film reported in [Bank et al., 1985], showed the spatial variations in u, J_n and J_p by using a color code. In each time frame u was placed at the top of the screen, J_n in the middle, and J_p at the bottom, so that they all shared the x variable. However, it was rather difficult to view the film and mentally form the correlations between two variables along the y direction. This is where the technique presented in Section 4.1 proved useful. It was used to portray any two variables (from among u, J_n and J_p) by superimposing them spatially and using depth alone to separate them in the manner of Figure 5. It was found impractical to superimpose all the variables because the viewers were not able to attend to the episode at three different depth planes; thus we were limited to showing only two of the profiles simultaneously.

The latchup episode spans a total of 16 time frames. For display purposes, each of the three variables is defined over a two-dimensional array of 320 rows by 420 columns of pixels, each with sixteen levels of intensities, which are converted into brightness levels for our stereo pairs. The total size of the input data file is thus $16 \times 3 \times (320 \times 420) \times 4$ bits = 3.225 MBytes. A second format used more frames (64), but with lower spatial resolution for each frame (59×250 pixels), thus resulting in a dataset of 1.416 MBytes.

We used two modes of display: In the first one, which we termed the grey-level mode, each of the pixels was assigned a brightness level proportional to the value of the variable at that point. The two images were then combined in the manner of Equation (8). Reasonable values for w_1 and w_2 were in the range 0.4 to 0.6. An example of this display mode is shown in Figure 8a, which is the 15th frame of an animation sequence with J_n and J_p occupying the frontal and rear planes, respectively, and with $w_1 = 0.55$ and $w_2 = 0.45$.

In the second mode, the contour mode, only the edges between regions of equal magnitude are shown. Figure 8b shows the same frame as that of Figure 8a, rendered in contour mode. It proved easier for novice viewers to discern the two variables of the episode at different depth planes in the contour mode, which is thus useful in training users before they advance to the grey-level mode. The latter contains more visual information since, in addition to the edges, it also shows the relative magnitude of the variables.

In both modes, we introduced the feature of delaying the sequence of either variable with respect to the other by a fixed number of frames. In this way, we provided the expert viewer with the capability of investigating cause-and-effect relationships that may involve delayed responses. On the perceptual side, we showed the 64-frame episode in 21.33 seconds by repeating each frame 10 times on a frame buffer system that runs at 30 Hz. This represents a slow down by a factor of about 10^9, since the actual episode lasted only a few nanoseconds.

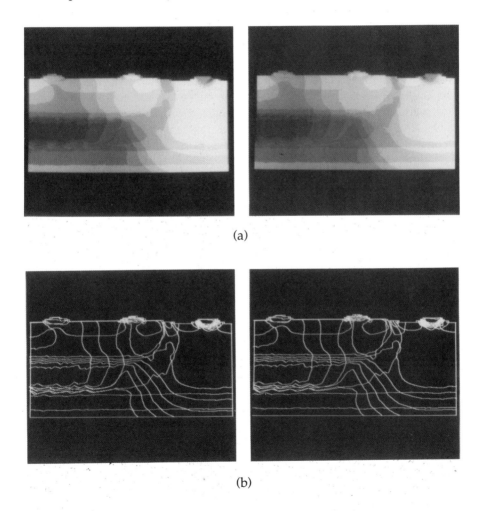

Figure 8. (a) Grey-level mode for displaying electron (frontal plane) and hole (rear plane) densities. (b) Contour mode depiction for the same frame as that of Figure 8a.

4.4 Experiments with CT-Scanned Data

The objective of these experiments, still in progress, is to recover depth information from two-dimensional computer tomographic (CT) scans of the human body. Computer tomography provides image data which are ideally suited for our technique for many reasons. First, each image is a "slice," providing a cross-section picture of the bone and muscle structure. When these slices are stacked up one behind the other, they naturally form our model of Figure 5a. Second, each slice is visually familiar to the specialist, thus facilitating the task of separating the images. Lastly, when these slices are stacked up in depth, adjacent images are correlated and they provide us with visual hints for reconstructing the three-dimensional (3-D) structure.

The problem of obtaining a 3-D model from a series of 2-D CT-scanned images has been addressed by computer graphics experts [Farrell, 1983; Vannier and Marsh, 1985; Farrell, Yang, and Zappulla, 1985; MacNicol, 1986]. Most of the techniques that have been developed attempt to identify corresponding edges and objects in contiguous slices and then connect these by interpolating curves or surfaces in 3-D space. Our method produces a sequence of images for fast previewing by radiologists and relies on the human visual machinery to perform the 3-D reconstruction.

Our CT data came from forty contiguous scans, or slices, s_0 s_1, $s_2, \ldots, s_{39}$ of the human torso and pelvis. Each scan was 256×256 pixels with eight bits per pixel for a total of 256 grey-levels. Thus the size of the dataset is 40×256×256 bytes (approximately 2.5 MBytes). The slices were arranged so that the twentieth one was assigned zero parallax so that it would be perceived at the plane of the screen. The eyes of the observer

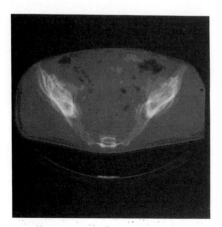
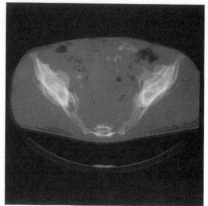

Figure 9. Multiple CT scans displayed simultaneously in depth in the manner suggested by Figure 5 and Equation (8).

are assumed to be on the torso side and the line bisecting the two eyes' lines of sight is normal to the slices. Thus, slices 0-19 have cross, i.e. negative, horizontal disparities and slices 21-39 have uncrossed, i.e. positive disparities. The values of disparity form a monotonically increasing series as we go from slice 0 to 39.

Our experience to date indicates that the maximum number of slices that can be combined in the manner of Figure 5 and still result in meaningful stereo percepts is no greater than five. Let this number be N; then, one way to assign $w_1, w_2, \ldots, w_N$ to the set of slices, say, $s_1, s_2, \ldots, s_N$, is to try to imitate physical reality, in which objects become dimmer as they recede from the viewer. Thus, we implemented a scheme in which the values for $w_1, w_2, \ldots, w_N$ form a linearly or exponentially decreasing sequence. Figure 9 shows an example of such a stereo pair with $N = 4$ and $w_1 = 100/358$, $w_2 = 93/358$, $w_3 = 86/358$ and $w_4 = 79/358$ (linearly decreasing).

Since it is impossible to view all forty slices simultaneously, we devised a scheme which uses animation to display these static data. The scheme is to show the first N slices in the first frame starting with slice s_0, and then, for frame k, to show N contiguous slices, starting at slice s_{k-1}. One way to visualize this is to think of a similar fictitious situation, where we move through a tunnel, 40 meters long, holding a flashlight which can only illuminate N meters ahead. The parallelism is strengthened if we imagine shooting one photograph per second and moving with a speed of one meter per second. Again, if we animate the resulting series of pictures, we will manage to portray the static structure of the tunnel through animation. We are currently experimenting with this scheme.

Acknowledgements

We thank Mr. James A. Schiavone for collaborating with us in the meteorological project; Messrs. Allan R. Wilks and Andrew M. Odlyzko for valuable advice on the mathematics of fractals; Mr. Eric H. Grosse for providing the VLSI simulation output data, and Ms. Sarah E. Chodrow for developing and implementing the code for the VLSI stereo animation; Mr. Michael Potmesil for providing the CT-scan data; Ms. Martina Bose for word processing, and Messrs. Richard S. Kennedy and Steve Cullerton for photography.

REFERENCES

Bank, R. E., Coughran, W. M. Jr., Fichtner, W., Grosse, E. H., Rose, D. J., and Smith, R. K. (1985). "Transient simulation of silicon devices and circuits," *IEEE Transactions on Electron Devices*, **ED-32**: 1992-2007.

Blinn, J. F. (1982). "Light reflection functions for simulation of clouds and dusty surfaces," *Computer Graphics, SIGGRAPH '82 Conference Proceedings*, **16**(3): 21-29. New York: Association for Computing Machinery.

Chambers, J. M., Cleveland, W. S., Kleiner, B., and Tukey, P. A. (1983). *Graphical Methods for Data Analysis*. Monterey, CA: Wadsworth.

Cleveland, W. S. (1985). *The Elements of Graphing Data*. Monterey, CA: Wadsworth.

Dewdney, A. K. (1985). "Exploring the Mandelbrot set," *Scientific American*, **253**(1): 16-20.

Farrell, E. J. (1983). "Color display and interactive interpretation of three-dimensional data," *IBM Journal of Research and Development*, **27**: 356-366.

Farrell, E. J., Yang, W. C., and Zappulla, R. A. (1985). "Animated 3D CT imaging," *IEEE Computer Graphics and Applications*, **5**(12): 49-52.

Foley, J. D. and van Dam, A. (1982). *Fundamentals of Interactive Computer Graphics*. Reading, MA: Addison-Wesley. p. 599

Gardner, G. Y. (1985). "Visual simulation of clouds," *Computer Graphics, SIGGRAPH '85 Conference Proceedings* **19**(3): 297-303. New York: Association for Computing Machinery.

Grotjahn, R. and Chervin, R. M. (1984). "Animated graphics in meteorological research and presentations," *Bulletin of the American Meteorological Society*, **65**: 1201-1208.

Hasler, A. F., des Jardins, M., and Negri, A. J. (1981). "Artificial stereo presentation of meteorological data fields," *Bulletin of the American Meteorological Society*, **62**: 970-973.

Hasler, A. F., Pierce, H., Morris, K. R., and Dodge, J. (1985). "Meteorological data fields in perspective," *Bulletin of the American Meteorological Society*, **66**: 795-801.

Hausman, B. A., Kroehl, H. W., and Kamide, Y. (1985). "Computer movie of ionospheric dynamics," *Proceedings of the International Conference on Interactive Information and Processing Systems for Meteorology, Oceanography, and Hydrology,*: 183-186. Boston, MA: American Meteorological Society.

Hibbard, W., Krauss, R. J., and Young, J. T. (1985). "3-d weather displays using McIDAS," *Proceedings of the International Conference on Interactive Information and Processing Systems for Meteorology, Oceanography, and Hydrology,*: 153-156. Boston, MA: American Meteorological Society.

Julesz, B. (1960). "Binocular depth perception of computer-generated patterns," *Bell System Technical Journal*, **39**: 1125-1162.

Julesz, B. (1971). *Foundations of Cyclopean Perception*. Chicago: University of Chicago Press.

Julesz, B. and Johnson, S. C. (1968). "Mental holography: stereograms portraying ambiguously perceived surfaces," *Bell System Technical Journal*, **49**: 2075-2083.

Julesz, B. and Oswald, H.-P. (1978). "Binocular utilization of monocular cues that are undetectable monocularly," *Perception*, **7**: 315-322.

Julesz, B. and Papathomas, T. V. (1985). "Independent movement channels in stereopsis," supplement to *Investigative Ophthalmology and Visual Science*, **26**: 242.

Klemp, J. and Rotunno, R. (1985). "Taking a good look at data," *NCAR Annual Report for 1984*, Report NCAR/AR-84: 46-49.

Kolers, P. A. (1969). "Voluntary attention switching between foresight and hindsight," *Quarterly Progress Reports*, No. 92, 381-385. Cambridge, MA: MIT, Research Laboratory of Electronics.

Kolers, P. A. (1972). *Aspects of Motion Perception*. New York: Pergamon Press.

Kreitzberg, C. W. and Perkey, D. J. (1976). "Release of potential instability: part I. a sequential plume model within a hydrostatic primitive equation model," *Journal of the Atmospheric Sciences*, **33**: 456-475.

MacNicol, G. (ed.) (1986). "SIGRAPH '85 technical slide set credits," *Computer Graphics*, **20**(1): relevant slides are by B. Drebin, p. 20, and W. E. Schmidt, p. 24.

Magnenat-Thalmann, N. and Thalmann, D. (1985). "Three-dimensional computer animation: more an evolution than a motion problem," *IEEE Computer Graphics and Applications*, **5**(10): 47-57.

Mandelbrot, B. B. (1983). *The Fractal Geometry of Nature*, New York: W. H. Freeman.

Neisser, U. and Becklen, R. (1975). "Selective looking: attending to visually specified events," *Cognitive Psychology*, **7**: 480-494.

Norton, A. (1982). "Generation and display of geometric fractals in 3-D," *Computer Graphics, SIGGRAPH '82 Conference Proceedings*, **16**(3): 61-67. New York: Association for Computing Machinery.

Papathomas, T. V. and Julesz, B. (1987). "Animation with fractals from variations on the Mandelbrot set," *The Visual Computer*, **3**: 23-26.

Reeves, W. T. (1983). "Particle systems — a technique for modeling a class of fuzzy objects," *Computer Graphics, SIGGRAPH '83 Conference Proceedings*, **17**(3): 359-376. New York: Association for Computing Machinery.

Robertson, P. K. and O'Callaghan, J. F. (1985). "The application of scene synthesis techniques to the display of multidimensional image data," *ACM Transactions on Graphics*, **4**: 247-275.

Schiavone, J. A., Papathomas, T. V., Julesz, B., Kreitzberg, C. W., and Perkey, D. J. (1986). "Anaglyphic stereo animation of meteorological fields," *Proceedings of the Second International Conference on Interactive Information and Processing Systems for Meteorology, Oceanography, and Hydrology*: 64-71. Boston, MA: American Meteorological Society.

Schlatter, T. W., Schultz, P., and Brown, J. M. (1985). "Forecasting convection with the PROFs system: comments on the Summer 1983 experiment," *Bulletin of the American Meteorological Society*, **66**: 802-809.

Shutz, H. G. (1961). "An evaluation of methods for presentation of graphic multiple trends — experiment III," *Human Factors*, **3**: 108-119.

Tufte, E. R. (1983). *The Visual Display of Quantitative Information*. Cheshire, CT: Graphics Press.

Tukey, J. W. (1977). *Exploratory Data Analysis*. Reading, MA: Addison-Wesley.

Vannier, M. W. and Marsh, J. L. (1985). "3D imaging aids skull surgeons," *Computer Graphics World*, **8**(12): 49-52.

Vonder Haar, T., Meade, A., Brubaker, T., and Craig, R. (1986). "Four-dimensional digital imaging for meteorological applications," *Proceedings of the Second International Conference on Interactive Information and Processing Systems for Meteorology, Oceanography, and Hydrology*: 357. Boston, MA: American Meteorological Society.

15

A HIGH PERFORMANCE COLOR GRAPHICS FACILITY FOR EXPLORING MULTIVARIATE DATA

C. Ming Wang
Harold W. Gugel
General Motors Research Laboratories

ABSTRACT

Graphical techniques have long been known as powerful aids for data analysis. Numerous papers can be found on computer generated static graphical methods for data representation. Due to advances in computing, recent research in this area has principally focused on highly interactive and dynamic methods. This paper describes a high-performance color graphics facility for exploring multivariate data. It was developed with commercially available software as a procedure under a major statistical system. Although designed to be device independent, it is most effective with high-end, intelligent graphic devices. Through simple movements and selections with a graphics locator the analyst can rapidly examine many different subregions of high-dimensional data. This facilitates the discovery of structure imbedded in complex data.

From *Dynamic Graphics for Statistics*, William S. Cleveland and Marylyn E. McGill, eds., copyright © 1988 by Wadsworth, Inc. All rights reserved.

1. INTRODUCTION

The advent of computer graphics spawned numerous developments in graphical techniques for data analysis. A detailed discussion of some of the more powerful static methods can be found in Chambers, et al. [1983]. Much of the recent research in statistical graphics has focused on applications of interactive and dynamic computer graphics. These operation modes are particularly useful for exploratory data analysis. They allow the human's talents for processing visual information and knowledge about data to be combined with the computer's ability to rapidly generate different graphic views of data to form a powerful environment for exploring complex data. Solutions to problems that are difficult to obtain by other computing means often become obvious with these interactive methods because of the relative ease of interjecting human intelligence.

Some dynamic graphics systems for data analysis, such as PRIM-9 [Fisherkeller, Friedman, and Tukey, 1974], PRIM-H [Donoho et al., 1982], and ORION [McDonald, 1982], used real time motion to display three-dimensional scatter plots. They could also be used to view an arbitrary three-dimensional subspace of higher dimensional data through rotation and projection [Huber, 1985] methods. These prototype systems required special and/or custom-made hardware and software. Thus, their availability was very limited.

With the increasing availability of intelligent graphics terminals and microcomputer workstations, high speed interactive graphics becomes a reality for every data analyst. Donoho, et al. [1985], provide powerful dynamic software on a Macintosh for three-dimensional point cloud rotation including numerous other capabilities such as animation, subsetting, and highlighting. Becker and Cleveland [1984, 1987] describe a dynamic graphics system for exploring high-dimensional data through "brushing" a scatter plot matrix. It executes under the control of the S statistical system [Becker and Chambers, 1984]. Special software drives a bitmap display of a monochrome Teletype 5620 intelligent terminal. The operational capabilities demonstrated by this system provided much of the impetus for developing a general, more widely accessible interactive color graphics facility at the GM Research Laboratories.

The GM facility was developed solely with commercially available, high-level software for use with commonly available graphics devices.[1] It is callable as a procedure within the SAS statistical system (SAS Institute, 1985). Although the procedure can be executed from any device which supports graphics input, it is far more effective when run on intelligent devices that support picture segmentation. To provide the rapid responses necessary for effective use, a graphic image must be created for

---

[1] The DI-3000 graphics system [Precision Visuals, 1984] was used to provide device independence and dynamic color graphics.

each component of a display which is to be individually selectable and/or changeable, and then stored as a retained segment in local memory. As selections are made, only those segments of the display which pertain to the selection are refreshed from local memory. Although the graphics may be adequate with certain non-intelligent devices, such as the IBM 3179 terminals, the response time with these devices can be so severely slowed, because of computational demands in creating refreshed displays from the host, that the effectiveness of the procedure is greatly impaired.

As in the Becker and Cleveland system, high-dimensional data in the GM facility is graphically presented on a two-dimensional display as a matrix of ordered pairwise scatter plots [Becker and Chambers, 1984; Becker and Cleveland, 1984]. This array of plots simultaneously displays the bivariate association between all the variables. To investigate higher order associations, the analyst picks, via a graphics locator, operations from a menu and applies them to selected elements of the data display. Through these operations he searches for interesting features and structure in the data such as patterns, high-order relationships, and outliers. He can easily subpartition the data by either zooming in on subsets of the observations or selecting a subset of the variables. Color is used primarily to assist in identifying high-dimensional characteristics of the data. It provides the necessary linkage across the matrix of plots to identify the points which comprise a multivariate observation.

The remainder of this paper describes the operations of the GM procedure. A reader interested primarily in an overview of its capabilities is advised to first read Section 4, which illustrates most of its features through color graphic displays using a sample case study. Section 2 describes the controls available to the user to configure the execution environment. Section 3 gives a discussion of the interactive operations available.

2. CONFIGURING THE INTERACTIVE ENVIRONMENT

Upon invoking the procedure, the display screen is formatted into five regions as shown in Figure 1. This format is used across all devices to simplify device independence. The central and major region is the "DATA DISPLAY" which, by default, is comprised of a "full" display of pairwise scatter plots. A "reduced" display can be requested which provides only the plots in the lower triangular region. The display of plots is formed to fill as much of the screen as possible. The individual plots can be forced to be square to prevent undesirable elongated plots. This can result whenever the sets of variables assigned to the horizontal and vertical axes are not identical. Spacing between plots can be specified and the total size of the display can be controlled to allow exact sizing for hardcopy recording for publications.

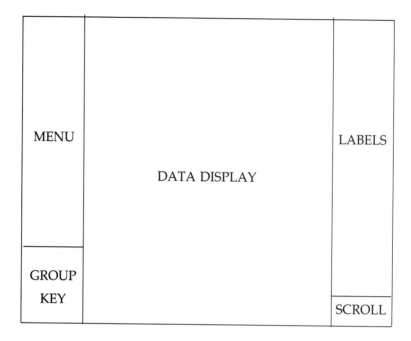

Figure 1. Display screen layout.

A variable can be supplied to provide identification labels for the observations. These labels are listed in the region on the right named "LABELS". If the labels exceed a full page, a "SCROLL" region in the lower right can be used to scroll through the labels.

If the data is naturally grouped, a grouping or classification variable can be indicated. (Additional groupings can be supplied through variables in the "DATA DISPLAY".) Group names and associated colors are given by a "GROUP KEY" region at the lower left. This color coding is used by nearly all operations to distinguish the data groupings and dynamically highlight selected observations.

A panel of selectable operators and options is given in a "MENU" region at the mid to upper left (or across the top for tall screens). These are discussed in the following section.

Other facilities include the ability to select the graphic input device and the symbol for the common data point markers. Categorical data can be jittered to more accurately portray repeated values (see Chambers et al., 1983, p. 106). A secondary classification variable can be designated to assign a specific marker symbol to each observation. Select variables can be assigned to the horizontal and vertical axes. (These controls are indicated through options as the facility is invoked from within the SAS system.

3. INTERACTIVE DISPLAY OPERATIONS

This section describes the interactive operation of the procedure as seen while running from a graphics display device. Frequent references will be made to figures in the next section which graphically illustrate the features described using a sample case study.

Recall that the display screen is formatted into five regions. The initial bivariate scatter plots in the "DATA DISPLAY" region, by themselves, generally contain a wealth of information (see Plate 5). One can scan a certain row and column and see one common variable plotted against all the other variables, the common variable changing role from a response to a predictor variable. This ensemble of plots allows the simultaneous study of pairwise associations between all the variables. In addition, data anomalies or outliers in one or two dimensions can be quickly spotted, and data clusters and interesting patterns can be detected. However, to effectively study higher order associations, operations provided through the interactive graphics facilities must be used. These operations include identifying, highlighting, and deleting sets of observations, zooming in on subsets of variables and observations, and fitting regression smoothers to the data.

All operations are indicated through a graphics locator in the form of cross-hairs or a cursor which is controlled by a user selected input device such as a mouse, tablet, or keyboard. These operations are triggered by either choosing items from the "MENU" or selecting certain pickable elements from the graphic display. The following describes the various operations.

Direct Locator Operations — Several operations can be performed directly on the display of observations. Using the locator, a specific observation can be identified in the "DATA DISPLAY" by selecting its label in the "LABELS" region. The points comprising the observation selected will be highlighted in the plots using the color associated with the observation group indicated by the "GROUP KEY". Similarly, any entry in the "GROUP KEY" can be selected with the locator, and all points for the observations in the selected group will be highlighted in all the plots using the designated color (see Plate 6). Unless otherwise instructed (see *Memory* below), as new selections are made from either the "LABELS" or the "GROUP KEY" regions, the highlighted points corresponding to the previous selections will be returned to their original state. Individual points in the "DATA DISPLAY" can also be directly selected with the locator to become the target for certain operators in the "MENU", as discussed below.

Window/Complmnt — The "Window" operator is used to define a rectangular region or window in the "DATA DISPLAY" to capture a set of points. Upon selecting "Window", two positions in the "DATA

DISPLAY" region must be indicated with the locator to define the opposite corners of a rectangle. The observations corresponding to the points enclosed by the window will then become the target for subsequent operations. Once a window is defined, it can be freely moved to any position by simply indicating with the locator the center for its new position. All subsequent operations selected will then be applied to the new observations until the window is again repositioned or it is deactivated. (See Plates 8 thru 11.)

"Complmnt" can similarly be used to define a window. However, the complement of the observations corresponding to the captured points, i.e., those outside the window, will become the target for the subsequent operations.

Highlight — The "Highlight" operator highlights those points in all the plots which are associated with the same observations as the points presently captured by the window. Highlighting is achieved by coloring the points according to their group color as shown in the "GROUP KEY". Observations can also be highlighted individually by simply selecting an associated point with the locator and activating "Highlight". Once "Highlight" is activated, additional observations can be rapidly highlighted by simply repositioning the locator. (See Plates 8 thru 10.)

Label — The "Label" operator displays in the "LABELS" region the identification labels of all observations captured by the current window. The labels are displayed in color giving their group association. If labels are not explicitly assigned, the observation sequence number will be used. (See Plates 8 and 11.)

Delete — The "Delete" operator temporarily deletes from view the observations corresponding to points captured by the current window. This can be used to bring into clearer focus selected subsets of the observations. Individual points can also be selected with the locator to delete observations one by one. (See Plates 7, 9 and 10.)

Smooth/CLASS — "Smooth/CLASS" fits smoothers through all the displayed scatter plots using the robust locally weighted regression technique due to Cleveland [1979]. Selecting "Smooth" will present the smoother for each plot, pooling all the data points within the plot. Selecting "CLASS" will present multiple smoothers within every plot, one for each group (class), and identified by the group color. The smoothers are computed at the time new displays are drawn. Thus, temporarily deleting points will not adjust the smoothers. (See Plates 7 and 11, top.)

Memory — "Memory" is a status option which applies to successive executions of operations. Without "Memory" activated, the results of operations are only shown for the most current selection. With "Memory" activated, previous results will be retained. (See Plates 6 and 9 thru 11.)

Value — "Value" is a status option which is used for displaying the values of observations. If "Value" is activated, the values of variables for selected observations will be listed in the open diagonal squares. (See Plate 8.)

Var Subset — The "Var Subset" operator is used for redisplaying a selected subset of the variables. Upon selecting "Var Subset", plots containing the desired variables must be picked with the locator. Selecting "Var Subset" again terminates variable picking and creates the new display. When picking an open diagonal square, the associated variable will be displayed along both the horizontal and vertical axes. (See Plates 7 and 11, top).

Stow Obs — The "Stow Obs" operator provides a method of graphically selecting subsets of observations to be returned to the host environment for further processing. Upon selecting "Stow Obs", the observations corresponding to the points captured by the current window are stored in a data set to be returned to SAS. Multiple selections can be accumulated.

Zoom — "Zoom" isolates a subset of the observations. A new display is created for the subset of observations corresponding to the points captured by the current window. The scales and regression smoothers are recomputed. (See Plate 11, bottom.)

Restore/Refresh — The "Restore" operator returns the original display created when the procedure was first invoked. "Refresh" only returns the current display to its original state.

End — The "End" operator terminates the procedure and returns the user to the host SAS environment.

4. A SAMPLE CASE STUDY

This section is intended to illustrate, through graphic displays, many of the capabilities provided by the GM facility. The figures present only a few "snapshots" in time of a large series of displays generated through interactive graphic operations applied to a set of data. They do not do justice in describing the full power of the facility; they merely provide an indication of the types of displays and results obtainable. An attempt is made to illustrate the function of each major operation rather than give a thorough analysis of the data set per se. Any statements made apply only to the current set of data.

The data used for this example is taken from Chambers, et al. [1983], pp. 352-355. The original source of the data is principally Consumer Reports and the EPA. It consists of selected characteristics for 71 automobiles in the 1979 model year sold in the U.S. Although the facility allows any number of variables, only six of the original fourteen variables are studied so that the individual bivariate plots remain large enough for data features to be discerned.

The initial display is shown in Plate 5. The car models identifying the observations are shown in the "LABELS" region and the five makes, color coded, are given in the "GROUP KEY" area. The discrete variables REPAIR77 and REPAIR78 (repair rating) were jittered to prevent severe overlapping of points. Despite the miniature size of the plots, a number of interesting features are still discernable.

- There are a number of apparent outliers (MPG row and plot WEIGHT*DISPLMNT).
- Note the strong correlation between WEIGHT and DISPLMNT and the close similarity between the plots of these two variables with all the other variables.
- Note how the variables correlate with MPG.
- There seem to be two relationships or clusters shown in both plots of PRICE with WEIGHT and with DISPLMNT.
- Plot REPAIR77*REPAIR78, as a result of jitter, shows the distribution of points; most points fall along the diagonal. Some points are considerably below the diagonal, indicating a significant improvement in repair rating from 1977 to 1978.

This collection of plots shows the bivariate relationships between the variables. The following discusses some additional structure exposed through use of the interactive facilities.

Plate 6 shows all the makes highlighted. This is obtained by enabling "Memory" and selecting each entry in the "GROUP KEY" area. Although from these reproduced color plots we cannot distinguish between all five of the individual makes, we can clearly differentiate foreign makes (reddish) from domestics (bluish). Note the clusters and the different patterns for foreign and domestic makes.

Plate 7 provides a clearer view of the relationships holding among the economy/size oriented variables. Here the differences between the five car makes are more distinguishable. This subset display was obtained from the previous display by using the "Var Subset" operator to select the four variables from the diagonal. All the data points were then "Deleted" and the "CLASS" operator was invoked to obtain the smoothers for each make.

Plate 8 identifies the low price, high mileage cars (see window in PRICE*MPG). "Highlight" was used to expose these cars in all the other plots. "Value" printed the values down the diagonal. These correspond, in order, to the models listed at the right which were obtained with "Label". Note that Chevette is the only "true" domestic make, DgColt and PlymCham, although domestics, are foreign made imports. Note that all are relatively high in repair ratings, low in weight, and except for Chevette, low in displacement.

Plates 9 and 10 isolate the low weight from the high weight cars. This required several operations. First, for each plate, the cars not shown were "Deleted". The remaining cars were "Highlighted" after capturing them in a "Window" (see plot WEIGHT*PRICE). First note that the low weight cars are predominantly foreign (reddish hue) while the heavier cars are mostly domestics (bluish hue). Note the different relationships holding for light and heavy cars in many of the plots, especially plots with PRICE and with DISPLMNT.

Plate 11 (top) studies in greater detail the plots of PRICE and MPG against WEIGHT. These were obtained by selecting them from the previous display using the "Var Subset" operator. Note that for a given weight, foreign cars appear to be higher in price but lower in mileage than domestics, except for the lowest weight makes, with European cars faring worst. The domestics appear quite tightly coupled showing only Ford to be generally lower in MPG across all weights. The upper plot identifies three high priced domestics. They are the high priced luxury models and appear to follow the PRICE*WEIGHT pattern of the foreign makes more so than the domestics.

Plate 11 (bottom) further examines the PRICE*WEIGHT plot by zooming in on only the lower weight cars. Note the recomputed scales. Two patterns are clearly visible; one primarily applies to foreign makes, the other to domestics. Also, note now the two clusters among the domestics. In the figure, the low weight, low price domestics (bluish) that also seem to fit the foreign pattern have been identified. Note that most of these are foreign made.

These are but a smattering of the types of displays obtainable with the GM facility. By using a number of options and operators in combination, all easily available through selections with a graphics locator, a virtually unlimited assortment of displays can be rapidly generated.

5. CONCLUDING REMARKS

The GM graphics facility described illustrates capabilities which can be readily developed today using off-the-shelf hardware and software. The facility was developed as a procedure callable from a major statistical system to provide needed capabilities in data management and extended statistical methods. ANSI standard Fortran was the development language. A commercially available, high-level graphics package was selected to provide color, abilities for high-performance interaction, and device independence.

Using the high-level graphics package simplified development efforts and provides far greater device accessibility, but at the expense of performance. Dynamics is sacrificed for incremental graphics. Every new position of the graphics locator must be signaled through some notification using a physical input device, such as a click of a mouse. Although the facility functions on most devices supporting graphics input, its operation is only acceptable on high-speed graphic devices which offer full support of picture segmentation.

Color is a key component of the GM facility. It provides the means to dynamically highlight selected observations and creates the linkage between the individual points in all the bivariate plots which comprise a multivariate observation.

Operations of the facility provide powerful means for exploring multivariate data. Through the highly interactive graphic capabilities, an analyst can rapidly explore many different subpartitions of high-dimensional data, concurrently looking at a number of dimensions. This allows him to more readily uncover interesting features and postulate hypotheses about the data. The synergism resulting by interactively combining high-speed computer graphics with human intelligence can expose structure in complex data which may be difficult to discover by other means.

REFERENCES

Becker, R. A. and Chambers, J. M. (1984). *S: An Interactive Environment for Data Analysis and Graphics*, Belmont, CA: Wadsworth.

Becker, R. A. and Cleveland, W. S. (1984). "Brushing a scatter plot matrix: high interaction graphical methods for analyzing multidimensional data," Technical Memorandum. Murray Hill, NJ: AT&T Bell Laboratories.

Becker, R. A. and Cleveland, W. S. (1987). "Brushing scatterplots," *Technometrics*, **29**: 127-142.

Chambers, J. M., Cleveland, W. S., Kleiner, B., and Tukey, P. A. (1983). *Graphical Methods for Data Analysis*. Boston, MA: Duxbury.

Cleveland, W. S. (1979). "Robust locally weighted regression and smoothing scatter plots," *Journal of the American Statistical Association*, **74**: 829-836.

Donoho, A. W., Donoho, D. L., and Gasko, M. (1985). *MACSPIN Graphical Data Analysis Software*, Austin, Texas: D^2 Software, Inc.

Donoho, D. L., Huber, P. J., Ramos, E., and Thoma, H. M. (1982). "Kinematic display of multivariate data," *Proceedings of the Third Annual Conference and Exposition of the National Computer Graphics Association*, **1**: 393-398. Fairfax, VA: National Computer Graphics Association.

Fisherkeller, M. A., Friedman, J. H., and Tukey, J. W. (1974). "PRIM-9: An interactive multidimensional data display and analysis system," SLAC-PUB-1408. Stanford, CA: Stanford Linear Accelerator Center.

Huber, P. J. (1985). "Projection pursuit," *Annals of Statistics*, **13**: 435-475.

McDonald, J. A. (1982). "Interactive graphics for data analysis," Technical Report Orion 11. Stanford, CA: Stanford University, Department of Statistics.

Precision Visuals, Inc. (1984). *DI-3000 User's Guide*. Boulder, CO.

SAS Institute Inc. (1985). *SAS User's Guide: Basics*, Version 5 Edition. Cary, NC: SAS Institute Inc.

16

DYNAMIC GRAPHICS FOR EXPLORING MULTIVARIATE DATA

Forrest W. Young
University of North Carolina

Douglas P. Kent
Warren F. Kuhfeld
SAS Institute Inc.

ABSTRACT

In this paper we discuss several dynamic graphics methods designed to help the data analyst visually discover and formulate hypotheses about structure in multivariate data. These methods include two new methods for seeing and manipulating multivariate data spaces, called *Hyperspace Interpolation* and *Hyperspace Residualization*, and a number of previously developed methods for manipulating 3D data spaces, including spinning, moving, brushing, subsetting and metamorphing. We also discuss a new software system, called VISUALS, that implements all of these methods.

We include an example of these methods being used to discover structure in multivariate data which are seven-dimensional. The structure cannot be seen with ordinary graphics as a spatial structure of objects, but can be seen with dynamic graphics as a structured pattern of movement. We also present two examples of these methods being used to compare the results of two different analyses of a set of multivariate data.

From *Dynamic Graphics for Statistics*, William S. Cleveland and Marylyn E. McGill, eds., copyright © 1988 by Wadsworth, Inc. All rights reserved.

INTRODUCTION

The course of scientific investigation can be classified into two broad stages: the exploratory stage, where the investigator forms hypotheses; and the confirmatory stage, where the investigator tests hypotheses. For many years statisticians focussed on developing and improving inferential methods to test hypotheses, to the neglect of exploratory methods to form hypotheses.

In recent years there have been many new developments in exploratory data analysis (EDA) methods. Tukey, in his landmark book *Exploratory Data Analysis* [1977, p. V] states that EDA

> "is about looking at data to see what it seems to say. It concentrates on simple arithmetic and easy-to-draw pictures. It regards whatever appearances we have recognized as partial descriptions, and tries to look beneath them for new insights. Its concern is with appearance, not with confirmation."

The work reported here falls under the general rubric of "exploratory data analysis", and is guided by the philosophy of "looking at data to see what it seems to say".

In the first section of this paper we briefly review a number of dynamic graphics methods for exploring multivariate data. Most of these methods, which we refer to collectively as VEDA (Visual EDA), are discussed more extensively by others in this volume. An excellent review and bibliography of this approach to understanding data is presented by Cleveland [1987] in his overview of a number of valuable contributions that appear in the same publication (see references elsewhere in the next section).

The focus of this paper is on two new VEDA methods called *hyperspace interpolation* and *hyperspace residualization*. These methods are designed to help data analysts understand multivariate data. We present these new methods in the context of VISUALS, a new dynamic graphics software system that implements the new VEDA methods as well as many of the others reviewed in the introduction.

1. VISUAL EXPLORATORY DATA ANALYSIS

VEDA differs from EDA in two major ways:

> > VEDA is entirely graphical, emphasizing describing data visually;

> VEDA is primarily multivariate, emphasizing the
the exploration of multivariate data.

VEDA methods capitalize on the pattern recognition power of human vision and the computational power of graphics workstations to help data analysts look for structure (form hypotheses) that may be hiding in their multivariate data. The goal of VEDA is to aid in forming hypotheses about the data's hyper-dimensional (hD) geometric structure, even though we can only see in 3D. To do this, the graphical representation must

> respect the data's hyper-dimensional geometry,

> respect the user's three-dimensional perception,

> respect the workstation's computational limits.

Of course, while we see in 3D, we can only draw in 2D, either on paper or on the computer screen. The problem that all VEDA methods tackle is presenting hD information in a 2D plane, such that our 3D perception can understand the hD geometry.

Presenting 3D in a 2D plane is not new, of course. Artists have done this for centuries, statisticians for over a century, and computers for two decades. Indeed, techniques for presenting hD in a 2D plane are not new. Tufte [1983, p. 40] presents a marvelous example, dating from 1861, of a 2D statistical graphic that incorporates 6 dimensions of information concerning the fate of Napoleon's army in Russia.

There are sophisticated computer techniques that create images that appear to genuinely occupy 3D volume. These techniques do a very convincing job of tricking us into "seeing 3D". Even ordinary computer generated 2D printer plots can have additional "dimensions" added by labeling the points in the 3D space. Kuhfeld [1986], for example, has developed a 2D printer plotting approach that incorporates several methods of maximizing the amount of information that can be presented in point labels. Stuetzle [1987] has developed a very flexible system for manipulating multiple 2D "plot windows" which helps in understanding multivariate structure. Also, with common computer generated color graphics more dimensions may be added by using various colors and shapes to distinguish the points on discrete dimensions. In addition, there are a variety of techniques for communicating 3D and hD on a 2D plane, including perspective and stereo projections, movement, dynamically changing object shapes, etc.

VEDA methods can be classified as being based on graphical methods which are either static or dynamic. Static graphics are snapshots of the data space, whereas dynamic graphics are movies of the shape.[1] Dynamic graphics can in turn be classified as passive or active. Active dynamic graphics require the user to interact with the computer to create the movie, whereas passive dynamic graphics only require the user to passively watch the movie that the computer creates. We briefly mention some examples of each kind of VEDA graphics in the next few paragraphs.

Tukey and Tukey [1981] and Inselberg [1985] have proposed static VEDA methods that use a computer to draw a picture that represents the hD space. Tukey and Tukey proposed the generalized draftsman's plot, a matrix where each "element" is a 2D scatterplot formed by plotting the row variable against the column variable. Recent developments of this approach are presented by Carr, Littlefield, Nicholson and Littlefield [1987]. Inselberg proposed a parallel coordinate system in place of the familiar orthogonal coordinate system: Dimensions are represented in a plane by parallel lines, and each multivariate "point" is represented by a polygonal line.

Gabriel and Odoroff [1986] have developed dynamic VEDA methods that have both passive and active aspects. The user can actively select, by typing alphanumeric statements, a view of a 3D space. This view may be in stereo. Once selected, their software can passively present dynamic views of the 3D space repeatedly rocking back and forth a small amount. This rocking motion contributes markedly to depth perception.

Fisherkeller, Friedman and Tukey [1974] developed an active dynamic VEDA system that constructs a 3D scatterplot that can be rotated by the user. The sophistication of this early approach has been greatly increased in later descendants [Donoho, et al., 1982; Friedman, McDonald, and Stuetzle, 1982; and Donoho, Donoho and Gasko, 1986]. With these systems, the user rotates the 3D scatterplot by using a mouse or the cursor keys. The user-controlled rotations create a reasonable sensation of depth. Huber [1987], an early developer of this approach, discusses his experiences with it. In addition, Donoho, et al. [1982] discuss an active dynamic 2D draftsman's plot.

Additional active dynamic VEDA methods have been developed to assist the data analyst in understanding the structure of data. Many of these methods involve forming subsets according to properties of the displayed objects or according to values of variables in the data set. Donoho, Donoho and Gasko [1986] implement "slicing" and "masking", methods for dynamically subsetting observations on a fourth dimension

---

[1] Writing about dynamic graphics is a very frustrating business. After all, writing is linear, static and non-graphical. What could be further from the subject we are addressing here, which is highly nonlinear, very dynamic, and decidedly graphical, let-alone high-dimensional! While some "intellectual" understanding can be gained from the words, equations, and static graphics presented in this paper, true insight about dynamic graphics can only be obtained by seeing the movie.

to determine where these observations are in the 3D scatterplot. Becker and Cleveland [1987] discuss "brushing", a way of dynamically subsetting observations simultaneously on two variables which helps the user determine where the observations fall on yet other variables. Many developers implement other subsetting methods that group observations according to variable values.

Subsetting is particularly powerful when combined with a tool we call "metamorphing", a VEDA method which allows the user to change the appearance of objects on the screen that represent subsets of observations. Various researchers have developed systems which allow objects to have attributes such as color, shape, labels, size, etc., and then allow these attributes to be metamorphed for subsets of observations.

Passive dynamic VEDA methods have been developed to search through the hD data space and to display smoothly changing, dynamic projections of this search. Friedman and Tukey [1974] developed a passive dynamic VEDA method they call Projection Pursuit for looking through the hD data space to find "interesting" 2D views. This approach is discussed extensively by Huber [1985]. Nicholson and Carr (1985) have developed another passive dynamic method that presents rocking views of objects whose dynamic shape represents the changing values of two additional variables. With this system, the rocking takes place on all five dimensions simultaneously. Asimov [1985] has proposed the Grand Tour, another passive dynamic VEDA method for displaying smoothly changing 2D projections of hD data space.

2. OVERVIEW OF VISUALS

2.1 Dynamic Graphics Features

VISUALS is an active dynamic VEDA graphics system designed to help the data analyst visually discover and formulate hypotheses about structure hidden in multivariate data. VISUALS does this by implementing many of the dynamic graphics methods discussed in the introduction, including:

> *3D motion* - including *rotation, translation* and *rocking*, performed on one or two 3D spaces;
> *brushing* - highlighting or displaying character variable values or observation numbers of objects in either space according to their position in one of the spaces;
> *subsetting* - creating subsets of objects according to their location (in either space), their attributes, their observation numbers, or their character variable values;

> *metamorphing* - dynamically changing the labels, colors
  and shapes of object subsets;
> *recording* - actively recording the graphics session
  to passively replay it at a later time.

VISUALS also implements three methods for visually understanding multivariate structure. We believe these methods are new contributions to the dynamic graphics literature [see Young, Edds, Kent, and Kuhfeld, 1986; and Young, Kent and Kuhfeld, 1986a; 1986b, 1987]:

> *interpolation* - dynamically calculating and displaying a
  smoothly changing sequence of interpolations between
  the two 3D spaces: in effect moving a 3D space
  through 6D space;
> *residualization* - redefining the two 3D spaces as linear
  combinations of 6 or more variables, allowing the data
  analyst to move the 3D space into a hD space;
> *refinement* - improving the way objects are displayed from
  wire frame drawings to smoothly shaded solid objects in
  order to enhance the 3D effect.

We discuss all eight of these features in Sections 3 and 4 of this paper.

2.2 3D Features

The most important aspect of VISUALS is that it presents the data analyst with a dynamic color picture of a 3D space. The picture contains a cloud of objects representing the observations in a data set. The three axes of the space correspond to three variables in the data set (or to three orthogonal linear combinations of the variables). The data analyst can interact with the observation cloud, making it spin (rotate) and/or move (translate). The movements take place in real time, producing a dynamic picture of the data cloud's structure. To enhance the 3D effect, the 3D space is projected in perspective onto the 2D graphics screen. A number of additional features, such as brushing, subsetting, metamorphing, refinement, and recording, are available to enhance the data analyst's understanding of the data cloud's structure. All of these features are discussed in Section 3 of this paper.

2.3 hD Features

There are two features of VISUALS which are original to our research. They are both based on the fact that the user can define a second 3D space. One of these new features is the ability to interpolate between the two spaces: If you imagine that the first 3D space is the first frame of a movie, and the second 3D space is the last frame, then you can interact with VISUALS to create other frames which lead smoothly from the beginning to the end of the movie. The second new feature of VISUALS is that for any frame of the movie VISUALS can compute the 3D space that is orthogonal to the 3D space displayed in the frame and which has the longest dimensions. This new space, plus the one displayed in the frame, can now serve as the beginning and end of a new interpolation movie. This second feature can be invoked any number of times. In addition, the user can use any of the 3D features (including moving or spinning) with any space in any frame of the movie. Together, these features provide the user with the potential of seeing and interacting with all 3D projections of the hD data space. We discuss the two new features in Section 4.

2.4 High-Interaction Interface

VISUALS is a high-interaction graphics system: Most of its dynamic behavior is controlled by single key-stroke "commands" whose effects are immediately visible. One of the primary design philosophies of VISUALS is that the user need not be concerned with any of the mathematical details concerning rotations, translations, interpolations or residualization. Instead of using equations, numbers, or programming statements, the user controls all dynamic features by single key-stroke commands. Numbers, equations and statements need never be seen or used. These aspects of VISUALS are discussed in Section 5.

2.5 Software and Hardware

There are two versions of VISUALS. One is a stand-alone program[1], and the other is an experimental Version 6 SAS[2] procedure. Both run on an IBM (or compatible) PC XT or AT with IBM (or compatible) Professional Graphics Adapter (PGA) and Professional Graphics Display (PGD)[3]. More details are given in Kent, Edds, Kuhfeld and Young [1986] and Young, Edds, Kent and Kuhfeld [1986] about the implementation and hardware.

---

[1] The stand-alone version is available from the first author at the following address: UNC Psychometrics, CB-3270 Davie Hall, Chapel Hill, NC, 27599-3270. (Bitnet: ULURU1@TUCC).

[2] Both versions of VISUALS were written at the University of North Carolina under a contract from SAS Institute Inc., to the University. There is currently no commitment by the Institute to support or distribute VISUALS, nor if it does that it will conform to the description given in this paper.

[3] Efforts are underway to make the stand-alone version of VISUALS available on the IBM PS/2 and MAC-II microcomputers. Please contact the first author concerning the current status of these and other developments.

The abilities of the PGA board (but not the PC host) are important for the 3D features described in Section 3, since these are all performed by the PGA board. There are PGA boards made by several manufacturers, some of which generate 8-10 images per second with 50-75 unlabeled simple objects, and as many as 6-8 images per second with considerably more objects. Thus, these boards produce smooth animation which effectively communicates the 3D structure of the object cloud.

The hyper-dimensional features described in Section 4 are performed by the central processor of the computer, thus, the specific PC host determines how quickly these features are performed. A math coprocessor improves performance. The IBM AT with coprocessor generates two or three interpolations/second for 50 observations.

3. USING VISUALS TO UNDERSTAND 3D STRUCTURE

In this section we describe several aspects of VISUALS which are designed to provide insight into 3D structure. These aspects include the attributes of the objects and the spaces they are in, 3D motion, subsetting, metamorphing, brushing, recording, and successive refinement. Hyperspace interpolation and residualization are described in the next section.

3.1 Invoking VISUALS

VISUALS is invoked by the following statement:

> *visuals* (x1var y1var z1var)
> (x2var y2var z2var)
> (d1var d2var ... dnvar);

This statement allows the user to define two 3D spaces and an additional nD space. The first parenthesized list, which must be used, specifies those variables in the input data set that are used to construct the three axes of the 3D space that is initially displayed to the user. The values of the observations on these variables are used to position the objects in the cloud. (The list may include only one or two variables, in which case a 1D or 2D space is formed). The second parenthesized list contains up to three variable names, while the third list may contain any number of variables. These lists, which are both optional, define spaces which are used with the hyper-dimensional functions described in Section 4.

The *visuals* statement has several options. Three of the options define an input data set and two output data sets. One of the output datasets is used to create output that can be input to VISUALS in another session, whereas the other output dataset is used to create output that can be input to other software that creates graphical hardcopy. There are also options that specify which input variables are used to determine the attributes of the objects in the initial display, and which variables can be used for labels (object attributes and labels are discussed below).

3.2 Example

An example of an initial VISUALS display is shown in Figure 1. In this plate the "x1var" variable is horizontal and the "y1var" variable is vertical. The "z1var" variable points straight out of the screen and cannot be seen until movement occurs. This plate is the initial view of the results of our reanalysis of data gathered by Fisher, *et. al* [1985]. The data are means of judgement made by 32 male and 21 female judges. About half the judges are regular users of Psychoactive Mushrooms, the other half are nonusers. The judgments are of the frequency that the judge, the judge's mother, and the judge's father use Alcohol, Tobacco, Marijuana, Cocaine, Amphetamines, Hallucinogens, and Opiates. There are 12 observations (2 sexes × 2 user types × 3 judged people) and 7 variables. This mean frequency matrix was analyzed by the method developed by Young and Kuhfeld [1986] to obtain principal components under the assumption that the data are ordinal. The analysis, which monotonically transforms variables to maximize the variance accounted for by three principal components, accounted for 99.7% of the total variance. The plate is a 3D biplot [Gabriel, 1981] in which the observation scores and variable coefficients on the three principal components are used to locate the objects. The observations are represented by 3D crosses, and the variables by vectors.

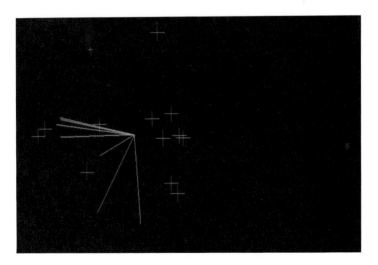

Figure 1

3.3 Attributes

VISUALS is an *object oriented* software system in which the user interacts with "objects" presented on the computer screen. One of these objects is the 3D space itself, while the others are objects representing the observations in the data set. These objects have attributes which the user can manipulate. We discuss these attributes in this section.

3.3.1 Observation Object Attributes

VISUALS represents each observation by a 3D object located according to the value of the observation on the three variables in the first parenthesized list of the *visuals* statement. The collection of objects forms the cloud of observation-objects. Each object has eight attributes: color, shape, size, label, label size, projection, vector and observation number. Except for observation number, the initial values for each attribute are determined by variables in the input data set. These variables have default names which may be overridden by using the options on the *visuals* statement. The attribute values may be changed while running VISUALS. Up to 16 colors and 7 shapes may be used. The shapes are null (invisible), pixel, 3D cross-hair, cube, pyramid, tetrahedron, and octahedron. The size of an object may vary. Lines may be drawn from any object orthogonally to any of the three planes formed by the axes of the space. A line may also be drawn from the object to the origin. Each object may have a character string label determined from variables in the input data set, from the observation number, or entered while VISUALS is running. These labels may vary in size. In Figure 1 there are twelve 3D cross-hair objects, and 7 "vector" objects (a vector object has the attributes "invisible" and "line to origin").

3.3.2 Space Attributes

The 3D space has many modifiable attributes, including axes, axis tick marks, axis labels, planes, perspective, fast environment, colors, etc. "Axes" are lines representing the x, y, and z axes of the space. These axes may or may not be displayed, and they may have tick marks and be labeled. "Planes" are grids representing the three planes formed by pairs of axes. The "perspective" attribute determines whether the 3D space is projected onto the 2D display screen in perspective or in parallel. "Fast environment" causes the axes and planes to be disappear when animated action modes (see below), so that the animation is as smooth as possible. In Figure 2 axes and planes are shown, and the figure is in perspective projection, although the visual effect is minimal. The axes are labeled, as are the vectors representing the drugs. Some of the drug labels cannot be read due to the orientation of the space.

3.4 3D Motion

The 3D motion commands allow the user to direct the motion of the cloud of objects. There are commands to *move* (translate) and *spin* (rotate) the 3D cloud of objects displayed on the screen. It is important to mention that the move and spin commands are *not* statements like the *visuals* statement: They do not involve typing a statement named *move* followed by information that specifies the nature of the move. Rather, these commands are single-keystroke, high-interaction commands. (The commands and their user interface are discussed in Section 5.)

Figure 3 presents an example of the effect of the move commands on the space shown in Figure 2. In this figure, the data analyst has moved the space closer, as well as up and to the right. From this viewpoint the effect of perspective projection on the drawing of the horizontal and vertical planes can be clearly seen. This effect can be seen on the z-axis of several of the 3D-cross objects, particularly for those near the upper left and right corners of the space.

Plate 12a shows the space in the previous figures oriented by the data analyst with move and spin commands so that the names of all the drugs are legible, and so that an interesting aspect of the drug structure (discussed below) can be seen. To further show this structure, the crosses (representing judges) have been made invisible.

In fact, the most striking information provided by VISUALS for these data is shown in Plate 12a. We see that five of the drugs (marijuana,

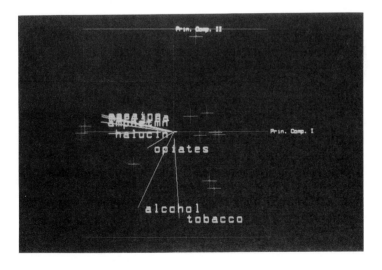

Figure 2

cocaine, amphetamines, hallucinogens and opiates) are essentially coplanar, whereas the other two (tobacco and alcohol) are orthogonal to the plane formed by the first five. This structure corresponds nicely with the legality of these drugs. Note that this structure cannot be seen in the ordinary 2D biplot of the first two principal components (Figure 1) nor from scatter plots formed from other pairs of components. However, the data analyst easily discovered this structure with VISUALS: He simply spun the 3D biplot.

The 3D motion may be "stop-action" (one movement per key stroke) or "animated" (repeated movements per key stroke). (While the structure of the drug data was first seen with animated moves and spins, the particularly revealing view shown in Plate 12a was fine tuned by stop-action commands.) All animated moves and spins may be combined together in any fashion, so that simultaneous rotation and translation is possible on all three axes. Also, the rates of each of the moves and spins are separately controlled. Thus, any arbitrary type of animated movement is possible, not just the canonical moves and spins. Finally, animated movements may be reversed by a single key-stroke, yielding rocking motion.

3.5 Other Features

In addition to the 3D motion features just discussed, there are a number of other features to aid in the understanding of 3D structure. These features are discussed in this section.

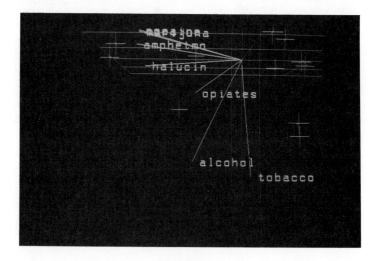

Figure 3

3.5.1 Subsetting

Subsets of the observations may be formed, saved, and used to change attribute values of objects in the subset. Multiple subsets may be formed. Subsets may be formed typographically, according to logical combinations of object attribute or character variable values, or graphically, by picking specific individual objects or by picking a group of objects occupying a rectangular region of the screen. Typographical and graphical methods may be logically combined. One of the subsets can be designated the "current" subject.

3.5.2 Metamorphing

The objects in the current subset of objects may be "metamorphed": The user may change their attributes. Figures 1-2 and Plate 12a show examples of a subset of labeled objects, and another of unlabeled objects. The labeling is a metamorphing of the objects from their appearance in Figure 1. Also, the crosses in Figures 1-3 have been metamorphed to "nulls" in Plate 12a. Other attributes, such as color, size, projections, etc., may also be metamorphed.

3.5.3 Brushing

A rectangular "brush" is available for "brushing" the objects. The size and location of the brush can be controlled by the user. An observation is brushed if it is inside the rectangular brush. Observations which are brushed in one space can be highlighted in either of the 3D spaces (defined by the two parenthesized lists), or can have hidden attributes revealed in either space. For those observations whose objects are inside the brush the user can display values of all character variables as well as observation numbers. Plate 12b shows the brush revealing that the two objects at the left of the space are "users", and Plate 12c shows that these are "self" reports about their own drug use. This fits nicely with the interpretation of the location of the variables in the space: The "user-self" objects represent psychoactive mushroom user's reports about their own drug use, and the "user-self" objects are in the part of the space toward which the illegal drug vectors point, accurately indicating that these objects represent respondents who frequently use illegal drugs (see Gabriel, 1981, for more information on interpreting biplots).

3.5.4 Refinement

Objects are represented as wire-frame drawings. While this increases animation speed, it reduces the 3D effect, particularly in static pictures appropriate for publication. Thus, when desired and when there is no movement on the screen, the objects may be refined from their wire-frame representation to solid objects, and then to shaded solid objects. (This is a design feature which is not yet fully implemented).

3.5.5 Recording

Everything that happens with VISUALS may be recorded may be recorded and played back. Recorded material may be overwritten with new material, if desired. The recording commands permit recording multiple frames, or only the currently displayed frame. The recording "head" can be moved to previous portions of the recording, thus allowing previously recorded material to be erased. The playback commands permit playing back the entire recording, portions of the recording, or only a single frame. Note that the recorder and player may be on simultaneously, permitting "tape loops" that can be repeated automatically an indefinite number of times. This can be combined with the reverse motion key to observe repeated rocking motion, an effective 3D cue. (This is a design feature which is not yet fully implemented).

4. USING VISUALS TO UNDERSTAND hD STRUCTURE

The central innovation in VISUALS is the way it tries to help the data analyst understand the structure of multivariate data. In addition to the features discussed in the previous section, which are useful for understanding hD structure but are predominantly oriented towards understanding 3D structure, there are two features in VISUALS which are specifically designed to help understand hD structure. These features are *hyperspace interpolation*, which dynamically displays a series of smoothly changing 3D projections of the hD data space, and *residualization*, which redefines the series of projections displayed by the hyperspace interpolation method. Together, these two features provide the user with the potential of seeing all 3D portions of the hD space. In this section we discuss these two new features.

4.1 An Analogy

As discussed in Section 3, the *visuals* statement allows the user to define one or two 3D spaces. If there is only one parenthesized list of variables on the statement, then one 3D space is defined. If there are two lists, then two 3D spaces are defined and the user can create a sequence of interpolations that lead from one of the two spaces to the other.

The sequence of interpolations can be thought of as a movie: The first frame of the movie is one of the two spaces, while the last frame is the other. Between the first and last frames VISUALS interpolates other frames which lead smoothly from the beginning to the end of the movie. The movie of the interpolation sequence is then shown to the viewer, being seen as a dynamic picture of a smoothly changing 3D data space.

The movie analogy breaks down, however, because the movie is not actually "taken" and then "shown". VISUALS does not compute all frames in a predetermined sequence and then display the sequence after it is computed. Rather, what VISUALS actually does is compute one frame at a time, displaying each frame before the next one is computed. This method has the characteristic that the user can interrupt the sequence at any time, changing the details of the interpolation. In fact, the movie can be displayed in *stop-action* mode, where the data analyst must interact with VISUALS to see the next frame, or in *animation* mode, where each frame in the sequence is seen until the user interrupts. At all times, the data analyst can actively controls the movie, rather than simply sitting passively and watching it occur. For example, the user can control which (and how many) axes of the spaces are subject to interpolation, and whether the interpolation proceeds all the way from one space to the other, or only part way.

4.2 Dynamic Hyper-Dimensional Graphics

What the user sees, in the movie that he or she tells VISUALS to create, is a 3D space containing moving objects. In this space, called the *interpolation* space, each object moves along it's own unique path, since the interpolation affects each observation uniquely. If, for example, the user decides to interpolate only between the two x-axes of the X_1 and X_2 spaces, then the objects move along paths that parallel the x-axis of the object cloud in the interpolation space. The movement is in either the positive or negative direction of the interpolation space's x-axis, with some objects moving rapidly, and others slowly. They move in the positive direction if the object's observation has a larger positive value on the x-axis in the space towards which the interpolation is directed. Negative movements result in the opposite case. Movements are fast when the difference in the object's values in the two spaces is large. Small differences create slow movement.

Animated interpolation may take place on two or three axes simultaneously. When interpolation involves two axes, then each object moves along a straight line in a plane parallel to the plane of interpolation. The direction and rate of movement of each point depends on where they are located when projected into the interpolation plane. Interpolation involving all three axes moves each object along the straight line between its two positions.

Note that the motion of each object is along its own unique path in the object cloud. This is a different kind of motion than that imparted on the objects by spins and moves where all objects move as a cloud (either all moving in the same direction, or all spinning around the same center). Thus, when interpolations, spins, and moves are combined it is possible to see the separate effects of each cause of motion on each object.

4.3 Examples

In this section we present two examples, one of VISUALS being used to explore a single high-dimensional space, and the other of VISUALS being used to compare two high-dimensional spaces.

4.3.1 *Exploring One hD Space*

In this section we present an example that illustrates looking for structure in high-dimensional data. It is based on a set of 7 dimensional data reporting the crime rate per 100,000 population in seven major crime categories in each of the 50 states of the United States. These data were submitted to a principal components analysis. VISUALS was then used to investigate the structure of the principal component scores and coefficients. The data and the principal components analysis are reported by SAS Institute, Inc. (1985), pp. 630-636.

In Plate 13a we show the initial biplot display constructed by VISUALS when it is initiated with the statement

visuals (pc1 pc2 pc3) (pc4 pc5 pc6) (pc7),

where the variables pc1, pc2, etc. contain the scores and coefficients on each principal component. What we see, initially, is the plane formed by the first two principal components (the principal plane). The first component is displayed horizontally, and the second vertically (the third cannot be seen because it is orthogonal to the screen).

In the display there are 50 crosses for the principal component scores of the 50 observations, and 7 vector-objects for the principal component coefficients of the 7 variables. The vectors are labeled. (Variables have been input to VISUALS to determine the object and label for each observation). The crosses represent the location of each state as projected into the principal plane, whereas the vectors represent the seven variables as projected into the principal plane. The length of a vector represents how close the variable is to the principal plane (the amount of variance in the variable that is associated with the plane).

Previous experience with these data (Young and Sarle, 1983) indicates that there is a cluster, at least in this plane, of southern states at the bottom of the plate. By brushing the display we can locate and label these states, as shown in Plate 13b (we have also metamorphed the objects for the other states into points and objects for variables into nulls). We can use VISUALS to investigate whether this cluster is a cluster in all seven dimensions, only in the two we see, or in some other dimensionality.

We first spin the space to see if the cluster is at least three dimensional. Looking at the spinning space confirms that the cluster is at least three dimensional (hiding all other objects in the space so that we only see the states in question helps reveal this).

Next, we use interpolation to see if the states form a cluster on more than the first three principal components. Preparing for the interpolation, we return the space to its original orientation, we move the space somewhat to the left and display only the eight southern states in question. To further emphasize the interpolation we display the grid representing the principal plane and the line that projects each object onto the principal component. (These last two actions are not needed when viewing the animated interpolation, but are helpful for the static plates shown in this paper.)

We then perform the interpolation. Four frames of the interpolation movie are shown in Plate 13c and 14, Plate 13c being the first frame and Plate 14c the last. Plate 14a is approximately 1/3 of the way into the movie, and 14b is about 2/3's of the way through it. At the right of each frame is information summarizing the current status of the move, spin and interpolation commands, as well as some other information (this is optional information which is shown in this plates to clarify details of the interpolation, but is not usually needed when it is viewed dynamically). The lowest gauge, consisting of three horizontal lines inside a box labeled "interpolation", indicates the amount of interpolation for each axis. The top line inside the box informs about the x-axis, the middle about the y-axis, and the lower about the z-axis. The heavy bar on each line moves back and forth along the line indicating the amount of interpolation between the space defined by the first parenthesized list of variables (left end of line) and the space defined by the second list (right end of line).

In Plates 13c and 14 the interpolation gauge shows that the interpolation is only taking place on the "x" and "y" dimensions (the two that can be seen in these plates). In Plate 13c we see that the observation-objects are located relative to each other in the same way as in Plates 13a and 13b, and the interpolation gauge indicates the interpolation space is exclusively the first space. In Plate 14a the objects have moved somewhat, and the gauge indicates the interpolation is 1/3 of the way from the first space towards the second space on dimensions "x" and "y". Then, in Plate 14b, the objects have moved further and the interpolation has proceeded 2/3'rds of the way on these dimensions. Finally, in Plate 14c the objects have moved to their locations in the plane defined by the first two variables in the second parenthesized list, and the interpolation gauge indicates the interpolation space now consists exclusively of the second space on dimensions "x" and "y" and exclusively the first space on dimension "z".

This interpolation sequence allows us to see that these states do not all form a single cluster in the six-dimensional space formed by the first six principal components. This can be seen most clearly by looking at Plate 14c, where we see that the two Carolinas are separated from the other southern states. But note, that we could have seen this by simply toggling to the second space (Plate 14c). Thus it is not necessary to use interpolation to discover the structure that we have just seen. However, we present this example because it is justified pedagogically, with the lesson to be learned being that one uses hyperspace interpolation to discover and understand the structure of multivariate data by looking at patterns of movement in the dynamic graphics.

Of greatest importance, though, is that the interpolation process that we just presented reveals additional structure in the eight southern states that cannot be seen simply by toggling between the two spaces. This structure is demonstrated in Plates 15a-15d, where we have metamorphed the objects to emphasize, in these static graphics, what is more obvious in dynamic graphics. In these plates we can see that those states which are represented by the same objects move along essentially parallel paths, and that these paths are not parallel with those for states represented by a different type of object. This suggests that these eight states fall into *three* clusters in six-dimensional space, not two.

Note that we cannot see these three clusters as spatially separated clusters in any of the four spaces shown in Plates 15a-15d (or in any other interpolation that we could show). Rather, we see these clusters as *movement clusters*, not spatial clusters. Indeed, it may be the case that we cannot see a structure that actually exists in hD in any of the infinity of different static 3D projections of that space. For our example there may be no 3D projection that shows these three groups of states as spatially separate clusters. However, the movement clusters imply that spatial clusters do exist in 6D. Thus, VISUALS gives us a way of "seeing" the structure of high-dimensional space by encoding the structure as movement in three-dimensional space.

Thus, in our example, VISUALS uses dynamic graphics to let us see that there are three six-dimensional clusters of states among the eight southern states, not the single cluster that we can see in a static graphic of the first two dimensions, or the two clusters we can see by looking at static graphics of other dimensions. One of the three clusters is the Carolinas cluster that can be seen in the static Plates 14c and 15d. However, the other two clusters, one consisting of Mississippi, Alabama and Louisiana, and the other of Georgia, Arkansas and Tennessee, are not obvious in any single picture, but can only truly be seen dynamically. Studying these frames carefully shows that one of the states - Arkansas - moves along a path which is somewhat like that of the Carolinas as well as like Georgia and Tennessee. However, when the space is spun during interpolation on all three axes, it is clear that Arkansas moves in a pattern like Georgia and Tennessee.

4.3.2 Comparing Two hD Spaces

Another use of the hD Dynamic Graphics features implemented in VISUALS is to compare two multivariate data spaces. These spaces may be of any dimensionality. They may have come from two alternative analyses of the same multivariate data, or from two (analyses of) two different sets of multivariate data. If the two data spaces come from two separate sets of multivariate data, the data must have been obtained from the same units of observation.

For example, we have used VISUALS to compare two analyses of the very well known iris data published by Fisher [1936]. In these data, the sepal length, sepal width, petal length and petal width were measured in millimeters on 50 specimens from each of three iris species, *Iris setosa*, *I. versicolor*, and *I. virginica*. These data have been widely used for examples in cluster analysis.

We analyzed them with the SAS clustering procedure FASTCLUS.[1] We then used VISUALS to compare the cluster analysis with a principal components analysis of the same data. Specifically, we performed a canonical discriminant analysis using the cluster membership as the classification variable, and compared the structure revealed in the plane formed by the two canonical variables (see Plate 15e) with the structure revealed in the plane formed by the first two principal components (see Plate 15f). Note that these plates present two different, but each in their own way optimal, 2D views of the 4D data space (4D due to the four variables). Plate 15e shows the 2D view that best reveals the clusters, whereas Plate 15f shows the 2D view that has the most variance. Indeed, in Plate 15e we can see that the clusters are more compact than Plate 15f, while in Plate 15f we see more spread overall. However, we also see that the plates are overall remarkably similar. Again, static graphics do not reveal this as clearly as dynamic graphics: Using VISUALS to interpolate from one space to the other, we can see that there is relatively little movement of individual observations between the two spaces. Thus, we conclude that the two analyses give us similar results, even though the analyses are very different.

As a second illustration, we use VISUALS to compare two 2D spaces resulting from two Correspondence Analyses reported by Greenacre [1984, pp. 280-291]. These spaces, which are shown in Plates 16a and 16b, result from two separate analyses of the pattern of protein consumption in Western and Eastern European countries. The first analysis (Plate 16a) showed Greenacre that Portugal is an outlier, having undue influence on the results. Thus, he reanalyzed these data with Portugal

---

[1] The only option used in the analysis specified that we wished to obtain three clusters. This cluster analysis, and the associated canonical discriminant analysis are reported by SAS Institute, Inc. [1985], pp. 387-393. A similar analysis, complete with VISUALS displays, is reported by Larus [1987].

removed from the dataset. The result of this second analysis is presented in Plate 16b. (See Greenacre and Hastie, 1987, for information on interpreting Correspondence Analysis results).

VISUALS shows that the pattern of protein consumption in Portugal differs from the patterns in other European countries. This is the straight forward observation that Portugal is an outlier in Plate 16a. Of much more interest, however, is that VISUALS shows us that the presence of Portugal in the analysis heavily influences the overall structure revealed by Correspondence Analysis. In particular, in these plates we can see that the vectors representing the protein variables have moved a great deal, indicating a radically different weighting scheme for these variables when Portugal is removed.

There are other differences that are immediately obvious in the dynamic graphics, differences concerning the location of certain countries. In Plates 16c and 16d we show the two spaces again, this time with the spaces translated closer, with the variables removed (to reduce the clutter) and with certain countries labeled. We have labeled Spain, because it is the country that moves the most between the two analyses, and we have labeled Albania, Yugoslavia, Rumania and Bulgaria because they are a cluster which doesn't move. It is interesting that Spain moves the most since Spain and Portugal obtain much of their protein from fish. It is also interesting that Albania, Yugoslavia, Rumania and Bulgaria, which are all culturally similar countries that obtain the vast majority of their protein from cereals, are a tight cluster of countries that move little between the two analyses. Note that we discovered this cluster by first watching the animated hyperspace interpolation, and then brushing the cluster to see which countries they were. We did not, for example, first ask the question: "Where are the Eastern European countries"? The dynamic graphics features in VISUALS have again helped to discover structure in multivariate data.

4.4 Algebra

It is important to repeat at this point that the user need not be aware of the algebraic details underlying interpolation. Rather, what the user sees is a graphical display of "dynamic algebra" that occurs behind the scenes. Indeed, the main design philosophy of VISUALS is to hide from the user all of the algebra that underlies the dynamic graphics. However, it is important for developers of dynamic graphics to completely explicate the algebraic details, a task which we address in this section.

4.4.1 Interpolation

To understand the algebraic details of the interpolation process, we must introduce some notation. We denote the two 3D spaces by the matrices X_1 and X_2. The matrix X_1 is defined by the first list, and X_2 is defined by the second. Both matrices have n rows containing the coordinates of the n observation points in the 3D space. The columns of the two matrices correspond to the variables specified in each list, and to the axes of each space. Both matrices have 3 columns, even if fewer than three variables are included in the list, in which case the appropriate column(s) contains zeros. For any given frame of the movie, VISUALS computes a 3D space whose three axes are linear combinations of the corresponding pairs of axes in the two spaces X_1, and X_2. This space, called the *interpolation* space, is represented by the $(n \times 3)$ matrix X, which contains the coordinates of the n points in the 3D interpolation space. The space is defined by the simple interpolation equation

$$X = X_1 A + X_2 (I - A), \, 0 \leq a_{ii} \leq 1$$

where A is a (3×3) diagonal matrix with constrained diagonal. Initially,

$$A = I,$$

so that the interpolation space is initially just X_1. The frames of the movie differ in the specific values of a_{ii}. These values are incremented or decremented (within the range constraints) for each successive frame according to the equation

$$a_{ii}^+ = a_{ii} + d_i,$$

where a_{ii}^+, indicates the value a_{ii} takes on for computing the next frame. The value of d_i, is determined by the high-interaction commands discussed in Section 5.

4.4.2 Residualization

Up to this point, we have been discussing *hyperspace interpolation*, which is only one of the two new VISUALS methods which are designed for understanding the structure of multivariate data. The second method provided by VISUALS is *residualization*, a method which extends the power of interpolation. In this section we discuss residualization.

The two parenthesized lists of variable names on the *visuals* statement define a 6D data space whose dimensions are the six variables. (Note: We have been using "axis" for the basis dimensions of the two 3D spaces. Here we use "dimension" for the basis dimensions of the abstract hD data space.) In addition, there is a third parenthesized list of variable names which can appear on the *visuals* statement. When the third parenthesized list is used the variables listed in it form additional dimensions of the data space. The data space is represented by the $(n \times h)$ matrix X_d, where there are a total of h variables mentioned in the three parenthesized lists.

Hyperspace residualization replaces the X_1 space with the current interpolation space X and replaces the X_2 space with the three longest orthogonal dimensions of the h-dimensional data space X_d that are also orthogonal to the interpolation space. The data space is related to the interpolation space by

$$X_d = XB + R,$$

where R is a $(n \times h)$ matrix of residuals from fit between the two spaces, and B is a $(3 \times h)$ matrix of coefficients of three orthogonal linear combinations of the h variables, determined by the equation

$$B = X - X_d,$$

where $X-$ is the generalized inverse of X. Then

$$R = X_d - \mathbf{X} - X_d$$

can be decomposed into

$$R = UGV'$$

using a singular value decomposition. We then redefine the X_1 space as the old interpolation space X and X_2 space as the first three columns of UG. Notice that residualization does not change X_d.

After residualization the user has two new 3D spaces. One of these was the interpolation space that appeared on the screen at the moment of residualization, while the other is the space that is orthogonal to the first space which also has the longest dimensions. These two spaces can serve as the basis for further interpolation, now showing a movie of new portions of the multivariate data space. The residualization can be performed again any number of times, potentially opening up all 3D projections of the data space for viewing, and through the spin and move commands, manipulation. Note that residualization is always taking the data analyst back to a portion of the space that contains the most spread.

5. HIGH-INTERACTION INTERFACE

It is important to understand that the user is not required to type in complicated equations or programming statements. Rather, the spins, moves and interpolations (and all other features) are controlled by "high-interaction", immediately effective, single-keystroke commands. In this section we discuss the layout of the spin, move and interpolation commands on the keyboard, and the way key-presses implement these commands.

5.1 Command Layout

We have arranged the commands in a pattern on the keyboard like the pattern shown in Figure 4. In this figure we show the interpolation commands on the left and the move and spin commands on the right. (In the actual layout on the IBM PC/AT the interpolation commands are at the far left on the function keypad, and the move and spin commands are the far right on the numeric keypad.)

The move and spin commands are assigned to the same keys, thus these keys have a move or spin mode, controlled by a toggle key, as shown. In move mode, the indicated keys move the cloud of objects up, left, right and down, at the indicated 45° angle, or towards or away from you. In spin mode the same keys spin the front of the cloud of objects up, down, left or right, at a 45° angle, or in a clockwise or counter-clockwise direction. Four additional keys to the right control related operations, including one which pauses animated moves or spins (to allow fine tuning), one which reverses animated moves or spins (to provide rocking), and two which increase or decrease the effect of one key press on the amount of move or spin.

The interpolation commands are arranged in a pattern like that shown on the left side of Figure 4. The top six keys control the interpolation itself. The top three keys on the left cause interpolation towards the "left" space (the one created by the first -- or left-hand -- of the two parenthesized lists on the *visuals* statement), and the three keys on the

Interpolate towards x_1	Interpolate towards x_2	Toggle: Menu On / Off	Toggle: Move/Spin	Toggle: Animation On / Off	Toggle: Pause On / Off
Interpolate towards y_1	Interpolate towards y_2	Up-Left	Up	Up-Right	Reverse
Interpolate towards z_1	Interpolate towards z_2	Left		Right	Decrease Move/Spin Increment
Decrease Interpolate Increment	Increase Interpolate Increment	Down-Left	Down	Down-Right	Increase Move/Spin Increment
Toggle: between Left/Interp	Toggle: between Right/Inter	Move: Away Spin: Clockwise		Move: Near Spin: Counter C W	

Figure 4

right causing interpolation towards the "right-hand" space. Of these six keys, the top pair cause interpolation along the "x" dimension, the middle pair cause interpolation on the "y" dimension, and the bottom pair cause interpolation on the "z" dimension. Thus, these six keys are arranged in a spatial pattern which reflects their function. The next two keys control the amount of interpolation increment that occurs on each press of the first six keys, with the left-hand key decreasing the increment and the right-hand key increasing it. Finally, the last two keys toggle between the three spaces, with the left key toggling between the "left" space and the interpolation space, and the right key toggling between the "right" space and the interpolation space. Only one space may be viewed at a time.

The arrangement of commands shown in Figure 4 corresponds closely to the way we have arranged the commands on the standard IBM PC/AT keyboard. This layout is particularly suited to this keyboard, although similar arrangements are possible on other keyboards. In the IBM PC/AT, the function keys and numeric keys are arranged rectangularly, just like those in the figure, which is particularly appropriate for the command layout. Furthermore, on this keyboard the cursor keys are also on the numeric keypad so that the move and spin commands for left, right, up and down correspond to the normal actions of the cursor keys. Finally, the increase and decrease increment commands correspond to the + and − keys. Switching between the move and spin modes is controlled by the "Num Lock" key, while switching between animate and stop-action modes is controlled by the "Scroll Lock" key.

This pattern feels very comfortable and is easily learned on this keyboard. This is partially due to the rectangular arrangement of the keys, to the IBM PC/AT's layout of the cursor keys on the numeric keypad, and to the natural correspondence of the cursor keys and the commands. But it is perhaps more due to the fact that the hands never have to leave the two keypads (even interacting with the menu system only requires using the cursor keys except when entering names for new subsets, variables or data sets). Even on keyboards that do not arrange the cursor keys on the numeric keypad, we recommend following this arrangement as closely as possible.

There has been much said in favor of using mice instead of keys for high-interaction control of dynamic graphics. It is our opinion that the important issue is not keys versus mice, but is whether the user can interact with the graphics without having to look away from them. The command layout we have developed permits this. Of course this layout can be represented graphically on the screen, and then interacted with through a mouse. We think this would be equally effective.

5.2 Algebra of the Commands

In this section we explain the algebra of the commands, showing how the commands effect the coefficients of the interpolation equation presented in Section 4.4.1. We do not discuss the move and spin commands since they work in the same fashion as the interpolation commands, and since we have only introduced the algebra underlying the interpolation commands (under the assumption that the reader is familiar with the algebra for translating and rotating 3D spaces).

In Section 4.4.1 we defined the equation which is used to change the interpolation coefficients. The equation is:

$$a_{ii}^+ = a_{ii} + d_i,$$

where a_{ii}^+ indicates the value that the interpolation coefficient a_{ii} takes on for computing the next frame (in stop-action mode) or the next sequence of frames (in animation mode). In this equation the value of d_i (which initially is zero) is incremented or decremented via the six interpolation commands in the top three rows of the interpolation keypad of Figure 4. The amount that d_i is incremented is a function of the mode (stop-action or animation), the number of key presses (in animation mode only), and the increment constant c (which is determined by the increment commands in the fourth row of the interpolation keypad). Thus, there are eight commands (six interpolation and two increment) commands which can be used to modify the coefficients of the interpolation equation.

As shown in Figure 4, the six interpolation commands are organized into three pairs. The top pair of commands, which control interpolation between the x-axes of the X_1 and X_2 spaces, change d_1 (and thus a_{11}). The second pair of commands provides similar control of interpolation along the y-axis (they modify d_2) and the third along the z-axis (through d_3). The commands in a pair effect the value of d_i via the following equation:

$$d_i = \begin{bmatrix} \text{sign}(p_i)\, c\,(2^{abs(pi)-1}) & \text{for } abs(p_i) = 1, 2, \ldots 7 \\ 0 & \text{for } p_i = 0 \end{bmatrix}$$

where initially

$$c = .01$$

and

$$p_i = 0.$$

The right-hand member of a pair of commands increments p_i by 1 each time the command is used, whereas the left-hand member decrements p_i by 1 each time.

The commands operate nearly identically in both stop-action and animation modes. In either mode, when a interpolate key is pressed the value of p_i is incremented or decremented by 1 (the right-hand interpolate keys increment p_i, the left-hand ones decrement p_i). If the absolute value of p_i exceeds 7 it is reset to 7. Also in either mode, when the increment key is pressed the value of c is doubled, but when the decrement key is pressed c is halved. After calculating the new value of d_i that results from the key-press, the picture sequence is computed. In stop-action mode this sequence consists of one picture, while in animation mode the sequence consists of an indefinite number of pictures. In stop-action mode, p_i is set back to 0 after the picture is computed, while this does not occur in animation mode.

If the system is in animation mode, all frames in the sequence are calculated using the same values of d_i until one of the following stopping rules is satisfied: 1) another interpolation command key (including the increment keys) is pressed; or 2) all three interpolation coefficients are at a boundary value. If d_i causes a_{ii}^+ to exceed its range constraints during the sequence, then a_{ii} is set to the nearest boundary, p_i is set to zero, and the sequence continues. Note that all three interpolation coefficients have to be at their boundary values before the sequence stops.

5.3 Stop-Action and Animation

In stop-action mode, the effect of repeated use of the six interpolation commands on the dynamic graphics display is straight forward: Each use displays a new picture which is a bit different from the previous one. However, it is more difficult to understand the effect of repeated use of the six interpolation commands in animation mode.

To gain this understanding, we must focus on the $2^{abs(p_i)}$ term in the equation above. Each time an interpolation command is used, this portion of the equation has the effect of changing the amount of interpolation by a power of 2. Thus, repeatedly using the same command doubles, then quadruples, etc., the amount of interpolation that takes place from one frame to the next.

For the interpolations, these six commands change the d_i values. The first use of one of the commands in a pair sets d_i to plus .01 (if c has not been changed), causing interpolation towards the axis of the X_2 space (increases a_{ii}). Since $d_i = .01$, each interpolated frame leads 1/100th of the way further towards X_2 (The first use of the other command in the pair causes an effort, since it causes the range constraints to be violated). If the same command is used again, then $d_i = .02$, causing the rate of interpolation to double. Another use sets $d_i = .04$, etc. At this point, if the other command in the pair is used, the rate of interpolation is halved. Another use further halves the interpolation rate. When these commands are used in animation mode, so that interpolations are repeatedly taking place, repeated use of the other command in the pair will continue halving the rate of interpolation until the interpolation stops. Then, further use of the command reverses the direction of interpolation towards the X_1 space, with each use doubling the rate of interpolation back towards the X_1 space.

The twenty-four move and spin commands operate identically, except that the move commands modify the coefficients of a translation equation and the spin commands effect the coefficients of a rotation equation. Thus, in animation mode, repeated use of, say, a spin command causes the cloud to spin ever faster. Then, repeated use of the opposite command gradually slows the spin rate until it stops, and then reverses direction, gradually picking up speed as the command continues to be used.

If VISUALS is in animation mode the first press of a key initiates the movement (interpolation, move or spin) in the indicated direction. Once animated motion is initiated it can be changed by continuing to enter other movement commands. As long as VISUALS is in animation mode, the effect of each key press accumulates with all of the previous key presses, even when you switch between move and spin modes (while still in animation mode). All twelve move commands and all twelve spin commands can be accumulated together, and can be combined with all eight interpolation commands to create virtually any kind of combined interpolation, moving and spinning. Also, note that repeated use of a key increases the particular action. If this key happens to be the opposite of the motion on the screen, repeated use gradually slows down the motion, and then creates an increasingly rapid counter motion. So, in animation mode all 32 move, spin and interpolate commands can be combined together to obtain virtually any combination of movements. Furthermore, stop-action motion of any one type (either moving, spinning or interpolation) can be combined with animated moves of the other types. This gives the data analyst great flexibility, a flexibility which can be confusing, but which is often quite useful. For example, simply combining spin commands means that the cloud can be spun in any fashion. While it is spinning, it can then be positioned with animated move commands to a more visually optimal location. Then, while the space is still spinning, its location can be fine tuned with stop-action move commands.

6. DISCUSSION

We repeat the purpose of VISUALS: to help the data analyst visually discover and formulate hypotheses about structure in multivariate data. VISUALS introduces interpolation, residualization, and refinement as new methods of discovering and observing such structure. Others researchers have developed other methods with the same goal in mind, as reviewed in the first section. Research should be done to see which, if any, of these methods accurately communicate structure of multivariate data.

6.1 Cues for the 3D Illusion

There are many cues that induce the 3D illusion. VISUALS emphasizes hand-eye interaction, perspective projection, and smooth motion (including rotation, translation and rocking). When possible, VISUALS also displays smoothly shaded solid objects to enhance the illusion. While it can be argued that these may be the most important cues, the 3D illusion could be improved by smoother motion, depth cuing, stereo projection, etc. (see Simkin and Hastie, 1987, for more on graphical perception).

Many of these cues have been used to induce the 3D illusion in other 3D statistical graphics systems. MacSpin [Donoho, Donoho and Gasko, 1986], for example, has hand-eye interaction, smooth motion and depth cuing, but only uses pixel objects and parallel projection. ANIMATE [Gabriel and Odoroff 1986] has smooth motion, stereo or perspective projection, and smoothly shaded solid objects, but only rocking motion that is not actively dynamic (i.e., does not involve hand-eye interaction).

Many of these software cues for inducing the 3D illusion can now be implemented on relatively inexpensive hardware. Alternatively, there exists specialized hardware that can present what appears to be a genuine, space filling display (as in 3D motion pictures). Unfortunately, such hardware is expensive. Furthermore, it is still the case that some of the software cue combinations (smooth shading with smooth motion, for example) require hardware that is expensive. Thus, an interesting empirical question is, which options are most cost effective in communicating 3D information?

Related to depth perception issues are issues concerning hand-eye interaction, and in particular, what kind of control devices are best for interfacing the active user with the 3D screen. VISUALS is controlled by cursor keys. Other possibilities include mice, track balls, sliders, pointers and joy sticks. Most of these devices exist in both 2D and 3D form, providing the interface designer with many possibilities.

Perhaps the fundamental hand-eye question is whether the distinction between active and passive dynamic systems made in the introduction is relevant to the strength of the 3D illusion: Do the hands contribute to the eyes' 3D perception? Our hunch is that active control of the motion is a strong cue in creating the illusion.[1]

Note, however, that a surprisingly large portion of the population do not perceive depth, and that for them, no matter how many cues are present there will never be a 3D illusion. It also seems that the popular distinction between "algebraists" and "geometers" is relevant here. There are many data analysts who would much rather look at tables of numbers and equations than at pictures of the numbers and equations, strange as that may seem to some of us. So, the endeavors reported in this paper, and others like them, should be viewed as a supplement to traditional data analysis methods. We hope that this supplement will prove useful to a fairly large segment of the data analysis community.

We are very pleased that VISUALS produces a reasonable 3D illusion on a relatively inexpensive, commonly available computer. We feel that it can help many data analysts form hypotheses about the structure of their data, a "hD illusion", if you will. In one or two years, when hardware costs have been further reduced and hardware capabilities further enhanced, the data analyst should be able to actively animate smoothly moving, smoothly shaded, solid object representations of multivariate data presented in either perspective or stereo projection. This should greatly enhance the 3D (and perhaps hD) effect over that obtained from the pixel objects that VISUALS can currently animate at 8-10 frames per second, or the wire-frame models that it can animate at only 3-4 frames per second.

6.2 Enhancements

A number of relatively simple enhancements to VISUALS would increase its power. The enhancement that might return the greatest payoff is stereo projection. Other enhancements could provide dynamic objects like those proposed by Nicholson and Carr [1984], and dynamic subsetting like the slicing and masking methods of Donoho, Donoho and Gasko [1986]. The dynamic subsetting notions could be implemented in 3D within VISUALS (by slicing, or masking X_2 and watching the effect on X_1).

---

[1] With passive dynamic graphics the user, as it were, is placed on a bus and given a tour of the data space. Active graphics is analogous to traveling through the data space instead of being given a tour. With active dynamic graphics the user is a traveler, who, on his or her own, decides what to see, rather than being a tourist at the whims of a tour guide. It has been said that a tourist goes somewhere to see what is there, whereas a traveler goes somewhere to be changed by what is there. We think that this saying applies to the difference between passive and active dynamic graphics.

Aside from these enhancements of VISUALS, there are steps that can be taken to create more powerful VEDA systems. We think the most important step is to "seamlessly" integrate: dynamic hyper-dimensional graphics for multiple 2D and 3D spaces; multivariate exploratory data analysis techniques; and a graphical data flow language for dynamic path diagrams.

6.3 Multiple Spaces

Multiple 3D spaces would support a number of uses, such as allowing interpolation between several 3D spaces. Graphical icons, such as regular polygons, could be used to control the interpolation between corresponding axes in several spaces. There would be three such icons, one for each axis. For example, in order to interpolate between the x-axes of three spaces, the user could move a pointer around inside a triangle, with different positions inside the triangle corresponding to different values of the interpolation coefficients. A square icon would permit interpolation between x-axes of four spaces.

Multiple 2D spaces would enable construction of spreadplots, our name for a high-interaction version of Tukey's generalized draftsman's plot (called a scatter plot matrix by Cleveland, 1985) for looking at all possible bivariate distributions. The spreadplot could effectively support brushing, slicing and masking. Spreadplots could also be constructed in 3D for looking at all possible trivariate distributions.

6.4 Multivariate Exploratory Data Analysis

Perhaps the greatest increase in power would come from integrating multiple space VEDA software with dimension reducing multivariate exploratory data analysis (MEDA) methods. Some of these methods, such as principal components, factor analysis, discriminant analysis, and canonical correlation analysis, are widely known in the statistical community. Others, including multidimensional scaling [Young and Hamer, 1987], correspondence analysis [Greenacre, 1984], projection pursuit [Huber, 1985], and the nonlinear multivariate analysis methods of de Leeuw [Gifi, 1981] and Young [1981], are becoming more well known.

Processing multivariate data to reduce its dimensionality can be a useful precursor to VEDA. Indeed, the axes of the plates displayed in this paper are principal components from a nonlinear principal components analysis performed by the SAS PRINQUAL procedure (Young and Kuhfeld, 1986). This analysis monotonically transforms the variables to maximize the proportion of variance in the transformed variables that is accounted for by a few principal components. The 3D space in the plates is actually a curved manifold that accounts for 97% of the variance of a 7D space. The plates are biplots [Gabriel, 1981] of principal components scores (plotted as 3D crosses) and coefficients (vectors).

Consider what could be done with these data with a VEDA system that could display multiple spaces. One 3D space could display the principal components, as was done above. In addition, another 3D space could be formed for each variable, by plotting the original untransformed variable against its transformed version and its least-squares prediction. Such a 3D plot displays, in three 2D views, the variable's nonlinear transformation and the residuals from both linear and nonlinear fit. In addition, 2D spreadplots could be formed by arranging multiple 2D spaces next to each other. Bivariate scatter plots of the untransformed variables could be arranged above the diagonal with the corresponding bivariate plots for the transformed variables below the diagonal (and transformations on the diagonal).

High interaction graphical techniques could be designed to allow the user to actively modify any part of the displayed information, with the system recalculating all or part of the remaining information. For example, the user could choose to modify one transformation. As the user does this, the system recalculates the principal components and/or transformations of other variables, displaying the changes in real time. Or, the user could decide to move one object in the principal component space to an intuitively more pleasing location, with the system making all of the compensatory changes in the transformations. This lets the user see the change in one portion of the data analysis model propagate throughout the rest of the model.

The ability to observe and interact with variable transformation plots, bivariate plots of untransformed variables, bivariate plots of transformed variables, and linear combinations of transformed variables (such as principal components), is particularly relevant to nonlinear multivariate analysis methods [Gifi, 1981; Young, 1981; Young and Kuhfeld, 1986; 1987]. As de Leeuw has recently shown [1985], these analysis methods attempt to find a system of transformations that linearizes all bivariate plots. An integrated MEDA/VEDA system could assist in obtaining these linearizing transformations.

The hyperspace interpolation feature can be used in a variety of new ways in a multiple 3D space VEDA system. One example is to interpolate between two trivariate scatter plots whose axes are formed from corresponding transformed and untransformed variables. Interpolation would dynamically display the nature of the transformation. Another example is to interpolate between the two canonical spaces resulting from canonical correlation analysis such as those performed by TRANSREG [Young and Kuhfeld, 1987]. Here, interpolation would dynamically display how well each observation is being fit by the two spaces. Finally, interpolation could be useful for comparing two 3D spaces resulting from two different analyses of the same set of data, highlighting the differences and similarities between the two analyses.

6.5 Graphical Data Flow Language

A highly interactive, highly integrated MEDA/VEDA system that incorporates all of the MEDA models and all of the multiple space VEDA notions mentioned above would indeed be a powerful data analysis tool. Powerful software tools must be properly interfaced with the user. Otherwise, they may remain unused or, even worse, they may be misused. Perhaps the ideal interface is a graphical data flow language that would implement, in essence, high interaction dynamic path diagrams. The user would manipulate 2D icons on the screen representing various portions of the multivariate data, connecting the icons together in ways representing the flow of the data through a specific MEDA model. When the icons are arranged and connected as desired, the user would request that the MEDA/VEDA system perform the analysis. The system would determine the specific MEDA model implied by the arrangement of icons, and would perform the analysis. The user could then "open" the icons, revealing 2D and 3D VEDA spaces that can be manipulated using the techniques described above. This graphical data flow language has been discussed by Young, Kuhfeld, and Smith [1987].

We are very optimistic that highly integrated, highly interactive, appropriately interfaced MEDA/VEDA systems will become very useful tools for exploring, understanding, and forming hypotheses about the structure of multivariate data. We believe that such systems will provide the data analyst with powerful new tools for "looking at data to see what it seems to say."

REFERENCES

Asimov, D. (1985). "The grand tour: a tool for viewing multidimensional data," *SIAM Journal on Scientific and Statistical Computing*, **6**: 128-143.

Becker, R. A. and Cleveland, W. S. (1987). "Brushing scatter plots," *Technometrics*, **29**: 127-142.

Carr, D. B., Littlefield, R. J., Nicholson, W. L., and Littlefield, J. S. (1987). "Scatter plot matrix techniques for large N," *Journal of the American Statistical Association*, **82**: 424-436.

Cleveland, W. S. (1985). *The Elements of Graphing Data*. Monterey, CA: Wadsworth.

Cleveland, W. S. (1987). "Research in statistical graphics," *Journal of the American Statistical Association*, **82**: 419-423.

de Leeuw, J. (1985). "Multivariate analysis with optimal scaling," International Conference on Advances in Multivariate Statistical Analysis, Indian Statistical Institute, Calcutta, India.

Donoho, A. W., Donoho, D. L., and Gasko, M. (1986). *MACSPIN: A tool for dynamic display of multivariate data*. Monterey, CA: Wadsworth & Brooks/Cole.

Donoho, D. L., Huber, P. J., Ramos, E., and Thoma, H. M. (1982). "Kinematic display of multivariate data," *Proceedings of the Third Annual Conference and Exposition of the National Computer Graphics Association*, **1**: 393-398. Fairfax, VA: National Computer Graphics Association.

Fisher, R. A. (1936). "The use of multiple measurements in taxonomic problems," *Annals of Eugenics*, **7**: 179-188.

Fisher, D. G., MacKinnon, D. P., Anglin, M. D., and Thompson, J. P. (1987). "Parental influences on substance use: gender differences and stage theory," *Journal of Drug Education*, **17**: 69-85.

Fisherkeller, M. A., Friedman, J. H., and Tukey, J. W. (1974). "PRIM-9: An interactive multidimensional data display and analysis system," SLAC-PUB-1408. Stanford, CA: Stanford Linear Accelerator Center.

Friedman, J. H., McDonald, J. A., and Stuetzle, W. (1982). "An Introduction to Real Time Graphics for Analyzing Multivariate Data," *Proceedings of the Third Annual Conference and Exposition of the National Computer Graphics Association*, **1**: 421-427. Fairfax, VA: National Computer Graphics Association.

Friedman, J. H. and Tukey, J. W. (1974). "A Projection Pursuit Algorithm for Exploratory Data Analysis," *IEEE Transactions on Computers*, **C-23**: 881-890.

Gabriel, K. R. (1981). "Biplot display of multivariate matrices for inspection of data and diagnosis," *Interpreting Multivariate Data*, V. Barnett, ed., 147-173. Chichester: Wiley.

Gabriel, K. R. and Odoroff, C. L. (1986). "ANIMATE: An Interactive Color Statistical Graphics System for Three-Dimensional Displays," *Proceedings of the Seventh Annual Conference and Exposition of the National Computer Graphics Association*, **3**: 723-731. Fairfax, VA: National Computer Graphics Association.

Gifi, A. (1981). "Non-Linear Multivariate Analysis," Department of Data Theory. Leiden, The Netherlands: University of Leiden.

Greenacre, M. J. (1984). *Theory and Application of Correspondence Analysis*. London: Academic Press.

Greenacre, M. J. and Hastie, T. (1987). "The geometric interpretation of correspondence analysis," *Journal of the American Statistical Association*, **82**: 437-447.

Huber, P. J. (1985). "Projection pursuit," *Annals of Statistics*, **13**: 435-475.

Huber, P. J. (1987). "Experiences with three-dimensional scatter plots." *Journal of the American Statistical Association*. **82**: 448-453.

Inselberg, A. (1985). "The plane with parallel coordinates," *The Visual Computer*, **1**: 69-97.

Kent, D. P., Edds, T., Kuhfeld, W. F., and Young, F. W. (1986). "PROC VISUALS: Experimental Software for Animated 3D graphics using the IBM Professional Graphics Device," *SAS Users Group International Proceedings*, **11**: 208-211.

Kuhfeld, W. F. (1986). "Metric and Nonmetric Plotting Models," *Psychometrika*, **51**: 155-161.

Kuhfeld, W. F., Edds, T. C., Kent, D. P., and Young, F. W. (1986). "New Developments in Alternating Least-Squares SAS Procedures," *SAS Users Group International Proceedings*, **11**: 846-851.

Kuhfeld, W. F., Sarle, W. S., and Young, F. W. (1985). "Methods for Generating Model Estimates in the PRINQUAL Macro," *SAS Users Group International Proceedings*, **10**: 962-971.

McDonald, J. A. and Pedersen, J. (1985). "Computing environments for data analysis, (I: Introduction, II. Hardware)," *SIAM Journal on Scientific and Statistical Computing*, **6**: 1004-1012, 1013-1021.

Larus, J. M. (1987). "Graphical Displays of Nonhierarchical Cluster Analysis Solutions," *Proceedings, SAS Users Group International*, **12**: 1089-1095.

Nicholson, W. L. and Carr, D. B. (1984). "Looking at more than three dimensions," *Computer Science and Statistics: Proceedings of the 15th Symposium on the Interface*, 201-209. Amsterdam: North-Holland.

SAS Institute Inc. (1985). *SAS User's Guide: Statistics*, Version 5 Edition. Cary, NC: SAS Institute Inc.

Simkin, D. and Hastie, R. (1987). "An information-processing analysis of graph perception," *Journal of the American Statistical Association*, **82**: 454-465.

Stuetzle, W. (1987). "Plot windows," *Journal of the American Statistical Association*, **82**: 466-475.

Tufte, E. R. (1983). *The Visual Display of Quantitative Information*. Cheshire, CT: Graphics Press.

Tukey, J. W. (1977). *Exploratory Data Analysis*. Reading, MA: Addison-Wesley.

Tukey, J. W. and Tukey, P. A. (1981). "Graphical display of data sets in 3 or more dimensions," *Interpreting Multivariate Data*, V. Barnett, ed., 189-275. Chichester: Wiley.

Young, F. W. (1981). "Quantitative analysis of qualitative data," *Psychometrika*, **46**: 357-388.

Young, F. W., Edds, T. C., Kent, D. P., and Kuhfeld, W. F. (1986). "Visual exploratory data analysis: PROC VISUALS, a user-written experimental SAS procedure," *SAS Users Group International Proceedings*, **11**: 840-845.

Young, F. W. and Hamer, R. M. (1987). *Multidimensional Scaling: History, Theory and Applications*. Hillsdale, NJ: Erlbaum.

Young, F. W., Kent, D. P., and Kuhfeld, W. F. (1986). "PROC VISUALS: software for dynamic hyper-dimensional graphics," *American Statistical Association 1986 Proceedings of the Section on Statistical Graphics*, 69-74. Washington, DC: American Statistical Association.

Young, F. W., Kent, D. P., and Kuhfeld, W. F. (1987). "PROC VISUALS: Software for Dynamic Hyper-Dimensional Graphics," *SAS Users Group International Proceedings*, **12**: 551-560.

Young, F. W. and Kuhfeld, W. F. (1986a). *The CORRESP Procedure: Experimental Software for Correspondence Analysis*. Cary NC: SAS Institute Inc.

Young, F. W. and Kuhfeld, W. F. (1986b). *The PRINQUAL Procedure: Experimental Software for Principal Components of Qualitative Data*. Cary, NC: SAS Institute Inc.

Young, F. W. and Kuhfeld, W. F. (1987). *The TRANSREG Procedure: Experimental Software for Transformation Regression*. Cary, NC: SAS Institute Inc.

Young, F. W., Kuhfeld, W. F., and Smith, J. B. (1987). "A Structured Environment for Data Analysis, Dynamic Graphics and Writing," unpublished manuscript. Chapel Hill, NC: University of North Carolina, Psychometric Laboratory.

Young, F. W. and Sarle, W. S. (1983). *Exploratory Multivariate Data Analysis*. Cary, NC: SAS Institute Inc.